Real-Life Math

Real-
Life
Math

Volume 1: A–L

K. Lee Lerner & Brenda Wilmoth Lerner,
Editors

GERMANNA COMMUNITY COLLEGE
GERMANNA CENTER FOR ADVANCED TECHNOLOGY
18121 TECHNOLOGY DRIVE
CULPEPER, VA 22701

THOMSON

GALE

Detroit • New York • San Francisco • San Diego • New Haven, Conn. • Waterville, Maine • London • Munich

Real-Life Math

K. Lee Lerner and Brenda Wilmoth Lerner, Editors

Project Editor
Kimberley A. McGrath

Editorial
Luann Brennan, Meggin M. Condino,
Madeline Harris, Paul Lewon,
Elizabeth Manar

Editorial Support Services
Andrea Lopeman

Indexing
Factiva, a Dow Jones & Reuters Company

Rights and Acquisitions
Margaret Abendroth, Timothy Sisler

Imaging and Multimedia
Lezlie Light, Denay Wilding

Product Design
Pamela Galbreath, Tracey Rowens

Composition
Evi Seoud, Mary Beth Trimper

Manufacturing
Wendy Blurton, Dorothy Maki

LIBRARY OF CONGRESS CATALOGING-IN-PUBLICATION DATA

Real-life math / K. Lee Lerner and Brenda Wilmoth Lerner, editors.
p. cm.
Includes bibliographical references and index.
ISBN 0-7876-9422-3 (set : hardcover: alk. paper)—
ISBN 0-7876-9423-1 (v. 1)—ISBN 0-7876-9424-X (v. 2)
1. Mathematics—Encyclopedias.
I. Lerner, K. Lee. II. Lerner, Brenda Wilmoth.

QA5.R36 2006
510'.3—dc22   2005013141

This title is also available as an e-book, ISBN 1414404999 (e-book set).
ISBN: 0-7876-9422-3 (set); 0-7876-9423-1 (v1); 0-7876-9424-X (v2)
Contact your Gale sales representative for ordering information.
Printed in the United States of America
10 9 8 7 6 5 4 3 2 1

Entries (With Areas of Discussion) vii

Introduction xix

List of Advisors and Contributors xxi

Entries 1

Table of Contents

Volume 1: A–L

Addition 1
Algebra 9
Algorithms 26
Architectural Math 33
Area 45
Average. 51
Base 59
Business Math 62
Calculator Math 69
Calculus 80
Calendars 97
Cartography 100
Charts 107
Computers and Mathematics 114
Conversions 122
Coordinate Systems 131
Decimals 138
Demographics 141
Discrete Mathematics 144
Division 149
Domain and Range 156
Elliptic Functions 159
Estimation 161
Exponents 167
Factoring 180
Financial Calculations, Personal 184
Fractals 198
Fractions 203
Functions 210
Game Math 215
Game Theory 225
Geometry. 232
Graphing 248
Imaging 262
Information Theory 269
Inverse 278
Iteration 284
Linear Mathematics 287
Logarithms 294
Logic 300

Volume 2: M–Z

Matrices and Determinants 303
Measurement 307
Medical Mathematics 314
Modeling 328
Multiplication 335
Music and Mathematics 343
Nature and Numbers 353
Negative Numbers 356
Number Theory 360
Odds 365
Percentages 372
Perimeter 385
Perspective 389
Photography Math 398
Plots and Diagrams 404
Powers 416
Prime Numbers 420
Probability 423
Proportion 430
Quadratic, Cubic, and Quartic Equations . . . 438
Ratio 441

Rounding 449
Rubric 453
Sampling 457
Scale 465
Scientific Math 473
Scientific Notation 484
Sequences, Sets, and Series 491
Sports Math 495
Square and Cube Roots 511
Statistics 516
Subtraction 529
Symmetry 537
Tables 543
Topology 553
Trigonometry 557
Vectors 568
Volume 575
Word Problems 583
Zero-sum Games 595

Glossary *599*

Field of Application Index *605*

General Index *609*

Addition

Financial Addition 4
Geometric Progression 6
Poker, Probability, and Other Uses of Addition . . 5
Sports and Fitness Addition 3
Using Addition to Predict and Entertain 6

Algebra

Art 21
Building Skyscrapers 19
Buying Light Bulbs 20
College Football 14
Crash Tests 18
Fingerprint Scanners 22
Flying an Airplane 16
Fundraising 19
Personal Finances 13
Population Dynamics 22
Private Space Travel 24
Skydiving 17
Teleportation 23
UPC Barcodes 15

Algorithms

Archeology 27
Artificial Intelligence 32
Computer Programming 27
Credit Card Fraud Detection 27
Cryptology 28
Data Mining 28
Digital Animation and Digital Model Creation . . 28
DNA or Genetic Analysis 28
Encryption and Encryption Devices 28
The Genetic Code 30
Imaging 29
Internet Data Transmission 29
Linguistics, the Study of Language 31
Mapping 30
Market or Sales Analysis 30
Operational Algorithms 27
Security Devices 31
Sports Standings and Seedings 31
Tax Returns 31

Architectural Math

Architectural Concepts in Wheels 43
Architectural Symmetry in Buildings 38

Architecture 36
Astronomy 43
Ergonomics 41
Geometry, Basic Forms and Shapes of 40
Golden Rectangle and Golden Ratio 38
Grids, Use of 37
Jewelry 41
Measurement 35
Proportion 34
Ratio 33
Ratio and Proportion, Use of 38
Scale Drawing 34
Space, Use Of 37
Sports 38
Symmetry 34
Symmetry in City Planning 41
Technology 41
Textile and Fabrics 43

Area

Area of a Rectangle 45
Areas of Common Shapes 46
Areas of Solid Objects 46
Buying by Area 47
Car Radiators 48
Cloud and Ice Area and Global Warming . . . 47
Drug Dosing 46
Filtering 47
Solar Panels 49
Surveying 48
Units of Area 45

Average

Arithmetic Mean 51
The "Average" Family 55
Average Lifespan 57
Averaging for Accuracy 55
Batting Averages 53
Evolution in Action 57
Geometric Mean 52
Grades. 54
How Many Galaxies? 55
Insurance. 57
Mean 52
Median 52
Space Shuttle Safety 56
Student Loan Consolidation 56
Weighted Averages in Business 54
Weighted Averages in Grading 54

Base

Base 2 and Computers 60

Business Math

Accounting 63
Budgets 63
Earnings 66
Interest 67
Payroll. 65
Profits 66

Calculator Math

Bridge Construction 76
Combinatorics 77
Compound Interest 74
Financial Transactions 73
Measurement Calculations 75
Nautical Navigation 73
Random Number Generator 75
Supercomputers 78
Understanding Weather 77

Calculus

Applications of Derivatives 86
Derivative 81
Functions 81
Fundamental Theorem of Calculus 85
Integral 83
Integrals, Applications 91
Maxima and Minima 85

Calendars

Gregorian Calendar 99
Islamic and Chinese Calendars. 99
Leap Year 99

Cartography

Coordinate Systems 103
GIS-Based Site Selection 105
GPS Navigation. 105
Map Projection 100
Natural Resources Evaluation and Protection . . 105
Scale 100
Topographic Maps. 104

Charts

Bar Charts 109
Basic Charts 107
Choosing the Right Type of Chart For the Data . .112
Clustered Column Charts 110
Column and Bar Charts 109
Line Charts 107
Pie Charts 110
Stacked Column Charts 110
Using the Computer to Create Charts 112
X-Y Scatter Graphs 109

Computers and Mathematics

Algorithms 115
Binary System 114
Bits 116
Bytes 116
Compression 118
Data Transmission 119
Encryption 120
IP Address 117
Pixels, Screen Size, and Resolution 117
Subnet Mask 118
Text Code 116

Conversions

Absolute Systems 127
Arbitrary Systems 128
Cooking or Baking Temperatures 127
Derived Units 124
English System 123
International System of Units (SI) 123
Metric Units 123
Units Based On Physical or "Natural"
         Phenomena 124
Weather Forecasting 126

Coordinate Systems

3-D Systems On Ordinance Survey Maps 136
Cartesian Coordinate Plane 132
Changing Between Coordinate Systems 132
Choosing the Best Coordinate System 132
Commercial Aviation 135
Coordinate Systems Used in Board Games . . .134
Coordinate Systems Used for Computer
         Animation 134
Dimensions of a Coordinate System 131

Longitude and John Harrison 135
Modern Navigation and GPS 135
Paper Maps of the World 134
Polar Coordinates 133
Radar Systems and Polar Coordinates 136
Vectors 132

Decimals

Grade Point Average Calculations 139
Measurement Systems 139
Science 139

Demographics

Census 142
Election Analysis 141
Geographic Information System Technology . . .143

Discrete Mathematics

Algorithms 145
Boolean Algebra 145
Combinatorial Chemistry 147
Combinatorics 145
Computer Design 146
Counting Jaguars Using Probability Theory . . .147
Cryptography 146
Finding New Drugs with Graph Theory 147
Graphs 146
Logic, Sets, and Functions 144
Looking Inside the Body With Matrices 147
Matrix Algebra 146
Number Theory 145
Probability Theory 145
Searching the Web 146
Shopping Online and Prime Numbers 147

Division

Averages 152
Division and Comparison 151
Division and Distribution 150
Division, Other Uses 153
Practical Uses of Division For Students 153

Domain and Range

Astronomers 157
Calculating Odds and Outcomes 157

Computer Control and Coordination 157
Computer Science 158
Engineering 157
Graphs, Charts, Maps 158
Physics 157

Elliptic Functions

The Age of the Universe 160
Conformal Maps 159
E-Money 160

Estimation

Buying a Used Car 162
Carbon Dating 165
Digital Imaging 164
Gumball Contest 163
Hubble Space Telescope 165
Population Sampling 164
Software Development 166

Exponents

Bases and Exponents 167
Body Proportions and Growth
    (Why Elephants Don't Have
    Skinny Legs) 179
Credit Card Meltdown 178
Expanding Universe 178
Exponential Functions 168
Exponential Growth 171
Exponents and Evolution 174
Integer Exponents 167
Interest and Inflation 177
Non-Integer Exponents 168
Radioactive Dating 177
Radioactive Decay 175
Rotting Leftovers 173
Scientific Notation 171

Factoring

Codes and Code Breaking 182
Distribution 182
Geometry and Approximation of Size 182
Identification of Patterns
    and Behaviors 181
Reducing Equations 181
Skill Transfer 182

Financial Calculations, Personal

Balancing a Checkbook 189
Budgets 188
Buying Music 184
Calculating a Tip 194
Car Purchasing and Payments 187
Choosing a Wireless Plan 187
Credit Cards 185
Currency Exchange 195
Investing 190
Retirement Investing 192
Social Security System 190
Understanding Income Taxes 189

Fractals

Astronomy 202
Building Fractals 199
Cell Phone and Radio Antenna 202
Computer Science 202
Fractals and Nature 200
Modeling Hurricanes and Tornadoes 201
Nonliving Systems 201
Similarity 199

Fractions

Algebra 205
Cooking and Baking 206
Fractions and Decimals 204
Fractions and Percentages 204
Fractions and Voting 208
Music 206
Overtime Pay 208
Radioactive Waste 206
Rules For Handling Fractions 204
Simple Probabilities 207
Tools and Construction 208
Types of Fractions 203
What Is a Fraction? 203

Functions

Body Mass Index 214
Finite-Element Models 212
Functions, Described 210
Functions and Relations 210
Guilloché Patterns 211
Making Airplanes Fly 211

The Million-Dollar Hypothesis 212
Nuclear Waste 213
Synths and Drums 213

Game Math

Basic Board Games. 220
Card Games 218
Magic Squares 221
Math Puzzles 223
Other Casino Games 219

Game Theory

Artificial Intelligence 230
Decision Theory 228
eBay and the Online Auction World 230
Economics 229
Economics and Game Theory 228
Evolution and Animal Behavior 229
General Equilibrium 229
Infectious Disease Therapy 230
Nash Equilibrium 229

Geometry

Architecture 237
Fireworks. 241
Fourth Dimension. 245
Global Positioning. 239
Honeycombs 239
Manipulating Sound 241
Pothole Covers 236
Robotic Surgery. 245
Rubik's Cube 243
Shooting an Arrow 244
Solar Systems 242
Stealth Technology. 244

Graphing

Aerodynamics and Hydrodynamics 259
Area Graphs 252
Bar Graphs 249
Biomedical Research 258
Bubble Graphs 257
Computer Network Design 259
Finding Oil 258
Gantt Graphs 254
Global Warming 257

GPS Surveying 258
Line Graphs 251
Physical Fitness 259
Picture Graphs 254
Pie Graphs 252
Radar Graphs 253
X-Y Graphs 254

Imaging

Altering Images. 263
Analyzing Images 263
Art 267
Compression 264
Creating Images 263
Dance 266
Forensic Digital Imaging. 266
Meat and Potatoes 266
Medical Imaging 264
Optics 264
Recognizing Faces: a Controversial
    Biometrics Application 264
Steganography and Digital Watermarks 266

Information Theory

Communications 273
Error Correction 275
Information and Meaning 273
Information Theory in Biology and Genetics . . .274
Quantum Computing 276
Unequally Likely Messages 271

Inverse

Anti-Sound 282
The Brain and the Inverted Image
    On the Eye 281
Cryptography 280
Definition of an Inverse 278
Fluid Mechanics and Nonlinear Design 281
Inverse Functions 279
The Multiplicative Inverse 278
Negatives Used in Photography 281
Operations Where the Inverse Does
    Not Exist 279
Operations With More Than One
    Inverse 279
Stealth Submarine Communications 282
Stereo 282

Iteration

Iteration and Business 285
Iteration and Computers 286
Iteration and Creativity 285
Iteration and Sports 284

Linear Mathematics

Earthquake Prediction 289
Linear Programming 291
Linear Reproduction of Music 292
Recovering Human Motion From Video 290
Virtual Tennis 291

Logarithms

Algebra of Powers of Logarithms 296
Computer Intensive Applications 297
Cryptography and Group Theory 299
Designing Radioactive Shielding
    for Equipment in Space 299
Developing Optical Equipment. 298
Estimating the Age of Organic Matter
    Using Carbon Dating 298
Log Tables 296
Logarithms to Other Bases Than 10 296
The Power of Mathematical Notation 295
Powers and Logs of Base 10 295
Powers and Their Relation
    to Logarithms 296
Supersonic and Hypersonic Flight 299
Use in Medical Equipment 298
Using a Logarithmic Scale to Measure
    Sound Intensity 297

Logic

Boolean Logic 300
Fuzzy Logic 300
Proposition and Conclusion 300
Reasoning 300

Matrices and Determinants

Designing Cars 305
Digital Images 304
Flying the Space Shuttle 305
Population Biology 305

Measurement

Accuracy in Measurement 309
Archaeology 310
Architecture 310
Blood Pressure 310
Chemistry 310
Computers 310
The Definition of a Second 310
Dimensions 308
Doctors and Medicine. 310
Engineering 309
Evaluating Errors in Measurement
    and Quality Control 309
Gravity 313
How Astronomers and NASA Measure
    Distances in Space 312
Measuring Distance 308
Measuring Mass 313
Measuring the Speed of Gravity 313
Measuring Speed, Space Travel, and Racing . .310
Measuring Time 310
Navigation 310
Nuclear Power Plants 310
Space Travel and Timekeeping 312
Speed of Light 312

Medical Mathematics

Calculation of Body Mass Index (BMI) 319
Clinical Trials 323
Genetic Risk Factors: the Inheritance of Disease . .321
Rate of Bacterial Growth 326
Standard Deviation and Variance for Use
    in Height and Weight Charts 319
Value of Diagnostic Tests 318

Modeling

Ecological Modeling 330
Military Modeling 331

Multiplication

Sports Multiplication: Calculating
    a Baseball ERA 338
Calculating Exponential Growth Rates 338
Calculating Miles Per Gallon 341
Electronic Timing 339
Exponents and Growth Rates 337
Investment Calculations 337

Measurement Systems 339
Multiplication in International Travel 339
Other Uses of Multiplication 340
Rate of Pay 339
Savings 341
SPAM and Email Communications 341

Music and Mathematics

Acoustic Design 348
Compressing Music 349
Computer-Generated Music 349
Digital Music. 348
Discordance of the Spheres 346
Electronic Instruments 347
Error Correction 349
Frequency of Concert A 351
Mathematical Analysis of Sound 347
Math-Rock 351
Medieval Monks 345
Pythagoras and Strings 343
Quantification of Music 345
Using Randomness 349
Well-Tempered Tones 346

Nature and Numbers

Fibonacci Numbers and the Golden Ratio 353
Mathematical Modeling of Nature 354
Specify Application Using Alphabetizable Title . . 355
Using Fractals to Represent Nature 355

Negative Numbers

Accounting Practice 357
Buildings. 359
Flood Control 358
The Mathematics of Bookkeeping 357
Sports 358
Temperature Measurement 357

Number Theory

Cryptography 362
Error Correcting Codes 363

Odds

Odds in Everyday Life 367
Odds in State Lotteries 368

Odds, Other Applications 369
Sports and Entertainment Odds 366

Percentages

Calculating a Tip 375
Compound Interest 376
Definitions and Basic Applications 372
Examples of Common Percentage
    Applications 374
Finding the Base Rate 374
Finding the Original Amount 375
Finding the Rate of Increase or Decrease . . . 375
Finding the Rate Percent 374
Important Percentage Applications 374
Percentage Change: Increase or Decrease . . . 375
Public Opinion Polls 379
Ratios, Proportions, and Percentages 373
Rebate Period and Cost 378
Rebates 377
Retail Sales: Price Discounts and Markups
    and Sales Tax 376
Sales Tax Calculation: In-Store Discount
    Versus Mail-In Rebate 377
Sales Tax Calculations 377
SAT Scores or Other Academic Testing 383
Sports Math 379
Tournaments and Championships 382
Understanding Percentages in the Media . . . 378
Using Percentages to Make Comparisons . . . 379

Perimeter

Bodies of Water. 386
Landscaping. 386
Military 387
Planetary Exploration. 388
Robotic Perimeter Detection Systems 388
Security Systems 386
Sporting Events 386

Perspective

Animation 392
Art 391
Computer Graphics 395
Film 393
Illustration 392
Interior Design 394
Landscaping. 395

Photography Math

The Camera 398
Depth of Field 400
Digital Image Processing 403
Digital Photography 401
Film Speed 398
Lens Aperture 400
Lens Focal Length 399
Photomicrography 403
Reciprocity 401
Shutter Speed 399
Sports and Wildlife Photography 402

Plots and Diagrams

Area Chart 406
Bar Graphs 406
Body Diagram 414
Box Plot 405
Circuit Diagram 414
Diagrams 404
Fishbone Diagram 406
Flow Chart 411
Gantt Charts 413
Line Graph 408
Maps 413
Organization Charts 413
Other Diagrams 414
Pie Graph 406
Polar Chart 406
Properties of Graphs 404
Scatter Graph 405
Stem and Leaf Plots 405
Street Signs 414
Three-Dimensional Graph 407
Tree Diagram 412
Triangular Graph 407
Weather Maps 414

Powers

Acids, Bases, and pH Level 418
Areas of Polygons and Volumes
    of Solid Figures 417
Astronomy and Brightness of Stars 418
Computer Science and Binary
    Logic 417
Earthquakes and the Richter Scale 417
The Powers of Nanotechnology 418

Prime Numbers

Biological Applications of Prime Numbers 421

Probability

Gambling and Probability Myths 425
Probability in Business and Industry 427
Probability, Other Uses 428
Probability in Sports and Entertainment 426
Security 424

Proportion

Architecture 432
Art, Sculpture, and Design 432
Chemistry 435
Diets 436
Direct Proportion 431
Engineering Design 435
Ergonomics 434
Inverse Proportion 431
Maps 434
Medicine 434
Musical Instruments 435
Proportion in Nature 436
Solving Ratios With Cross Products 430
Stock Market 436

Quadratic, Cubic, and Quartic Equations

Acceleration 439
Area and Volume 439
Car Tires 439
Guiding Weapons 440
Hospital Size 440
Just in Time Manufacturing 440

Ratio

Age of Earth 446
Automobile Performance 445
Cleaning Water 446
Cooking 446
Cost of Gas 443
Determination of the Origination of the Moon . .447
Genetic Traits 443
Healthy Living 446
Length of a Trip 443
Music 445

Optimizing Livestock Production 447
Sports 445
Stem Cell Research. 446
Student-Teacher Ratio 445

Rounding

Accounting 451
Bulk Purchases 450
Decimals 450
Energy Consumption 451
Length and Weight 450
Lunar Cycles 451
Mileage 452
Pi 450
Population 451
Precision 452
Time 452
Weight Determination 451
Whole Numbers. 449

Rubric

Analytic Rubrics and Holistic Rubrics 455
General Rubrics and Task-Specific
        Rubrics 455
Scoring Rubrics 453

Sampling

Agriculture 459
Archeology 463
Astronomy 462
Demographic Surveys 462
Drug Manufacturing 460
Environmental Studies 462
Market Assessment 463
Marketing 463
Non-Probability Sampling 458
Plant Analysis 460
Probability Sampling 457
Scientific Research 460
Soil Sampling 460
Weather Forecasts 461

Scale

Architecture 468
Atmospheric Pressure Using Barometer 469
The Calendar 469

Expanse of Scale From the Sub-Atomic
        to the Universe 471
Interval Scale 466
Linear Scale 465
Logarithmic Scale 465
Map Scale 467
Measuring Wind Strength 469
The Metric System of Measurement 472
Music 471
Nominal Scale 467
Ordinal Scale 467
Ratio Scale 466
The Richter Scale 470
Sampling 472
Technology and Imaging 469
Toys 471
Weighing Scale 468

Scientific Math

Aviation and Flights 478
Bridging Chasms 478
Discrete Math 474
Earthquakes and Logarithms 482
Equations and Graphs 476
Estimating Data Used for Assessing Weather . . .477
Functions and Measurements 473
Genetics 483
Logarithms 475
Matrices and Arrays 475
Medical Imaging 480
Rocket Launch 480
Ships 482
Simple Carpentry 479
Trigonometry and the Pythagorean Theorem . . .474
Weather Prediction 476
Wind Chill in Cold Weather 476

Scientific Notation

Absolute Dating 489
Chemistry 486
Computer Science 487
Cosmology 487
Earth Science 489
Electrical Circuits 486
Electronics 489
Engineering 487
Environmental Science 488
Forensic Science 488
Geologic Time Scale and Geology. 488

Light Years, the Speed of Light,
and Astronomy 486
Medicine 488
Nanotechnology 490
Proteins and Biology 490

Sequences, Sets, and Series

Genetics 493
Operating On Sets 492
Ordering Things 493
Sequences 491
Series 492
Sets 491
Using Sequences 493

Sports Math

Baseball 498
Basketball 499
Cycling—Gear Ratios and How They
Work 505
Football—How Far was the Pass Thrown? . . .507
Football Tactics—Math as a
Decision-Making Tool 501
Golf Technology 506
Math and the Science of Sport 504
Math and Sports Wagering 508
Math to Understand Sports Performance . . .497
Mathematics and the Judging of Sports 504
Math to Understand Sports Performance—Capology 101 . . .507
Money in Sport—Capology 101 507
North American Football 499
Pascal's Triangle and Predicting a Coin Toss . . .500
Predicting the Future: Calling the Coin Toss . . .500
Ratings Percentage Index (RPI) 503
Rules Math 496
Soccer—Free Kicks and the Trajectory
of the Ball 506
Understanding the Sports Media Expert 502

Square and Cube Roots

Architecture 513
Global Economics 515
Hiopasus's Fatal Discovery 513
Names and Conventions 512
Navigation 514
Pythagorean Theorem 513
Sports 514
Stock Markets 515

Statistics

Analysis of Variance 522
Average Values 519
Confidence Intervals 522
Correlation and Curve Fitting 521
Cumulative Frequencies and Quantiles 521
Geostatistics 525
Measures of Dispersion 520
Minimum, Maximum, and Range 518
Populations and Samples 516
Probability 517
Public Opinion Polls 527
Quality Assurance 526
Statistical Hypothesis Testing 522
Using Statistics to Deceive 523

Subtraction

Subtraction in Entertainment
and Recreation 533
Subtraction in Financial Calculations 531
Subtraction in Politics and Industry 535
Tax Deductions 532

Symmetry

Architecture 541
Exploring Symmetries 539
Fractal Symmetries 541
Imperfect Symmetries 542
Symmetries in Nature 542

Tables

Converting Measurements 545
Daily Use 549
Educational Tables 545
Finance 546
Health 548
Math Skills 544
Travel 549

Topology

Computer Networking 555
I.Q. Tests 555
Möbius Strip 555
Visual Analysis 554
Visual Representation 555

Trigonometry

Chemical Analysis 566
Computer Graphics 566
Law of Sines 561
Measuring Angles 557
Navigation 562
Pythagorean Theorem. 559
Surveying, Geodesy, and Mapping 564
Trigonometric Functions 560
Types of Triangles 558
Vectors, Forces, and Velocities 563

Vectors

3-D Computer Graphics 572
Drag Racing 572
Land Mine Detection 572
The Magnitude of a Vector 569
Sports Injuries 573
Three-Dimensional Vectors 569
Two-Dimensional Vectors 568
Vector Algebra 570
Vectors in Linear Algebra 571

Volume

Biometric Measurements. 581
Building and Architecture 579
Compression Ratios in Engines 579
Glowing Bubbles: Sonoluminescence . . .579
Medical Applications 578
Misleading Graphics 581
Pricing 577
Runoff 582
Sea Level Changes 580
Swimming Pool Maintenance 581
Units of Volume 575
Volume of a Box 575
Volumes of Common Solids 575
Why Thermometers Work 580

Word Problems

Accounts and VAT 592
Archaeology 585
Architecture 590
Average Height? 593
Banks, Interest Rates, and Introductory Rates . .591
Bearings and Directions of Travel 592
Comparisons 586

Computer Programming 584
Cooking Instructions 591
Creative Design 584
Cryptography 585
Decorating 594
Disease Control. 591
Ecology 587
Efficient Packing and Organization 590
Engineering 585
Exchange Rates 586
Finance 591
Geology 591
Global Warming 594
Graph Theory 588
Hypothesis Testing 585
Insurance 584
Linear Programming 588
Lotteries and Gambling 591
Measuring Height of a Well 594
Medicine and Cures 585
MMR Immunization and Autism 594
Navigation 587
Opinion Polls 593
Percentages 586
Phone Companies 586
Postman 589
Proportion and Inverse Proportion 586
Quality Control 592
Ranking Test Scores 589
Recipes 591
ROTA and Timetables 589
Searching in an Index 590
Seeding in Tournaments 590
Shortest Links to Establish Electricity
    to a Whole Town 589
Software Design 584
Stock Keeping 592
Store Assistants 592
Surveying 592
Teachers 584
Throwing a Ball 593
Translation 587
Travel and Racing 586
Traveling Salesperson 589
Weather 593

Zero-Sum Games

Currency, Futures, and Stock Markets 597
Experimental Gaming 597
Gambling 596
War 597

Real-Life Math takes an international perspective in exploring the role of mathematics in everyday life and is intended for high school age readers. As *Real-Life Math* (*RLM*) is intended for a younger and less mathematically experienced audience, the authors and editors faced unique challenges in selecting and preparing entries.

The articles in the book are meant to be understandable by anyone with a curiosity about mathematical topics. *Real-Life Math* is intended to serve all students of math such that an 8th- or 9th-grade student just beginning their study of higher maths can at least partially comprehend and appreciate the value of courses to be taken in future years. Accordingly, articles were constructed to contain material that might serve all students. For example, the article, "Calculus" is intended to be able to serve students taking calculus, students finished with prerequisites and about to undertake their study of calculus, and students in basic math or algebra who might have an interest in the practical utility of a far-off study of calculus. Readers should anticipate that they might be able to read and reread articles several times over the course of their studies in maths. *Real-Life Math* challenges students on multiple levels and is designed to facilitate critical thinking and reading-in-context skills. The beginning student is not expected to understand more mathematically complex text dealing, for example, with the techniques for calculus, and so should be content to skim through these sections as they read about the practical applications. As students progress through math studies, they will naturally appreciate greater portions of more advanced sections designed to serve more advanced students.

To be of maximum utility to students and teachers, most of the 80 topics found herein—arranged alphabetically by theory or principle—were predesigned to correspond to commonly studied fundamental mathematical concepts as stated in high school level curriculum objectives. However, as high school level maths generally teach concepts designed to develop skills toward higher maths of greater utility, this format sometimes presented a challenge with regard to articulating understandable or direct practical applications for fundamental skills without introducing additional concepts to be studied in more advanced math classes. It was sometimes difficult to isolate practical applications for fundamental concepts because it often required more complex mathematical concepts to most accurately convey the true relationship of mathematics to our advancing technology. Both the authors and editors of the project made exceptional efforts to smoothly and seamlessly incorporate the concepts necessary (and at an accessible level) within the text.

Although the authors of *Real-Life Math* include math teachers and professors, the bulk of the writers are

Introduction

practicing engineers and scientists who use math on a daily basis. However, *RLM* is not intended to be a book about real-life applications as used by mathematicians and scientists but rather, wherever possible, to illustrate and discuss applications within the experience—and that are understandable and interesting—to younger readers.

RLM is intended to maximize readability and accessibility by minimizing the use of equations, example problems, proofs, etc. Accordingly, *RLM* is not a math textbook, nor is it designed to fully explain the mathematics involved in each concept. Rather, *RLM* is intended to compliment the mathematics curriculum by serving a general reader for maths by remaining focused on fundamental math concepts as opposed to the history of math, biographies of mathematicians, or simply interesting applications. To be sure, there are inherent difficulties in presenting mathematical concepts without the use of mathematical notation, but the authors and editors of *RLM* sought to use descriptions and concepts instead of mathematical notation, problems, and proofs whenever possible.

To the extent that *RLM* meets these challenges it becomes a valuable resource to students and teachers of mathematics.

The editors modestly hope that *Real-Life Math* serves to help students appreciate the scope of the importance and influence of math on everyday life. *RLM* will achieve its highest purposes if it intrigues and inspires students to continue their studies in maths and so advance their understanding of the both the utility and elegance of mathematics.

"[The universe] cannot be read until we have learnt the language and become familiar with the characters in which it is written. It is written in mathematical language, and the letters are triangles, circles, and other geometrical figures, without which means it is humanly impossible to comprehend a single word." Galilei, Galileo (1564–1642)

K. Lee Lerner and Brenda Wilmoth Lerner, Editors

In compiling this edition, we have been fortunate in being able to rely upon the expertise and contributions of the following scholars who served as contributing advisors or authors for *Real-Life Math*, and to them we would like to express our sincere appreciation for their efforts:

William Arthur Atkins

Mr. Atkins holds a BS in physics and mathematics as well as an MBA. He lives at writes in Perkin, Illinois.

Juli M. Berwald, PhD

In addition to her graduate degree in ocean sciences, Dr. Berwald holds a BA in mathematics from Amherst College, Amherst, Massachusetts. She currently lives and writes in Chicago, Illinois.

Bennett Brooks

Mr. Brooks is a PhD graduate student in mathematics. He holds a BS in mathematics, with departmental honors, from University of Redlands, Redlands, California, and currently works as a writer based in Beaumont, California.

Rory Clarke, PhD

Dr. Clark is a British physicist conducting research in the area of high-energy physics at the University of Bucharest, Romania. He holds a PhD in high energy particle physics from the University of Birmingham, an MSc in theoretical physics from Imperial College, and a BSc degree in physics from the University of London.

Raymond C. Cole

Mr. Cole is an investment banking financial analyst who lives in New York. He holds an MBA from the Baruch Zicklin School of Business and a BS in business administration from Fordham University.

Bryan Thomas Davies

Mr. Davies holds a Bachelor of Laws (LLB) from the University of Western Ontario and has served as a criminal prosecutor in the Ontario Ministry of the Attorney General. In addition to his legal experience, Mr. Davies is a nationally certified basketball coach.

John F. Engle

Mr. Engle is a medical student at Tulane University Medical School in New Orleans, Louisiana.

List of Advisors and Contributors

William J. Engle

Mr. Engle is a retired petroleum engineer who lives in Slidell, Louisiana.

Paul Fellows

Dr. Fellows is a physicist and mathematician who lives in London, England.

Renata A. Ficek

Ms. Ficek is a graduate mathematics student at the University of Queensland, Australia.

Larry Gilman, PhD

Dr. Gilman holds a PhD in electrical engineering from Dartmouth College and an MA in English literature from Northwestern University. He lives in Sharon, Vermont.

Amit Gupta

Mr. Gupta holds an MS in information systems and is managing director of Agarwal Management Consultants P. Ltd., in Ahmedabad, India.

William C. Haneberg, PhD

Dr. Haneberg is a professional geologist and writer based in Seattle, Washington.

Bryan D. Hoyle, PhD

Dr. Hoyle is a microbiologist and science writer who lives in Halifax, Nova Scotia, Canada.

Kenneth T. LaPensee, PhD

In addition to professional research in epidemiology, Dr. LaPensee directs Skylands Healthcare Consulting located in Hampton, New Jersey.

Holly F. McBain

Ms. McBain is a science and math writer who lives near New Braunfels, Texas.

Mark H. Phillips, PhD

Dr. Phillips serves as an assistant professor of management at Abilene Christian University, located in Abilene, Texas.

Nephele Tempest

Ms. Tempest is a writer based in Los Angeles, California.

David Tulloch

Mr. Tulloch holds a BSc in physics and an MS in the history of science. In addition to research and writing he serves as a radio broadcaster in Ngaio, Wellington, New Zealand.

James A. Yates

Mr. Yates holds a MMath degree from Oxford University and is a teacher of maths in Skegnes, England.

ACKNOWLEDGMENTS

The editors would like to extend special thanks to Connie Clyde for her assistance in copyediting. The editors also wish to especially acknowledge Dr. Larry Gilman for his articles on calculus and exponents as well as his skilled corrections of the entire text. The editors are profoundly grateful to their assistant editors and proofreaders, including Lynn Nettles and Bill Engle, who read and corrected articles under the additional pressures created by evacuations mandated by Hurricane Katrina. The final editing of this book was interrupted as Katrina damaged the Gulf Coast homes and offices of several authors, assistant editors, and the editors of *RLM* just as the book was being prepared for press. Quite literally, many pages were read and corrected by light produced by emergency generators—and in some cases, pages were corrected from evacuation shelters. The editors are forever grateful for the patience and kind assistance of many fellow scholars and colleagues during this time.

The editors gratefully acknowledge the assistance of many at Thompson Gale for their help in preparing *Real-Life Math*. The editors wish to specifically thank Ms. Meggin Condino for her help and keen insights while launching this project. The deepest thanks are also offered to Gale Senior Editor Kim McGrath for her tireless, skilled, good-natured, and intelligent guidance.

Addition

Overview

Addition is the process of combining two or more numbers to create a new value, and is generally considered the simplest form of mathematics. Despite its simplicity, the ability to perform basic addition is the foundation of most advanced mathematics, and simple addition, repeated millions of times per second, actually underlies much of the processing performed within the most advanced electronic computers on earth. Despite its elementary nature, the process of adding numbers together remains one of the most useful mathematical operations available, as well as perhaps the most common type of calculation performed on a daily basis by most adults.

Fundamental Mathematical Concepts and Terms

An addition equation requires only two terms to describe its component parts. When asked to name the simplest equation possible, most adults would respond with $1 + 1 = 2$, probably the first math operation they learned. In this simple equation, the two 1s are termed addends, while the result of this or any other addition equation is known as the sum, in this case the value 2. Because this final value is called a sum, it is also correct, though less common, to describe the process of adding as summing, as in the expression, "Sum the five daily values to find the total attendance for the week." While the addition sign is properly called a plus sign, one does not ever refer to the process of addition as "plus-ing" two values.

A Brief History of Discovery and Development

Because the basic process of addition is so simple, its exact origins are impossible to identify. Near the beginning of recorded history, a variety of endeavors including commerce, warfare, and agriculture required the ability to add numbers; for some lines of work, addition was such a routine operation that specific tools became necessary in order to streamline the process. The most basic counting tools consisted of a small bag of stones or other small objects that could be used to tally an inventory of goods. In the case of shipping, a merchant counting sacks of grain as they were loaded onto his ship would move one small stone aside for each sack loaded, providing both a running total and a simple method to double-check the final count. Upon arrival, this same collection of stones would serve as the ship's manifest, allowing a

7 2 3 0 1 8 9
NUMBER REPRESENTED

The Chinese abacus was one of the earliest tools for everyday addition. CORBIS-BETTMANN. REPRODUCED BY PERMISSION.

The earliest known example of what we today recognize as the hand-held abacus was invented in China approximately 5,000 years ago. Consisting of wood and moveable beads, this counting tool did not actually perform calculations, but instead assisted its human operator by keeping a running total of items added. The Chinese abacus was recognized as an exceptionally useful tool, and progressively spread throughout the world. Modern examples of the abacus are little changed from these ancient models, and are still used in some parts of the world, where an expert user can often solve lengthy addition problems as quickly as someone using an electronic calculator.

As technology advanced, users sought ways to add more quickly and more accurately. In 1642, a French mathematician Blaise Pascal (1623–1662) invented the first mechanical adding machine. This device, a complex contraption operated by gears and wheels, allowed the user to type in his equation using a series of keys, with the results of the calculation displayed in a row of windows. Pascal's invention was revolutionary, specifically because it could carry digits from one column to another. Mechanical calculators, the distant descendents of Pascal's design, remained popular well into the twentieth century; more advanced electrically operated versions were used well into the 1960s and 1970s, when they were replaced by electronic models and spreadsheet software.

In a strange case of history repeating itself, the introduction of the first high-priced electronic calculators in the 1970s was coincidentally accompanied by television commercials offering training in a seemingly revolutionary method of adding called Chisenbop. Chisenbop allowed one to use only his fingers to add long columns of numbers very quickly, and television shows of that era featured young experts out-performing calculator-wielding adults. Chisenbop uses a variety of finger combinations to represent different values, with the right hand tallying values from zero to nine, and the left hand handling values from ten and up. The rapid drop in calculator prices during this era, as well as the potential stigma associated with counting on one's fingers, probably led to the method's demise. Despite its seemingly revolutionary nature, this counting scheme is actually quite old, and may in fact predate the abacus, which functions in a similar manner by allowing the operator to tally values as they are added. Multiple online tutorials today teach the technique, which has gradually faded back into obscurity.

running count of the shipment as it was unloaded. In the case of warfare, a general might number his horses using this same method of having each object represented by a stone, a small seashell, or some other token. The key principle in this type of system was a one-to-one relationship between the items being counted and the smaller symbolic items used to maintain the tally.

Over time, these sets of counting stones gradually evolved into large counting tables, known as abaci, or in the singular form, an abacus. These tables often featured grooves or other placement aids designed to insure accuracy in the calculations being made, and tallies were made by placing markers in the proper locations to symbolize ones, tens, and hundreds. The counting tables developed in numerous cultures, and ancient examples survive from Japan, Greece, China, and the Roman Empire. Once these tables came into wide use, a natural evolution, much like that seen in modern computer systems, occurred, with the bulky, fixed tables gradually morphing into smaller, more portable devices. These smaller versions were actually the earliest precursors of today's personal calculator.

While the complex calculations performed by today's sophisticated computers might appear to lie far beyond anything achieved by Pascal's original adding machine, the remnants of Pascal's simple additions can still be found deep inside every microprocessor (as well as in a simple programming language which bears his name in honor of his pioneering work). Modern computers offer user-friendly graphic interfaces and require little or no math or programming knowledge on the part of the average user. But at the lowest functional level, even a cutting edge processor relies on simple operations performed in its arithmetic logic unit, or ALU. When this basic processing unit receives an instruction, that instruction has typically been broken down into a series of simple processes which are then completed one at a time. Ironically, though the ALU is the mathematical heart of a modern computer, a typical ALU performs only four functions, the same add, subtract, multiply, and divide found on the earliest electronic calculators of the 1970s. By performing these simple operations millions of times each second, and leveraging this power through modern operating systems and applications software, even a process as simple as addition can produce startling results.

Real-life Applications

SPORTS AND FITNESS ADDITION

Many aspects of popular sports require the use of addition. For example, some of the best-known records tracked in most sports are found by simply adding one success to another. Records for the most homeruns, the most 3-point shots made, the most touchdown passes completed, and the most major golf tournaments won in a career are nothing more than the result of lengthy addition problems stretched out over an entire career. On the business side of sports are other addition applications, including such routine tasks as calculating the number of fans at a ballgame or the number of hotdogs sold, both of which are found by simply adding one more person or sausage to the running total.

Many sports competitions are scored on the basis of elapsed time, which is found by simply adding fractions of a second to a total until the event ends, at which time the smallest total is determined to be the winning score. In the case of motor sports, racers compete for the chance to start the actual race near the front of the field, and these qualifying attempts are often separated by mere hundredths or even thousandths of a second. Track events such as the decathlon, which requires participants to attempt ten separate events including sprints, jumps, vaults, and throwing events over the course of two grueling days, are scored by adding the tallies from each separate event to determine a final score. In the same way, track team scores are found by adding the scores from

each individual event, relay, and field event to determine a total score.

Although the sport of bowling is scored using only addition, this popular game has one of the more unusual scoring systems in modern sports. Bowlers compete in games consisting of ten frames, each of which includes up to two attempts to knock down all ten bowling pins. Depending on a bowler's performance in one frame, he may be able to add some shots twice, significantly raising his total score. For example, a bowler who knocks down all ten pins in a single roll is awarded a strike, worth ten plus the total of the next two balls bowled in the following frames, while a bowler who knocks down all ten pins in two rolls is scored a spare and receives ten plus the next one ball rolled. Without this scoring system, the maximum bowling score would be earned by bowling ten, ten-point strikes in a row for a perfect game total of 100. But with bowling's bonus scoring system, each of the ten frames is potentially worth thirty points to a bowler who bowls a strike followed by two more strikes, creating a maximum possible game score of 300.

While many programs exist to help people lose weight, none is more basic, or less liked, than the straightforward process of counting calories. Calorie counting is based on a simple, immutable principle of physics: if a human body consumes more calories than it burns, it will store the excess calories as fat, and will become heavier. For this reason, most weight loss plans address, at least to some degree, the number of calories being consumed. A calorie is a measure of energy, and 3,500 calories are required to produce one pound of body weight. Using simple addition, it becomes clear that eating an extra 500 calories per day will add up to 3,500 calories, or one pound gained, per week.

While this use of addition allows one to calculate the waistline impact of an additional dessert or several soft drinks, a similar process defines the amount of exercise required to lose this same amount of weight. For example, over the course of a week, a man might engage in a variety of physical activities, including an hour of vigorous tennis, an hour of slow jogging, one hour of swimming, and one hour officiating a basketball game. Each of these activities burns calories at a different rate. Using a chart of calorie burn rates, we determine that tennis burns 563 calories per hour, jogging burns 493 calories per hour, swimming burns 704 calories per hour, and officiating a basketball game burns 512. Adding these values up we find that the man has exercised enough to burn a total of 2,272 calories over the course of the week. Depending on how many calories he consumes, this may be adequate to maintain his weight. However if he is consuming an extra 3,500 calories per week, he will need to burn an additional 1,228 calories to avoid storing these extra calories as fat. Over the course of a year, this excess of 1,228 calories will eventually add up to a net gain of more than 63,000 calories, or a weight gain of more than 18 pounds.

While healthy activities help prolong life, the same result can be achieved by reducing unhealthy activities. Cigarette smoking is one of the more common behaviors believed to reduce life expectancy. While most smokers believe they would be healthier if they quit, and cigarette companies openly admit the dangers of their product, placing a health value or cost on a single cigarette can be difficult. A recent study published in the British Medical Journal tried to estimate the actual cost, in terms of reduced life expectancy, of each cigarette smoked. While this calculation is admittedly crude, the study concluded that each cigarette smoked reduces average life-span by eleven minutes, meaning that a smoker who puffs through all 20 cigarettes in a typical pack can simply add up the minutes to find that he has reduced his life expectancy by 220 minutes, or almost four hours. Simple addition also tells him that his pack-a-day habit is costing him 110 hours of life for each month he continues, or about four and one-half days of life lost for each month of smoking. When added up over a lifetime, the study concluded that smokers typically die more than six years earlier than non-smokers, a result of adding up the seemingly small effects of each individual cigarette.

FINANCIAL ADDITION

One of the more common uses of addition is in the popular pastime of shopping. Most adults understand that the price listed on an item's price-tag is not always the full amount they will pay. For example, most states charge sales tax, meaning that a shopper with $20.00 to spend will need to add some set percentage to his item total in order to be sure he stays under budget and doesn't come up short at the checkout counter. Many people estimate this add-on unconsciously, and in most cases, the amount added is relatively small.

In the case of buying a car, however, various add-ons can quickly raise the total bill, as well as the monthly payments. While paying 7% sales tax on a $3.00 purchase adds only twenty-one cents to the total, paying this same flat rate on a $30,000 automobile adds $2,100 to the bill. In addition, a car purchased at a dealership will invariably include a lengthy list of additional items such as documentation fees, title fees, and delivery charges, which must be added to the sticker price to determine the actual cost to the buyer.

As of 1999, Americans spent almost 40 cents of every food dollar at the 300,000 fast food restaurants in the country. Because they are often in a hurry to order, many customers choose one of the so-called value meals offered at most outlets. But in some cases, simple addition demonstrates that the actual savings gained by ordering a value meal is only a few cents. By adding the separate costs of the individual items in the meal, the customer can compare this total to learn just how much he is saving. He can also use this simple addition to make other choices, such as substituting a smaller order of French fries for the enormous order usually included or choosing a small soda or water in place of a large drink. Because most customers order habitually, few actually know the value of what they are receiving in their value meals, and many could save money by buying *à la carte* (piece by piece).

Deciding whether to fly or to drive is often based on cost, such as when a family of six elects to drive to their vacation destination rather than purchasing six airline tickets. In other cases, such as when a couple in Los Angeles visits relatives in Connecticut over spring break, the choice is motivated by sheer distance. But in some situations, the question is less clear, and some simple addition may reveal that the seemingly obvious choice is not actually superior. Consider a student living in rural Oklahoma who wishes to visit his family in St. Louis. This student knows from experience that driving home will take him eight hours, so he is enthusiastic about cutting that time significantly by flying. But as he begins adding up the individual parts of the travel equation, he realizes the difference is not as large as he initially thought. The actual flight time from Tulsa to St. Louis is just over one hour, but the only flight with seats available stops in Kansas City, where he will have to layover for two hours, making his total trip time from Tulsa to St. Louis more than three hours. Added to this travel time is the one hour trip from his home to the Tulsa airport, the one hour early he is required to check in, the half hour he will spend in St. Louis collecting his baggage and walking to the car, and the hour he will spend driving in St. Louis traffic to his family's home. Assuming no weather delays occur and all his flight arrive on time, the student can expect to spend close to seven hours on his trip, a net savings of one hour over his expected driving time. Simple addition can help this student decide whether the price of the plane ticket is worth the one hour of time saved.

In the still-developing world of online commerce, many web pages use an ancient method of gauging popularity: counting attendance. At the bottom of many web pages is a web counter, sometimes informing the visitor, "You are guest number . . .". While computer gurus still hotly debate the accuracy of such counts, they are a common feature on websites, providing a simple assessment of how many guests visited a particular site.

In some cases, simple addition is used to make a political point. Because the United States government finances much of its operations using borrowed money, concerns are frequently raised about the rapidly rising level of the national debt. In 1989, New York businessman Seymour Durst decided to draw attention to the spiraling level of public debt by erecting a National Debt Clock one block from Times Square. This illuminated billboard provided a continuously updated total of the national debt, as well as a sub-heading detailing each family's individual share of the total. During most of the clock's lifetime, the national debt climbed so quickly that the last digits on the counter were simply a blur. The clock ran continuously from 1989 until the year 2000, when federal budget surpluses began to reduce the $5.5 trillion debt, and the clock was turned off. But two years later, with federal borrowing on the rise once again, Durst's son restarted the clock, which displayed a national debt of over $6 trillion. By early 2005, the national debt was approaching $8 trillion.

POKER, PROBABILITY, AND OTHER USES OF ADDITION

While predicting the future remains difficult even for professionals such as economists and meteorologists, addition provides a method to make educated guesses about which events are more or less likely to occur. Probability is the process of determining how likely an event is to transpire, given the total number of possible outcomes. A simple illustration involves the roll of a single die; the probability of rolling the value three is found by adding up all the possible outcomes, which in this case would be 1, 2, 3, 4, 5, or 6 for a total of six possible outcomes. By adding up all the possibilities, we are able to determine that the chance of rolling a three is one chance in six, meaning that over many rolls of the die, the value three would come up about 1/6 of the time. While this type of calculation is hardly useful for a process with only six possible outcomes, more complex systems lend themselves well to probabilistic analysis. Poker is a card game with an almost infinite number of variations in rules and procedures. But whichever set of rules is in play, the basic objective is simple: to take and discard cards such that a superior hand is created. Probability theory provides several insights into how poker strategy can be applied.

Consider a poker player who has three Jacks and is still to be dealt her final card. What chance does she have of receiving the last Jack? Probability theory will first add up the total number of cards still in the dealer's stack, which for this example is 40. Assuming the final Jack has not been dealt to another player and is actually in the stack, her chance of being dealt the card she wants is 1 in 40. Other situations require more complex calculations, but are based on the same process. For example, a player with two pair might wonder what his chance is of drawing a card to match either pair, producing a hand known as a full house. Since a card matching either pair would produce the full house, and since there are four cards in the stack which would produce this outcome, the odds of drawing one of the needed cards is now better than in the previous example. Once again assuming that 40 cards remain in the dealer's stack and that the four possible cards are all still available to be dealt, the odds now improve to 4 in 40, or 1 in 10. Experienced poker players have a solid grasp of the likelihood of completing any given hand, allowing them to wager accordingly.

Probability theory is frequently used to answer questions regarding death, specifically how likely one is to die due to a specific cause. Numerous studies have examined how and why humans die, with sometimes surprising findings. One study, published by the National Safety Council, compiled data collected by the National Center for Health Statistics and the U.S. Census Bureau to predict how likely an American is to die from one of several specific causes including accidents or injury. These statistics from 2001 offer some insight into how Americans sometimes die, as well as some reassurance regarding unlikely methods of meeting one's end.

Not surprisingly, many people die each year in transportation-related accidents, but some methods of transportation are much safer than others. For example, the lifetime odds of dying in an automobile accident are 1 in 247, while the odds of dying in a bus are far lower, around 1 in 99,000. In comparison, other types of accidents are actually far less likely; for instance, the odds of being killed in a fireworks-related accident are only 1 in 615,000, and the odds of dying due to dog bites is 1 in 147,000. Some types of accidents seem unlikely, but are actually far more probable than these. For example, more than 300 people die each year by drowning in the bathtub, making the lifetimes odds of this seemingly unlikely demise a surprising 1 in 11,000. Yet the odds of choking to death on something other than food are higher by a factor of ten, at 1 in 1,200, and about the same as the odds of dying in a structural fire (1 in 1,400) or being poisoned (1 in 1,300). Unfortunately, these odds are roughly equivalent to the lifetime chance of dying due to medical or surgical errors or complications, which is calculated at 1 in 1,200.

Geometric Progression

An ancient story illustrates the power of a geometric progression. This story has been retold in numerous versions and as taking place in many different locales, but the general plot is always the same. A king wishes to reward a man, and the man asks for a seemingly insignificant sum: taking a standard chessboard, he asks the king to give him one grain of rice on day one, two grains of rice on day two, and so on for 64 days. The king hastily agrees, not realizing that in order to provide the amount of rice required he will eventually bankrupt himself.

How much rice did the king's reward require? Assuming he could actually reach the final square of the board, he would be required to provide 9,223,372,036,854,775,808 grains of rice, which by one calculation could be grown only by planting the entire surface of the planet Earth with rice four times over. However it is doubtful the king would have moved far past the middle section of the chessboard before realizing the folly of his generosity. The legend does not record whether the king was impressed or angered by this demonstration of mathematical wisdom.

USING ADDITION TO PREDICT AND ENTERTAIN

Addition can be used to predict future events and outcomes, though in many cases the results are less accurate than one might hope. For example, many children wonder how tall they will eventually become. Although numerous factors such as nutrition and environment impact a person's adult height, a reasonable prediction is that a boy will grow to a height similar to that of his father, while a girl will approach the height of her mother. One formula which is sometimes used to predict adult height consists of the following: for men, add the father's height, the mother's height, and 5, then divide the sum by 2. For women, the formula is (father's height + mother's height − 5) / 2. In most cases, this formula will give the expected adult height within a few inches.

One peculiar application of addition involves taking a value and adding that value to itself, then repeating this operation with the result, and so forth. This process, which doubles the total at each step, is called a geometric progression, and beginning with a value of 1 would

appear as 1, 2, 4, 8, 16, 32 and so forth. Geometric progressions are unusual in that they increase very slowly at first, then more rapidly until in many cases, the system involved simply collapses under the weight of the total.

One peculiarity of a geometric progression is that at any point in the sequence, the most recent value is greater than the sum of all previous values; in the case of the simple progression 1, 2, 4, 8, 16, 32, 64, addition demonstrates that all the values through 32, when added, total only 63, a pattern which continues throughout the series. One seemingly useful application of this principle involves gambling games such as roulette. According to legend, an eighteenth century gambler devised a system for casino play which used a geometric progression. Recognizing that he could theoretically cover all his previous losses in a single play by doubling his next bet, he bragged widely to his friends about his method before setting out to fleece a casino. The gambler's system, known today as the Martingale, was theoretically perfect, assuming that he had adequate funds to continue doubling his bets indefinitely. But because the amount required to stay in the game climbs so rapidly, the gambler quickly found himself out of funds and deep in debt. While the story ends badly, the system is mathematically workable, assuming a gambler has enough resources to continue doubling his wagers. To prevent this, casinos today enforce table limits, which restrict the maximum amount of a bet at any given table.

Addition also allows one to interpret the cryptic-looking string of characters often seen at the end of series of motion picture credits, typically something like "Copyright MCMXXLI." While the modern Western numbering system is based on Arabic numerals (0–9), the Roman system used a completely different set of characters, as well as a different form of notation which requires addition in order to decipher a value. Roman numerals are written using only seven characters, listed here with their corresponding Arabic values: M (1,000), D (500), C (100), L (50), X (10), V (5), and I (1). Each of these values can be written alone or in combination, according to a set of specific rules. First, as long as characters are placed in descending order, they are simply added to find the total; examples include VI (5 + 1 = 6), MCL (1,000 + 100 + 50 = 1,150), and LIII (50 + 1 + 1 + 1 = 53). Second, no more than two of any symbol may appear consecutively, so values such as XXXX and MCCCCV would be incorrectly written.

Because these two rules are unable to produce certain values (such as 4 and 900), a third rule exists to handle these values: any symbol placed out of order in the descending sequence is not added, but is instead subtracted from the following value. In this way, the proper sequence for 4 may be written as IV (1 subtracted from 5), and the

An Iraqi election officer checks ballot boxes at a counting center in Amman, Jordan, 2005. Counting ballots was accomplished by adding ballots one at a time, by hand. AP/WIDE WORLD PHOTOS. REPRODUCED BY PERMISSION.

Roman numeral for 900 is written CM (100 subtracted from 1,000). While this process works well for shorter numbers, it becomes tedious for longer values such as 1997, which is written MCMXCVII (1,000 − 100 + 1,000 − 10 + 100 + 5 + 1 + 1). Adding and multiplying Roman numerals can also become difficult, and most ancient Romans were skilled at using an abacus for this purpose. Other limitations of the system include its lack of notation for fractions and its inability to represent values larger than 1,000,000, which was signified by an M with a horizontal bar over the top. For these and other reasons, Roman numerals are used today largely for ornamental purposes, such as on decorative clocks and diplomas.

Potential Applications

While addition as a process remains unchanged from the method used by the ancient Chinese, the mathematical tools and applications related to it continue to evolve. In particular, the exponential growth of computing power will continue to radically alter a variety of processes. Gordon Moore, a pioneer in microprocessor design, is credited with the observation that the number of transistors on a processor generally doubles every two years; in practice, this advance means that computer processing power also doubles. Because this trend follows the principle of the geometric progression, with its doubling of size at each step, expanding computer power will create unexpected changes in many fields. As an example, encryption schemes, which may use a key consisting of 100 or more digits to encode and protect data, could potentially become easily decipherable as computer power increases. The rapid growth of computing power also holds the potential to produce currently unimaginable applications in the relatively near future. If the consistent geometric progression of Moore's law holds true computers one decade in the future will be fully 32 times as powerful as today's fastest machines.

Where to Learn More

Books

Orkin, Mike. *What are the Odds? Chance in Everyday Life.* New York: W.H. Freedman and Company, 2000.

Seiter, Charles. *Everyday Math for Dummies.* Indianapolis: Wiley Publishing, 1995.

Walker, Roger S. *Understanding Computer Science.* Indianapolis: Howare W. Sams & Co., 1984.

Periodicals

Lin, B-H., E. Frazao, and J. Guthrie. "Away-From-Home Foods Increasingly Important to Quality of American Diet," *Agricultural Information Bulletin, U.S. Department of Agriculture and U.S. Department of Health and Human Services.* (1999).

Shaw, Mary, Richard Mitchell, and Danny Dorling. "Time for a smoke? One cigarette reduces your life by 11 minutes." *British Medical Journal.* (2000): 320 (53).

Web sites

Aetna Intellihealth. "Can We Predict Height?" February 11, 2003. <http://www.intelihealth.com/IH/ihtIH/WSIHW000/353 20/35323/360788.html?d=dmtHMSContent#bottom> (March 15, 2005).

Arabic 2000. "The Arabic Alphabet." <http://www.arabic2000 .com/arabic/alphabet.html> (March 15, 2005).

Brillig.com. "U.S. National Debt Clock." <http://www.brillig. com/debtclock> (March 15, 2005).

ComputerWorld. *"Inside a Microprocessor."* <http://www .computerworld.com/hardwaretopics/hardware/story/0,10 801,64489,00.html> (March 14, 2005).

DECA: The Decathlon Association. "The Decathlon Rules." <http://www.decathlonusa.org/rules.html> (March 15, 2005).

Gambling Gates. "The truth behind the limits: What maximum and minimum bets are about." <http://www.gambling gates.com/Tips/maximum_bets8448.html> (March 15, 2005).

Intel Research. "Moore's Law." <http://www.intel.com/research/ silicon/mooreslaw.htm> (March 15, 2005).

Mathematics Magazine. "Chisenbop Tutorial." <http:// www. mathematicsmagazine.com/5-2003/Chisenbop_ 5_2003.htm> (March 13, 2005).

National Safety Council. "What are the odds of dying?" <http://www.nsc.org/lrs/statinfo/odds.htm> (March 15, 2005).

Nutristrategy. "Calories Burned During Exercise." <http://www .nutristrategy.com/activitylist.htm> (March 15, 2005).

Sigma Educational Supply. "History of the Abacus." <http:// www.citivu.com/usa/sigmaed/> (March 14, 2005).

Tallahassee Democrat (AP). "National 'Debt Clock' Restarted." July 11, 2002. <http://www.tallahassee.com/mld/tallahassee/ business/3643411.htm> (March 15, 2005).

The Great Idea Finder. "Fascinating Facts About the Invention of the Abacus by Chinese in 3000 BC." Inventions. <http://www.ideafinder.com/history/inventions/abacus .htm> (March 14, 2005).

The Math Forum at Drexel. "How Are Roman Numerals Used today?" (March 15, 2005).

University of Maryland Physics Department. "A2-61: Exponential Increase - Chessboard and Rice." <http://www.physics .umd.edu/lecdem/services/demos/demosa2/a2-61.htm> (March 15, 2005).

Overview

Algebra is the study of mathematical procedures that combine basic arithmetic with a wide range of symbols in order to express quantitative concepts. Arithmetic refers to the study of the basic mathematical operations performed on numbers, including addition, subtraction, multiplication, and division, and is widely viewed as a separate field of mathematics because it must be taught to students before they can progress to higher studies; but arithmetic is basically algebra without the symbols and advanced operations. In this sense, algebra is often referred to as a generalization of arithmetic, which can be applied to more sophisticated ideas than numbers alone. From adding up the price of groceries and balancing a checkbook, to preparing medicines or launching humans into space, algebra enables almost any idea to be written in standard mathematical notation that can be utilized by people around the world. No matter how advanced the mathematics involved, algebraic rules and notations provide the instructions that dictate how to handle the various combinations of numbers and symbols.

Algebra

Fundamental Mathematical Concepts and Terms

Algebraic symbols can be classified into symbols for representing quantities (usually numbers and letters); symbols for representing operations (such as addition, subtraction, multiplication, division, exponents, and roots); symbols representing equality and inequality (equal to, approximately equal to, less than, greater than, less than or equal to, greater than or equal to, and not equal to); and symbols for separating and organizing terms, and determining the order of operations (typically parentheses and brackets).

Multiplication in an algebraic expression is often represented by a dot when written out by hand (e.g., $4 \cdot 5 = 20$), or an asterisk when using a computer or graphing calculator (e.g., $4*5 = 20$). Adjacent sets of parentheses also signify multiplication, as in $(4)(5) = 20$. A number or variable attached to the outside of a set of parentheses also signifies multiplication. That is, $60 \times t = (60)(t) = (60)t = 60(t)$. These notations are used instead of 4×5 and $60 \times t$ in order to avoid confusion between the multiplication sign and the commonly used variable x. The symbol for multiplication is often omitted from an equation altogether: aside from the notation, $60t$ is identical to $60 \times t$. When two numbers are multiplied together, there must always be some sort of symbol to

Mathematician Dr. Tasha Inniss corrects a factorization shown on blackboard. Can you spot the error? AP/WIDE WORLD PHOTOS. REPRODUCED BY PERMISSION.

indicate the multiplication in order to avoid confusion. For example, $(2)(3) = 2 \times 3 \neq 23$.

Repeated multiplication can be simplified using exponential notation. If the letter n is used to represent a generic number, then $n \times n = n^2$ (n squared), $n \times n \times n = n^3$ (n cubed or n to the 3rd power), and so on. For example, if $n = 3$, then $n^3 = 3^3 = 3 \times 3 \times 3 = 27$. In general, the value of n multiplied by itself y times can be expressed as n^y, read n to the power of y.

Performing operations in the proper order is essential to finding the correct solution to an equation. In general, the proper order of operations is as follows:

1. Using the following guidelines, always perform operations moving from left to right;
2. Perform operations within parentheses or brackets first;
3. Next, evaluate exponents;
4. Then perform multiplication and division operations;
5. Finally, perform addition and subtraction operations.

In algebraic equations, numbers are typically referred to as constants because their values do not change. Letters are most often used as variables, which represent either unknown values or placeholders that can be replaced with any value from a range of numbers. For example, if a car is traveling at a speed of 60 miles per hour, then the distance that the car has traveled can be represented as $d = 60t$, where d represents the distance in miles that the car has traveled and t represents the number of hours that the car has been moving. The variable t can be replaced with any nonnegative value (zero and the positive numbers); as time progresses, t increases, and as would be expected, the distance d increases.

An expression that involves variables, numbers, and operations is called a variable expression, or algebraic expression. For example, $x^2 + 3x$ is a variable expression. An equation, like $x^2 + 3x = 18$, is created when a variable expression is set equal to a number, variable, or another variable expression. An algebraic inequality is expressed when a variable expression is separated from a number, variable, or another variable expression by a greater than sign, less than sign, greater than or equal to sign, or less than or equal to sign. Inequalities can be used to determine upper or lower bounds for a possible range of values. For example, the idea that it takes less than 15 minutes to boil an egg can be expressed as $t < 15$, where t represents time measured in minutes.

The parts of an equation that are separated by the symbols of addition, subtraction, and equality (or inequality) are called the terms of the equation. In the equation $4x^2 + 3x = 76$, the three terms are $4x^2$, $3x$, and 76. The symbols of positive and negative can also be taken into account in the terms of the equation, so that the terms are only separated by the symbols of addition and equality. In the equation $8x^2 - 3x = 26$, the terms are $8x^2$, $-3x$, and 26, because $8x^2 - 3x$ can be written as $8x^2 + (-3x)$. In general, subtraction can be thought of as addition of a negative term.

When a constant and a variable are multiplied, the constant is called the coefficient of the term. In the variable expression $8x^2 - 3x$, the coefficient of the first term is 8 and the coefficient of the second term is -3.

A special type of equation or inequality in which there are an infinite number of solutions is known as an algebraic formula. Formulas are useful for performing repeated mathematical tasks. The previous equation for determining the distance that a car has traveled if traveling at 60 miles per hour for a given amount of time, $d = 60t$, is a formula because for every value of t, there is a new value for d. If the value for d is known, the value of t can be determined, and vice versa. This formula can be generalized to allow for different speeds as well. In the formula $d = st$, any speed can be substituted for the variable s. An equation like $2x^2 + x = 10$ is not a formula because only a finite number of values of x satisfy the equation.

Equations in which the highest power of any term is one are called linear equations (recall that $x^1 = x$). The equation $d = 60t$, for instance, is linear. Nonlinear equations are those that involve at least one term raised to a power greater than one. Equations in which the highest power of any term is two are referred to as quadratic equations. The equation $5x^2 + 3x = 2$ is an example of a quadratic equation. In general, a quadratic equation can be simplified into the form $ax^2 + bx + c = 0$, where a, b, and c are the coefficients of the terms. There are various methods for solving quadratic equations. One of the most common methods is the known as the quadratic formula, which states that

$$x = \frac{-b \pm \sqrt{b^2 - 4ac}}{2a}$$

For example, the equation $5x^2 + 3x = -2$ can be rewritten as $5x^2 + 3x + 2 = 0$ (by adding 2 to both sides of the equation); so the values of the coefficients are $a = 5$, $b = 3$, and $c = 2$. Substituting these values into the quadratic formula reveals the values of x that satisfy the equation:

$$x = \frac{-3 \pm \sqrt{3^2 - 4(5)(-2)}}{2(5)} = \frac{-3 \pm \sqrt{9 + 40}}{10} = \frac{-3 \pm 7}{10}$$

Therefore, the values of x that satisfy this equation are $-\frac{2}{5}$ and -1. These values can be substituted for x to verify that they satisfy the equation.

Equations in which the highest power of any term is three are called cubic equations. Equations involving higher powers are usually referred to as 4th-order equations, 5th-order equations, and so on.

The various methods and rules for simplifying the terms of an algebraic equation constitute the most important tools for working with any mathematical construction. For example, rules like the associative, commutative, and distributive properties dictate how terms can be added and multiplied to simplify and solve algebraic expressions.

Combining like terms is a useful method of simplification. To illustrate this method, consider the task of counting the number of boys and girls in a gymnasium. One way to simplify this problem is to ask all of the boys to move to one side of the room, and all the girls to move to the other side. Similar reasoning is used to simplify a messy algebraic equation like $3x^2 - 7x - x^2 + 9x + x^2 - 4x - 4x^2 + 2x^2 - 2 = 6$. Terms involving the same variable raised to the same power are called like terms and can be added and subtracted just like numbers. This equation involves three powers of the variable x, so by collecting like terms it can be simplified to an equation with only three terms. First, by grouping the like terms together, the equation becomes $3x^2 - x^2 + x^2 - 4x^2 + 2x^2 - 7x + 9x - 4x = 6 + 2$ (note that 2 was added to each side of the equation in order to group the constants on the right side). Next, by adding and subtracting like terms, the equation becomes less of any eyesore: $x^2 - 2x = 8$.

Factoring allows seemingly difficult equations to be expressed in different ways that can immediately reveal a solution. For example, finding the values of x that satisfy the equation $x^2 - 2x = 8$ may at first seem intimidating; but this equation can be rewritten as $(x - 4)(x + 2) = 0$, which reveals that $x = 4$ and $x = -2$ both satisfy the equation (if either of these values is substituted for x, then one of the two parenthetical expressions is equal to zero, so when it is multiplied with the other expression, the entire left side is equal to zero).

The endless rules and tricks for evaluating algebraic expressions permeate mathematics and science at all levels. Whether noticed not, the fundamental concepts of algebra can be found in daily activities, and can be used to explain many concepts in the universe.

A Brief History of Discovery and Development

The symbols and syntax of algebra have developed slowly over thousands of years to become what is now recognized as the fundamental language of mathematics. In ancient times, mathematical problems were often written out in the verbal language of the time. As similar problems repeatedly arose, people began to invent abbreviations, and eventually symbols, for common terms in the problems. As mathematical concepts progressed and great mathematicians continued to make breakthroughs based on the findings of earlier mathematicians, less words were used to describe problems and the language of mathematics was continuously refined and adapted to the pressing problems of the various times and civilizations. Eventually, almost any problem or arithmetical fact that humans found could be expressed using the mathematical symbols of algebra.

In general, algebra refers to mathematical operations involving unknown values represented by some sort of symbols, where other symbols are used as shorthand for commands. In this sense, algebra was studied extensively in ancient Egypt, possibly as early as 2000 B.C. However, algebra in the form that is recognized today (even the word algebra) would not be discovered for thousands of years after the Egyptians began using these concepts. In ancient Egypt, and in many civilizations prior to the development of the current conception of algebra, algebraic equations were not generalized as ideas that could be applied to other types of problems. Individual practical problems of the time were studied and documented. Once a problem was solved, the writings could be used to solve a similar problem using different values for the unknowns; but mathematical language was seldom shared between different types of problems. For example, the method for finding the optimal amount of fertilizer to place on a crop was not seen as related mathematically to the method for figuring how much grain would fit in a storage structure, even though both procedures involve the operations now known as addition and multiplication.

The Greek mathematician Diophantus made great progress in generalizing algebraic symbolism in his writings.

Little is known of Diophantus' life, but it is commonly held that his most important works took place about A.D. 250. He discovered a general method for solving equations and finding values of unknowns. Diophantus is attributed as the first mathematician to use abbreviations for unknowns (variables) and powers of unknowns, and abbreviating words such as the Greek word meaning "is equal to". The use of these abbreviations was a major step toward the sophisticated algebraic symbolism (e.g., using letters to represent variables) found in modern mathematics. However, Diophantus did not use notation that could represent two or more unknowns and resorted to using words to describe multiple unknowns.

Like the earlier Egyptians, Diophantus viewed his mathematical ideas not as theories in the workings of numbers, but as a means for solving common problems of his day. His main work, *Arithmetica*, is a collection of pertinent problems described using numerical solutions of mathematical structures, the predecessors of algebraic equations. Diophantus' original works did not compensate for abstract ideas such as negative numbers. The idea of a negative number, or an equation like $x + 20 = 2$, was not explored because the idea of a negative quantity, a negative stone or book for example, was not needed in his society. Nonetheless, an essential branch of algebraic analysis that deals with solving certain types of rational equations (equations that allow for fractions and roots in addition to whole numbers) has been named Diophantine analysis (or analysis of Diophantine equations) in celebration of his work.

An Arab mathematician named Abu Abdullah Muhammad ibn Musa al-Khwarizmi contributed greatly to the language and concepts of algebra. Like Diophantus, little is known about the life of Khawarizmi, but it is commonly accepted that most of his important works took place around A.D. 820. In addition to his resounding developments in various fields of mathematics, he also contributed greatly to astronomy, geography, the inner workings of clocks, and the degree measurements of angles. Khawarizmi's writings on arithmetic and algebra have had resounding effects on the fundamental ideas of modern mathematics.

Based on knowledge documented by Greek mathematicians and the innovative notation for numbers and mathematical operations proposed by Hindu contemporaries in India, Khawarizmi developed the basis for modern arithmetical notation. His writings introduced and developed several fundamental arithmetic procedures, including operations performed on fractions. He was the first to spread the decimal number system (now commonly referred to as Arabic numerals) and the idea of the number zero outside of India, introducing it directly to Arabs, and later to Europe when his writings were translated to Latin and other European languages. Khawarizmi's original book on arithmetic was lost, leaving only translations.

Another of Khawarizmi's books, *Kitab al-Jabr w'al-Muqabala*, sparked the analysis of algebra as a well-organized form of mathematics. The title of the book has been interpreted in various ways, including "Rules of Reintegration and Reduction" and "The book of summary concerning calculating by transposition and reduction." The word algebra is derived from the term al-jabr in the title of the book, which can be taken to mean "reunion of broken parts," "reduction," "connection," or "completion." The rest of the title loosely translates to "to set equal to" or "to balance." The title of the book relates to the fundamental procedures involved in solving algebraic problems, such as shifting terms from one side of an equation to the other and combining like terms. The methods described in Khawarizmi's books have been built upon ever since; and the word algebra evolved for centuries before it was spelled and used as it is today.

At the beginning of the thirteenth century, Leonardo da Pisa (also known as Leonardo Fibonacci), an Italian mathematician, traveler, and tradesman, discovered that the potential of algebraic computations using the Hindu (Arabic) notation for numbers far exceeded the capacities of the Roman numeral system that was standard in Europe at that time. In his writings on algebra, he discussed the superiority of the symbols and concepts borne in distant lands. His writings included little original discoveries and were intended to illuminate pertinent ideas and problems found in his culture at that time. Unfortunately, his proposals were generally viewed as nothing more than interesting, and the ideas that he attempted to spread would not catch on in Europe for almost 300 years.

In the late fifteenth century, an Italian named Lucas Paciolus (Lucas de Burgo) authored multiple works on arithmetic, geometry, and algebra. Though most of the mathematical elements are taken from earlier writings, his algebraic writings were integral in the development of algebraic methods because of his efficient use of symbols. Due to the invention of the printing press earlier in the century, Paciolus' writings were among the first widely distributed algebraic texts, at long last effectively introducing the benefits of algebraic reasoning and Arabic numerals.

In the sixteenth century, algebra began to be used in a purely mathematical sense, with symbols and numbers completely representing general quantitative ideas. Robert Recorde—an English mathematician and originator of the symbol = for representing equality—is attributed with the first use of the term algebra in a strictly mathematical sense. François Viete made much progress in the use of symbols for representing generic numbers, which enabled mathematic ideas to be represented in a more general manner and ultimately led to the view of algebra as generalized arithmetic.

The recognition and understanding of negative values, irrational numbers, and negative roots of quadratic equations were crucial developments in the progression of algebraic theories because they opened doors to more advanced mathematical concepts. The discovery of negative numbers is often attributed to Albert Girard in the early seventeenth century. Unfortunately for Girard, the work of René Descartes—another great mathematician of the time—overshadowed his findings.

In the field of algebra, the most notable accomplishment of Descartes was the discovery of relationships between geometric measurements and algebraic methods, now referred to as analytic geometry (geometry analyzed using algebra). Using this analytic method of describing measurements such as lengths and angles, Descartes showed that algebraic manipulations (e.g., addition, multiplication, extraction of roots, and representation of negative values) could be represented by investigating related geometric shapes. Descartes' fusion of algebra and geometry elucidated both mathematical fields.

In the more than two centuries following Descartes discoveries, mathematicians have continued to refine algebraic notation and analyze the properties of more sophisticated aspects of mathematics. Many algebraic advances enable mathematicians and scientists to investigate and understand real-world phenomena that were previously thought impossible or unnecessary to analyze. In the twenty-first century, it seems that there are as many types of algebra as there are problems to be solved, but all of them depend wholly on the concepts of basic algebra.

Real-life Applications

PERSONAL FINANCES

Many people use a checkbook registry or financial software to track their income and expenses in order to make sure that they are making enough money to pay their bills and accomplish their financial goals. The registry in a checkbook is basically a form that helps to perform the algebraic operations necessary to track expenses. A checkbook registry includes columns for describing transactions (including the dates on which they take place), and columns for the recording amount of each transaction. There is usually a column labeled "deposits" and another column labeled "debits" so that transactions that add money and transactions that subtract money can be kept separate for quick and easy analysis. For example, when birthday money or a paycheck is received, it is logged in the deposits column. Things like groceries, rent, utilities, and car payments are recorded in the debits column. A "balance" column is provided for calculating the amount of money present in the bank account after each transaction. The process of recording transactions and balances in a checkbook registry basically involves performing a large, highly descriptive, ongoing algebraic equation. When a transaction is recorded in the column for deposits, a positive term is appended to the equation. When a transaction is recorded in the debits column, a negative term is appended to the equation. The balance column represents the other side of the equation. As the terms are appended to the equation, the balance column may be updated immediately, or the various transactions can be recorded and the total can be found later; but either way, the balance is always the result of the addition and subtraction of the terms represented by the values in the debits and deposits columns.

Every April, millions of United States citizens must analyze their financial income from the previous year in order to determine how much income tax they are required pay to federal and state government offices. Government taxes help pay for many social benefits, such as healthcare and social security. The Internal Revenue Service (IRS) provides various forms with step by step instructions for performing algebraic operations to calculate subtotals, and ultimately the amount of money that must be sent to the government. The various items on a tax form include the amount of taxes that are withheld from each paycheck; the amount of money taken home from each check after estimated taxes are deducted; items of personal worth such as savings, investments, and major possessions; and work-related expenses such as company lunches, office supplies, and utility bills. Many people receive money back from the government because the items representing expenses and taxes throughout the year add up to more than the total taxes due for the year. Some people end up owing taxes at the end of the year. Other people, such as self-employed workers, may or may not have taxes withheld from each paycheck. These people generally use a different type of IRS form and need to save money throughout the year to pay their taxes come tax time. Whatever form is used, the various items are added, subtracted, multiplied and divided just like the terms of an algebraic equation. In essence, an IRS tax form is an expanded algebraic equation, with the terms

and operations written out as explicit, intuitive instructions. The variables are described with words and a blank line or box is provided for filling in the value of each variable.

COLLEGE FOOTBALL

Unlike other college sports, National Collegiate Athletic Association (NCAA) football does not hold a national tournament at the end of the season to determine which team is the year's national champion. Instead, a total of 25 bowl games are held throughout the country, pitting teams with winning records against each other. The Bowl Championship Series (BCS) consists of four of these bowl games: the Orange Bowl, the Fiesta Bowl, the Sugar Bowl, and the Rose Bowl. These four bowl games feature eight of the highest rated teams of the year, and each year a different bowl game is designated as the national championship game. An invitation to any BCS game guarantees that a school will receive a hefty sum of money at the end of the year. Winning a BCS game could bring in millions of dollars.

The mathematical formula used to figure out which teams make it to the BCS games (and which two teams will fight to be crowned the national champions) turns out to be a rather complicated application of statistical analysis; but algebra provides the backbone of the entire operation. Across the country every week, each team's BCS ratings are updated according to four major factors: Computer rankings, the difficulty of the team's schedule, opinion polls, and the team's total number of losses. Each of these four components yields a numerical value.

The computer rankings, for one, are determined by complex computer programs created by statisticians. Computer ranking programs crunch an enormous amount of statistical data, including numerical values representing a multitude of factors ranging from the score of the game, the number of turnovers, and each team's total yardage, to the location of the game and the effects of weather.

The difficulty of a team's schedule is also determined by algebraic equations with terms accounting for the difficulty of the team's own schedule and the difficulty of the schedules of the teams that they will play throughout the season.

There are two separate opinions polls: one involving national sports writers and broadcasters, and one involving a select group of football coaches. Each poll results in a numerical ranking for all of the teams. For each team, an average of a these two rankings determines their national opinion poll ranking.

A team's number of losses is the most straightforward factor. The number of losses is figured directly into the general mathematical model, and each loss throughout the season has a large effect on a team's overall ranking.

The four numerical values are added together to calculate the team's national ranking. The top two teams at the end of the regular season are invited to the national championship BCS game. However, the selection of the six teams that are invited to the other three BCS games is not as straight forward. These other six teams are selected from the top 12 teams across the nation (excluding the top two that are automatically invited to the championship game). How these 12 teams are narrowed to six depends mainly on which teams are expected to attract the most attention and, therefore, create the most profits for the hosting institution, the television and radio stations that broadcast the game, and the various sponsors. These financial considerations are also modeled using algebraic formulas.

UPC BARCODES

Universal Product Code (UPC) barcodes are attached to almost all items purchased from mass merchandisers, such as department stores and grocery stores. These barcodes were originally used in grocery stores to help track inventory and speed up transactions, but shortly thereafter, UPC barcodes were appearing on all types of retail products.

UPC barcodes have two components: the barcode consisting of vertical lines that can be read by special scanning devices, and a set of numbers that can be read by humans (see Figure 1). Each component represents the same 12-digit number in a different language. That is, the barcode is simply the number below it represented in the language that can be read by the barcode scanner. The language of barcode scanners is based on vertical lines of two different colors (usually black and white) and four different sizes (the skinniest lines, and lines that are two, three, and four times as thick).

The UPC numbers for all items throughout the world are created and maintained by a central group called the Uniform Code Council (UCC). The first six numbers of a product's UPC number identify the manufacturer. Any manufacturer that wants to use UPC barcodes must submit an application to the UCC and pay an annual fee. Every barcode found on products sold by the same manufacturer will start with the same six digits. The first digit of the manufacturer number (the first digit in the entire UPC number) organizes all manufacturers into different categories. For example, the UPC numbers for

Figure 1: UPC bar code. KELLY QUIN. REPRODUCED BY PERMISSION.

pharmaceutical items, such as medicines and soaps, begin with 3. Some numbers at the beginning of UPC numbers are reserved for special items like coupons and gift certificates.

The second set of five digits represents the product itself. This five-digit product code is unique on every different product sold by a manufacturer, even different sizes of the same product. Some larger manufacturers have secured choice manufacturer codes and product codes that contain consecutive zeros. In certain configurations, consecutive zeros can be left out so that the UPC barcode can be squeezed onto small products, such as packs of chewing gum. There are ways to determine the positions of missing zeros when less than 12 numbers appear; but regardless, the barcode represents all 12 digits so that a quick swipe in front of a scanner determines all of the necessary information.

In any store, the price of each item is stored in a separate computer, which is attached to all of the checkout registers and provides the price for each item scanned. The prices of items are not indicated on barcodes because different stores charge different prices and all stores need to be able to change prices quickly.

The final digit of a UPC number is called the check digit and is used to minimize mistakes made by barcode scanners. The final digit can be calculated from the preceding 11 digits using a standard set algebraic operations. Following is an explanation of the algebra involved in calculating the check digit of the UPC number 43938200039, which has a check digit of 9:

1. Starting with the first digit, add together all of the digits in every other position (skipping every other number): $4 + 9 + 8 + 0 + 0 + 9 = 30$. In a sense, this sum is a variable in the equation for calculating the check digit because it represents values that can be changed.

2. Then multiply that value by 3: 3*30 = 90. The number 3 is a constant value in the check digit equation, and can be thought of as the coefficient of the variable in the previous step. Together, this coefficient and the variable sum in the previous step form a term in the equation for calculating the check digit.

3. Next, add up all of the other digits in the UPC number (starting with the second digit and skipping every other digit): 3 + 3 + 2 + 0 + 3 = 11. This sum is another term of the equation used to calculate the check digit. This variable value is not multiplied by a constant, so there is no coefficient of this term.

4. Now add the values of these two terms together: 90 + 11 = 101.

5. Finally, determine the smallest number that, when added to the value found in the previous step, results in a multiple of ten. That is, find the smallest number that can be added to the number in the previous step such that the sum divided by ten does not yield a remainder. In this case, that number is 9: 101 + 9 = 110. Using the mathematical concept of remainders, this value can be represented in an algebraic equation as well.

6. Compare the number found in the previous step with the final digit in the UPC number. The fact that this number matches the check digit in the UPC number confirms that the previous 11 digits were read correctly.

The entire calculation of the check digit can be represented by a single equation. A barcode scanner performs these calculations almost instantaneously every time a barcode is scanned. The actual check digit is represented at the end of the barcode (just as it appears at the end of the UPC number that humans can read). If the check digit calculated using the first 11 digits does not match the actual check digit, the scanner communicates to the store clerk—usually by making a loud beep and displaying a message on the screen of the cash register—that an error has occurred and the item needs to be rescanned.

FLYING AN AIRPLANE

In order for an 870,000-pound (394,625-kg) 747 jumbo jet to fly thousands of miles, it must be built according to strict specifications to create and balance the forces needed to carry this huge collection of metal through the air.

Thrust is the force that an airplane creates by moving forward very fast and causing air to move quickly past its wings. Airplanes use powerful propellers, jet engines, or rockets to create enough thrust to drive the airplane forward. When an airplane moves through the air, it also creates drag, a force that acts in the opposite direction of thrust and slows the plane down. When a hand is sticking out of a moving car, it creates similar drag. The faster the car is moving, the more the passing air acts on the hand, causing it to move backward with respect to the movement of the car. An airplane must create enough thrust to counteract the drag forces. This is why large planes must be traveling at hundreds of miles per hour in order to get off the ground. After a plane takes off, the landing gear is retracted because, much like a hand sticking out of a car, the landing gear creates drag. In fact, the drag created by the landing gear of a large airplane would most likely rip the landing gear off, leaving the pilots and passengers in a terrible predicament.

There are two other important forces that act on an airplane in motion. Gravity is pulling the airplane toward Earth so the weight (mass) of the airplane is an important factor. In order to raise the weight of the airplane upward, the airplane must create another force, called lift. As an over-simplified explanation of these four forces: the airplane must create enough thrust to move the plane quickly forward and counteract drag; and the airplane must be designed in such a way that when the plane moves forward fast enough, sufficient lift is created in order to counteract the forces of gravity acting on the mass of the plane. The wings play the biggest role in creating lift. The details of how an airplane creates lift involve advanced concepts of physics (including the idea that air is a fluid and acts much like water); but the calculations involved in planning and executing the safe operation of any airplane involve an immense amount of algebra. In algebraic equations that model the lift that an airplane produces, for instance, variables represent the factors that affect lift, including the density of the air, the speed of the airplane, the shape and surface area of the wings, and the angle at which the wings meet the oncoming air.

In addition to the calculations that must be checked and rechecked to ensure that an airplane can create sufficient lift, each trip involves a variety of important algebraic formulas. For example, deciding how much fuel to load into an airplane for each trip involves factors including the desired distanced to be traveled, the weather, and the effect that the weight of the fuel has on take off and landing procedures. Obviously, enough fuel must be present in the airplane to keep the engines running for a longer amount of time than the airplane will be flying. But this amount can be affected by strong winds and changes in air pressure, which must be predicted and taken into account in the formulas used to decide on an amount of fuel. Surprisingly, the maximum weight of an airplane on take off is a higher value than the maximum weight of the airplane during the landing sequence. When flying a 747 jumbo jet, for example, the maximum weight that the airplane can manage to get off of the ground is about 870,000 pounds (394,625 kg). But the maximum weight of the aircraft that will enable a safe landing is about 630,000 pounds (285,793 kg). All loss of weight is due to the burning of fuel, and it is essential that enough fuel is used during the flight to bring the weight of the airplane down below the safe landing weight. Therefore, unless an airplane will be traveling the longest distance possible, the fuel tanks are rarely filled to their maximum capacity. In an emergency landing, the pilot must usually dump some of the fuel from the aircraft in order to lower the weight below the maximum safe landing weight. All of the factors that determine how much fuel to load into an airplane are calculated and rechecked using standard mathematical formulas that require a solid understanding of algebra.

SKYDIVING

In addition to a good deal of courage, the act of jumping out of an airplane involves a lot of algebra. In addition to the important calculations involved in any flight of an airplane, algebra is used by all skydivers to ensure that the plummet to Earth is as controlled as possible. For example, careful calculations are performed and rechecked in order to ensure that proper of size parachute is packed according to each diver's body weight.

Algebra is also needed to analyze the speed and acceleration of a diver. In turn, this analysis is critical to calculating the amount of time that a diver should wait to deploy the parachute after jumping out of the airplane. In a typical skydiving session, the pilot takes the airplane to an altitude of about 10,000 feet (3,048 m). After jumping, the average diver accelerates to a top speed (known as terminal velocity) of about 120 miles per hour (193 km/h). This gives a diver about 45 seconds of free fall (falling at terminal velocity), at which time the diver will be approximately 2,500 feet (762 m) above the ground. At this height, the diver must deploy a small parachute, called a drogue chute. The drogue chute is attached to the main parachute, and the main parachute is held in its container until the diver pulls a cord that allows the drogue chute to pull the main parachute out.

For the first jump, a diver is usually strapped to an instructor who makes sure that everything goes smoothly. This is known as a tandem jump, and requires different calculations to plan the jump safely. The small drogue chute is deployed almost immediately after exiting the airplane in order to slow the pair of divers for the duration of the free fall. If the drogue shoot were not open, the extra weight would cause the two divers to accelerate to a terminal velocity of up to 200 miles per hour (322 km/h), making tandem jumps inconsistent and unsafe. The main parachute remains in its container until the correct altitude is reached and one of the two divers pulls the release cord, allowing the drogue to open the main parachute. In a tandem jump, the main parachute must be much larger than the main parachute in a solo jump in order to stabilize the two bodies and slow them to a safe landing speed.

In another type of dive, called a high-altitude, low-open (HALO) jump, the diver jumps from an airplane traveling at a much higher altitude (often about 30,000 feet [9,144 m]) and does not open a parachute until reaching a significantly lower altitude than in a typical jump. In any jump higher than about 15,000 feet (4,572 m), divers must where oxygen masks because the air becomes too thin at higher altitudes. In a HALO jump, free fall can last for up to three minutes and the diver can reach speeds of over 200 miles per hour (322 km/h). This type of jump requires more training and preparation. Algebra provides the essential tools for performing the many calculations required to plan all of these different types of jumps.

Several algebraic formulas are used to analyze the effects that changes in the materials and shape of the main parachute have on a dive. Most parachutes are made of materials that allow no air to pass through them, making the parachute more effective for slowing the fall of the diver. However, if the parachute opens too quickly, it can slow the diver too quickly, possibly causing serious physical injury or damage to the parachute and other gear. To prevent the parachute from opening all at once, a mechanism is attached to the cords that hold the parachute to the diver. This mechanism slides slowly down the cords and controls the speed at which the cords can separate from each other, and in turn controlling how quickly the parachute opens after it is deployed. Algebraic formulas are essential for finding a safe speed at which the parachute should open, and for properly manufacturing the device that controls this speed.

Many older parachutes (and some still made for special purposes) are round, a shape that allows the diver to fall straight down in the absence of wind. Standard contemporary parachutes are rectangular in shape. These rectangular parachutes cause the diver to move forward while falling. The main benefit of a rectangular parachute is that it allows for much more directional control while falling, enabling smoother and more accurate landings.

To ensure safe, reliable operation, the dimensions of a parachute must be as close to perfect as possible. Luckily the specifications of all parachutes are determined and tested according to in-depth mathematical models. While these models are rather advanced applications of mathematics and physics, they rely heavily on algebraic reasoning.

Most divers employ an automatic activation device (AAD), a small computation device that performs constant calculations in order to deploy a reserve parachute if something goes wrong. The AAD unit is turned on when the diver is on the ground, and from then on it constantly monitors the altitude of the diver. When the diver jumps out of the airplane, the AAD senses that it is falling quickly, and is programmed to recognize this as the beginning of the fall. If the diver falls past a certain altitude without deploying the main parachute, the AAD shoots a piece of metal into the cord that holds the reserve parachute in place, deploying it automatically. As long as the reserve parachute opens correctly, the AAD will most likely save the life of a diver that is distracted or has lost consciousness. To make things more complicated, the AAD must also be programmed to differentiate a loss in altitude during free fall from a loss in altitude due to other events, such as the plane landing before the diver ever jumps out. This ensures that the AAD will only activate the reserve parachute if the diver is free falling.

Every aspect of skydiving—from the altitude and timing to the lengths of all the cords and the computer assistance of the AAD—involves the addition, subtraction, multiplication, and division of terms and expressions that represent many factors. An enormous amount of algebraic formulas helps divers, instructors, pilots, and equipment manufacturers understand the multitude of factors that must be controlled in every skydiving session.

CRASH TESTS

Every year, hundreds of new model cars, trucks, vans, and sports utility vehicles (SUVs) are purposely involved in controlled crashes in order to analyze the safety features of each model of automobile. The various components of these crash tests involve a seemingly endless amount of calculations. The slightest miscalculation can result in the recall of an entire model (which costs the manufacturer a substantial amount of money), and much worse, injury or death of people involved in real crashes. Therefore, the calculations involved in crash tests are checked multiple times under various conditions.

The design of crash test dummies, the main focus of all crash tests, involves a great deal of algebraic calculations. To ensure consistent results, all official crash tests use the same type of crash test dummy, belonging to the Hybrid III family of dummies. Various Hybrid III dummies are used to simulate different ages and body types for both genders. For each dummy, characteristics including height and weight are measured and factored in to the mathematical formulas used to analyze the amount of damage done to the dummy during an accident.

Crash test dummies must possess rather complex structures in order to simulate all of the parts of a human body that are usually affected in car crashes. For example, an elaborate spine consisting of metal discs connected by rubber cushions is attached to sensors that collect data used to analyze the damage done to the simulated spine. Sensors for measuring how quickly different parts of the body speed up and slow down are present throughout the body of a dummy. These sensors collect data that help to analyze the potential injury sustained due to the sudden decrease in speed caused by a crash (e.g., whiplash). Other sensors are placed throughout the dummy to measure the amount of impact endured by body parts (e.g., how hard the dummy's arm slams into the dashboard). Different colors of paint are applied to a dummy's various body parts so that, when an impact is detected by these sensors during in a crash test, researchers can determine which parts of the body collided with which parts of the car or airbag. A single sensor in a dummy's chest measures how much the chest is compressed due to the forces applied by the seatbelt and airbag.

All of the information collected by these sensors is injected into mathematical formulas in order to test and improve the timing and power of the seatbelts and airbags. For example, modern seatbelts sense abrupt decreases in an automobile's speed, immediately lock up but allow for a small amount of movement forward, then quickly increase the tension to bring the body to a stop, and finally decrease the tension so that the seatbelt does not cause injuries. In this way, the body slows down more gradually than it would if strapped in by a constantly stiff seatbelt. If the seatbelt stopped the body from moving forward all at once, the seatbelt itself could cause substantial injuries. Most cars now include airbags to supplement seatbelts in absorbing the forward force of the body and keeping it from slamming into anything solid. In order to effectively supplement a seatbelt, an airbag must deploy with perfect timing immediately after the seatbelt begins to lock up.

Algebraic operations are integral to the mathematical models used to analyze the various factors in a car crash. The wide range of problems solved using the mathematical models found in a crash test include the realistic design of dummies and the analysis of data collected from their sensors; determining the effects of modifying the weight of the automobile and the load it carries; selection of the speed at which to hurl the automobile toward a concrete wall (frontal impact tests), or how fast to slam an object into the side of the automobile (side impact tests); analysis of the deliberate crunching of the materials that make up the automobile, which helps to absorb much of the impact; and calculation of the odds of surviving such a crash in real life. The rating systems used to indicate the effectiveness of an automobile's safety features also rely on algebraic formulas.

FUNDRAISING

In any fundraiser, the planners must be sure that enough money is brought in to cover the costs of the event and meet their fundraising goals. For example, to raise money for new equipment at a local hospital, the hospital's fundraising committee decides to sell raffle tickets for a new $30,000 car. The committee needs to raise at least $20,000 to be able to pay for the new equipment. To ensure that at least this amount of money is available after the paying for the car and the various components of the fundraiser, the committee decides to set up a mathematical model. To analyze the financial details of the event and decide the price of the raffle tickets and the minimum number of tickets that need to be sold, the committee prepares an algebraic formula. The formula will take into account the price of the car, the number of tickets sold, the price to be charged for each ticket, and the ultimate financial goal of raising $20,000. The formula they come up with is $G = TP - C - E$, where the variable G represents the amount of money to be raised, T is the number of tickets sold, P is the price of each ticket, C is the cost of the car, and E is the cost of the event. This equation states that the amount of money raised will be equal to the number of tickets sold multiplied by the price of each ticket, minus the cost, the car, and the event itself.

Because the purpose of this equation is to determine how many tickets to sell and at what price, the committee rewrites the equation with the term TP (representing the number of tickets multiplied by the price of each ticket) alone on the left side of the equation. By subtracting TP from both sides, the equation becomes $G - TP = -C - E$. Next, subtracting G from both sides gives $-TP = -G - C - E$. The term TP is now alone on the left side of the equation, but notice that all of the terms are negative. By multiplying both sides of the equation by -1, all of the negative terms become positive to yield $TP = G + C + E$. This equation now states that the amount of money that will be taken in from the sales of the raffle tickets is equal

to the financial goal of the event plus the cost of the car plus the cost of the event itself. This equation allows the committee to substitute the values for the financial goal and the costs of the car and the event in order to determine the number of tickets that need to be sold, and at what price. However, the committee does not necessarily want the money brought in from the ticket sales to be exactly equal to the costs and fundraising goals; the money brought in needs to be greater than or equal to the money spent. Thus the committee makes this equation into an algebraic inequality by replacing the equal sign with the greater than or equal to symbol to get $TP \geq G + C + E$.

Next, the committee begins to plug numbers into their algebraic inequality. The committee's financial goal for the event is to raise $20,000, so $G = 20,000$. The car costs $30,000, so $C = 30,000$. The costs of the eventflyers, food, musical entertainment, renting a venue, and so onare estimated at $5,000, so $E = 5,000$. By substituting these values into the equation $TP \geq G + C + E$, the committee finds that $TP \geq 20,000 + 30,000 + 5,000 = 55,000$. Since $TP \geq 55,000$, the committee knows that the sale of tickets must amount to at least $55,000.

At a similar fundraising event in the previous year, a little over 12,000 raffle tickets were sold. To be safe, the committee decides to predict an underestimate of 11,000 raffle tickets sold at this year's event. Substituting 11,000 for the variable T, the inequality becomes $11,000P \geq 55,000$. Dividing through by 11,000 gives $P \geq 5$; so the committee needs to charge at least $5.00 for each ticket in order to pay for the car and the event, and have enough left over to pay for the new equipment. Since 11,000 was an underestimate for the number of tickets sold, the committee decides that it is safe to charge $5.00 for the tickets.

BUILDING SKYSCRAPERS

A massive amount of calculations are involved in all phases of skyscraper construction, from determining the amounts of time, manpower, money, concrete, steel, wiring, pipes, and paint needed to build the skyscraper, to determining the number of exits, bathrooms, and electrical outlets needed to serve the maximum capacity of people in the building. The calculations used in the actual creation of the structure are dependent on basic algebra; but long before construction can begin, more sophisticated mathematical formulas are developed to design a building that meets strict safety guidelines.

A skyscraper towering high above a city is susceptible to many unpredictable forces and must be able to withstand a wide range of punishing forces. These forces include large changes in weight due to people coming and going and precipitation collecting on the outside of the

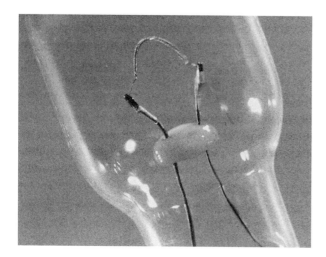

Figure 2: Light bulb filament. ROYALTY-FREE/CORBIS.

building, fluctuations in air pressure and wind, and seismic activity (earthquakes). A skyscraper must even be able to endure a sizeable fire or other direct damage to the structure of the building. For example, in 1945, a United States Army B-25 bomber, whose pilot had been disoriented by dense fog, crashed into the side of the Empire State Building, tearing gigantic holes in the walls and support beams, and igniting fires on five floors. However, the nearly 1,500-foot (457 m) tall skyscraper (the tallest in the world at the time) stood and the damage was repaired. If even small miscalculations had taken place in the planning of the Empire State Building, the crash might have caused the entire building to topple.

In-depth architectural specifications used to make a building visually pleasing and functionally efficient require countless algebraic systems. Formulas for modeling the various safety issues involved in constructing such a tall building take into account all of the factors that can compromise the structure of a skyscraper. The biggest problem to overcome when attempting to design a safe skyscraper is to make the structure stable enough to withstand wind and other forces. A skyscraper cannot be perfectly rigid. The structure must be allowed to sway slightly in all directions or its own weight would cause the structure to snap like a dry stick when acted on by forces like wind and earthquakes. On the other hand, if the skyscraper were allowed to sway too far from perfectly vertical, the building would fall over. Under normal conditions, the movement of a skyscraper is undetectable by the human eye, and unnoticed by occupants. The amount of flexibility in the structure must be controlled perfectly by the structure of each floor. Modeling the nature of a skyscraper's flexible components involves the use of the imaginary number, i, where $i^2 = -1$.

In the study of basic algebra, the value of i is not logical because multiplying any real number by itself results in a positive number, e.g., $2^2 = (-2)^2 = 4$. Multiples of i, such as $2i$ and $-3i$, are called imaginary numbers or complex numbers. An entire field of mathematics, known as complex analysis, is devoted to the study of the properties of imaginary numbers. Although imaginary numbers do not follow the rules of basic algebra, they are often used to simplify enormous, intricate polynomial equations—like those used to model the stability of skyscrapers—into more manageable equations. After an equation is solved using imaginary numbers, the solution can often be transformed back into real numbers. The use of imaginary numbers enables mathematicians and scientists to solve problems that would otherwise be unsolvable. For example, by assuming that i exists and using it in algebraic expressions, mathematicians, physicists, chemists, statisticians, and engineers are able to model and simplify complicated phenomena. In addition to modeling the slight swaying of a skyscraper, imaginary numbers can be used to model the behavior of electrical circuits, the springs that absorb shock in automobiles, and sophisticated economic systems.

BUYING LIGHT BULBS

Incandescent light bulbs produce light by passing electricity through a thin metal coil, called a filament (see Figure 2). When electricity is passed through the filament, it glows and illuminates the light bulb. The electricity also produces heat as it passes through the filament. In fact, special bulbs, called heat lamps, are intended to produce heat for purposes such as heating food and drying plants; but in most light bulbs, heat is an undesired (and unavoidable) side effect which eventually causes the filament to burn out. The amount of time that a light bulb can be turned on for before it is expected to burn out is printed on most packages so that shoppers can compare the life expectancy of the various available bulbs. It turns out that the amount of electricity that passes through the filament when the light bulb is turned on is all that is needed to predict how long a light bulb will last.

By logging and analyzing the results of various experiments to test the life of light bulbs under different conditions, the life expectancy of a light bulb has been found to be inversely proportional to the voltage that is applied to the filament. That is, the life expectancy is equal to some number divided by the number of volts raised to a power; when two values are inversely proportional, decreasing one value causes an increase in the other value. The life expectancy of most incandescent light bulbs is inversely proportional to the 12th power of the applied voltage and can be expressed as $L = a/V^{12}$, where L represents the life expectancy, a is a constant, and V represents the applied voltage. This expression indicates that using a lower value for the variable V results in the constant a being divided by a smaller number as long as V is >1, so that the life expectancy L is equal to a larger number. This means that lowering the voltage that is allowed to surge through the filament increases the life expectancy of the light bulb.

On the other hand, less light is produced if less electricity is allowed to pass through the filament. The amount of light that is produced is dependent on the voltage and can be expressed as $X = bV^{3.3}$ where X represents the amount of light produced by the light bulb, b is a constant, and V (which is raised to a power of 3.3) represents the voltage. In contrast to the relationship between voltage and life expectancy, it is said that the amount of light produced is directly proportional to the voltage, meaning that the amount of light is equal to some number multiplied by the voltage raised to a power. In this type of relationship, increasing the value of one variable also increases the value of the other variable. As can be deduced by examining these two equations, lowering the voltage has a much smaller effect on the amount of light produced than it does on the life expectancy of the light bulb. In other words, a small decrease in voltage increases the life expectancy by a relatively substantial amount, but decreases the amount of light produced only slightly. Therefore, buying light bulbs with lower voltage values usually increases the amount of time before light bulbs need to be replaced without resulting in a marked decrease in illumination.

ART

Though art is often seen as an pure expression of creativity, artists cannot help but use mathematics in the creation of any piece, whether the artist realizes it or not. The size and dimensions of a canvas and frame are carefully chosen by a painter; and though the choice is often dependent only on the artist's preference, the measurements of the rectangular canvas can be analyzed to determine which will be the most pleasing to the average person, or which will help to effect the emotions that the

Figure 3: Fractals in art. BILL ROSS/CORBIS.

artist hopes convey in the painting. The colors of the paints can also be examined to determine the best mixtures of the primary colors, or to analyze the emotions effected by different pigments and patterns.

In the age of computers, art has expanded its definition to encapsulate computer-generated art, including intricate and realistic images, and fractal images automatically created by computer programs using mathematical formulas (see Figure 3). Fractals are actually complex geometric shapes; but just as algebra can be used to define and analyze the characteristics of a circle or rectangle, algebraic sequences can be used to create and investigate fractal images. In a fractal, each part of the pattern has the same characteristics as every larger part. That is, when part of the fractal is magnified, it is generally undistinguishable from the original image. In many fractal images, a large part has small copies of itself sticking out of it, and these copies have small copies of themselves sticking out of them. For example, a triangle appears to have breaks in it, and smaller triangles fill the gaps so that the line is continuous. These smaller triangles have breaks in them, and even smaller copies of the triangle fill those gaps. Theoretically, this pattern can repeat infinitely so that no matter how many times the image is magnified, the same pattern will appear.

These infinite patterns are defined by special algebraic constructions, known as infinite series, which are determined by infinitely repeating numerical patterns. An infinite series repeats infinitely with, for example, each term raised to the power equal to the number of terms that precede the term in the series. Such an infinite series defines a fractal image similar to a snowflake, with the resulting image like a six-sided star, where each point can be thought of as a triangle missing one of its sides. The sides of each triangular point are cut at regular intervals and filled in with the next smaller size of triangles in the pattern. As the series continues, smaller and smaller triangles are added to the image. In theory, the series continues on forever and the image contains infinitely smaller and smaller triangles.

Though fractal images are frequently used to create beautiful artistic graphics, they have applications in other computer imaging projects as well. For example, computer generated maps use fractals to create realistic coastlines and mountain regions. No matter the use, fractal images are created by defining infinite series that involve algebra at every step.

POPULATION DYNAMICS

In any population—including bacteria, ants, fish, birds, and humans—many factors contribute to the number of individuals present at a given time and the rate at which the population increases or decreases. Some of the most common and important factors include the availability of food, the abundance of predators, and the inherent reproductive capacity and natural mortality rate of the species. In most investigations of population dynamics, researchers attempt to set up algebraic formulas using terms that represent all of the pertinent factors that determine the way that a population fluctuates. Basic population models are similar to $N = aZ + bY + cX + dW + eV + fU$, where N is the current number of individuals in the population. Each term on the other side of the equation represents a different factor in the life of the species. The variables Z, Y, X, W, V, and U represent different factors. These variables often take into account the number of individuals in the population in the immediate past (e.g., reproductive rates of are dependent on the number mature individuals). The constants a, b, c, d, e, and f are the coefficients of the terms and define the extent to which each represented factor affects the population. A large coefficient indicates that the term has a relatively significant effect, while a coefficient smaller than one indicates that the term has a relatively minimal effect. The coefficients for factors that decrease the number of living individuals (e.g., mortality rate and abundance of predators) have negative values, essentially subtracting individuals from the total. Positive coefficients are attached to terms representing factors such as reproductive rate and availability of food. This type of algebraic formula has helped to save many endangered species by facilitating important research of the affects that human developments have on wildlife populations around the world.

FINGERPRINT SCANNERS

Security is an essential consideration for many organizations, including police and military groups, financial institutions protecting money, and hospitals protecting sensitive medical records. All forms of security can be penetrated by an experienced attacker. Although many security devices attempt to give the appearance of being impenetrable, the true goal of a solid security system is to minimize the number of successful attacks by maximizing the time and energy required to circumvent the implemented security measures. The odds of an attacker successfully breaking a security system can be calculated using algebraic expressions that represent the various factors involved (e.g., the thickness of a safe door, or the odds of guessing a given password).

A form of identification steadily increasing in popularity, called biometrics, involves comparing an individual's unique physical characteristics with previously stored data about the individual. When someone first uses a biometric device, physical characteristics are measured, translated into mathematical formulas by a computing device, and stored for future comparison. The most widely developed biometric security devices include cornea and iris scanners that measure the characteristics of the parts of an individual's eyes; face scanners that can recognize major facial features; voice scanners that measure the frequencies in an individual's voice; and fingerprint scanners that read and interpret the unique curves and patterns of an individual's fingerprints.

Fingerprint scanners are widely accepted as one of the most effective forms of identification, and are becoming more common in all types of secure environments. The uses of fingerprint scanners range from the physical protection of a secured room to the protection of sensitive computer files. Many computer mice and keyboard manufacturers integrate fingerprint scanners into their products in hopes of replacing passwords as the most common form of identity verification for personal computers. An increasing number of automobile manufacturers have begun to incorporate fingerprint scanners into door locking mechanisms and ignition systems, so that the owner of a vehicle does not need a key to lock and unlock the car, or start the engine. Banking institutions are also beginning to look to fingerprint technology in hopes of replacing bank cards and personal identification numbers (PINs).

Like all biometric devices, fingerprint scanners map the unique characteristics of a fingerprint into mathematical formulas, which are used later to determine whether or not the fingerprint present on the scanner matches stored data. The size and relative location of the prominent features in each fingerprint are represented by the terms of mathematical formulas, so fingerprint scanners utilize massive amounts of algebraic operations during each security session.

Potential Applications

TELEPORTATION

Throughout history, humans have invented increasingly advanced methods of transportation—from the invention of the wheel to the first trip into space—in order to enable and expedite the process of traveling from one physical location to another. However, even with all of the advances in transportation, no vehicle can take passengers from one point to another without traveling across the space in between. Learning how to skip the intermediate locations is the goal of scientists who are attempting to invent and perfect teleportation devices. Similar to the

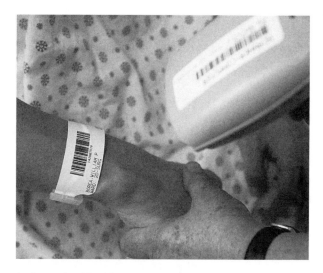

An bar-coded identification bracelet is scanned at Georgetown University Hospital in Washington, D.C. A/P WIDE WORLD. REPRODUCED BY PERMISSION.

fantastic idea first popularized in science-fiction, teleportation devices essentially collect information about an object, destroy the object, and send the information about the object to a different location, where the object is reconstructed. In a sense, this idea is similar to a fax machine, which translates a copy of a document into numerical information, and sends the information to another fax machine that uses the information to construct and print another copy. A teleportation device collects information about an object and translates it into numerical information according to mathematical formulas. Different formulas are then used to reconstruct the object elsewhere.

In 1998, a group of physicists performed the first successful teleportation experiment. In the experiment, information about a photon (a particle of energy that carries light) was collected, sent through a cable one meter in length, and used to construct a replica of the photon. When the replica was created, the original photon no longer existed, so the photon is considered to have traveled instantaneously to a different location.

It is uncertain whether or not this type of travel will ever be safe for living organisms. The idea of being destroyed and reconstructed elsewhere sounds rather foreign and frightening, and it will surely be difficult to find willing subjects for experiments. However, the ability to teleport energy will likely have profound effects on computer networks. Instead of using cables or airwaves to transfer information between computers, information will be instantaneously teleported from machine to machine, essentially eliminating delays in the transfer of information.

Figure 4: SpaceShipOne: the future of travel? JIM SUGAR/CORBIS.

PRIVATE SPACE TRAVEL

A new form of transportation will most likely revolutionize the way that humans think about space travel. SpaceShipOne (see Figure 4) is the first manned spacecraft project that does not depend on government funds. This privately owned and operated craft is intended to take anyone who can afford a ticket on a brief trip into space. In order to alleviate the most difficult part of any flight into space—the launch from the ground—SpaceShipOne is launched from a second aircraft, called White Night. While attached to White Night and during the launch into space, SpaceShipOne is in a contracted configuration. After the spacecraft launches and reaches its highest altitude, it spreads its wings in a configuration that slows its decent back into the inner atmosphere. Finally, the craft is reconfigured again to work much like an airplane, allowing the pilot to safely steer and land.

Because this project is not funded by the government, the company that designed and built SpaceShipOne and White Night had to create their innovative technology from scratch, a task that involves an unimaginable amount of calculations, all of which rely on the fundamental concepts of algebra. Developing a rocket propulsion system alone involves a multitude of mathematical formulas. Designing the three different configurations of SpaceShipOne also involves an enormous amount of mathematical models for determining the optimal size and shape of the various parts of the spacecraft. The idea of an average private citizen having the opportunity to travel into space on a regular basis is a shining example of the endless potential for using algebra to explore the real world.

Where to Learn More

Books

Johnson, Mildred. *How to Solve Word Problems in Algebra*, 2nd Ed. New York, NY: McGraw-Hill, 1999.

Ross, Debra Anne. *Master Math: Algebra*. Franklin Lakes, NJ: Career Press, 1996.

Periodicals

Backaitis, S.H., H.J. Mertz, "Hybrid III: The First Human-Like Crash Test Dummy." *Society of Automotive Engineers, International*. Vol. PT-44 (1994): 487–494.

Key Terms

Algebra: A collection of rules: rules for translating words into the symbolic notation of mathematics, rules for formulating mathematical statements using symbolic notation, and rules for rewriting mathematical statements in a manner that leaves their truth unchanged.

Arithmetic: The study of the basic mathematical operations performed on numbers.

Coefficient: A coefficient is any part of a term, except the whole, where term means an adding of an algebraic expression (taking addition to include subtraction as is usually done in algebra. Most commonly, however, the word coefficient refers to what is, strictly speaking, the numerical coefficient. Thus, the numerical coefficients of the expression $5xy^2 - 3x + 2y -$ are considered to be 5, -3, and $+2$. In many formulas, especially in statistics, certain numbers are considered coefficients, such as correlation coefficients.

Constant: A value that does not change.

Equation: A mathematical statement involving an equal sign.

Exponent: Also referred to as a power, a symbol written above and to the right of a quantity to indicate how many times the quantity is multiplied by itself.

Formula: A general fact, rule, or principle expressed using mathematical symbols.

Term: A number, variable, or product of numbers and variables, separated in an equation by the signs of addition and equality.

Variable: A symbol representing a quantity that may assume any value within a predefined range.

Web sites

How Stuff Works. "How Skyscrapers Work." Science Engineering Department. <http://science.howstuffworks.com/skyscraper.htm> (March 22, 2005).

NASA. "Beginner's Guide to Aerodynamics." Glenn Research Center. March 4, 2004. <http://www.grc.nasa.gov/WWW/K-12/airplane/bga.html> (May 27, 2005).

Algorithms

An algorithm is a set of instructions that indicate a method for accomplishing a task.

Algorithms—often used subconsciously—help us solve a wide range of problems in everyday life. Algorithms can be written to describe the method to tie a shoelace, bake a cake, or address an envelope. Algorithms are composed of the steps needed to accomplish a task and are written in such a way that no "judgment"—other than the fact that a particular step has been performed—is required to accomplish the overall task.

Fundamental Mathematical Concepts and Terms

In mathematics, an algorithm is a method for solving a mathematical problem by using a finite number of computations that repeat certain operations or steps. Not all algorithms lead to a single solution (deterministic algorithms), some algorithms can be designed that lead to multiple solutions (nondeterministic algorithms).

The length of time required to complete an algorithm is directly dependent on the number of steps involved and the speed with which the steps are completed. For example, a young child might take several minutes to add a long column of numbers—an algorithmic task performed by most computers in a fraction of a second.

A Brief History of Discovery and Development

The term algorithm is derived from the name of the ninth century Arabic mathematician and Tashlent cleric al-Khowarizmi, the mathematician most often credited with the early development of algebra.

With the rise of an industrial mechanized society, algorithms were developed to control a broad array of devices and procedures from traffic signals to the operation of production lines. Algorithms were used in almost every facet of communication and control (e.g., in routing aircraft at designated flight levels and speeds).

Microchip technology has increased the computational speed of computers so that, by using algorithms, they can quickly scan large arrays of data. For example, computers can use algorithms based upon the rules of chess to quickly evaluate the outcome of potential chess moves. Although the human brain is far more complex than even the most powerful supercomputers, high-speed supercomputers using well-designed algorithms have sometimes been able to defeat world chess champions in test matches.

Real-life Applications

OPERATIONAL ALGORITHMS

Probably the most commonly used algorithm is one used in the operation of addition. This algorithm is used everyday in countless ways, and is so basic to mathematics that most people do not realize that they are using an algorithm when adding numbers.

The addition algorithm relies upon Hindu-Arabic positional notation—the most commonly used notational system that imparts a value to the position of a numeral. Each position or column is 10 times larger than the column or position to its right (e.g., the number 3,456 is interpreted as the sum of 3 "thousands," 4 "hundreds," 5 "tens," and 6 "ones").

This positional notation and addition algorithm makes possible the easy addition of large numbers, and long columns of numbers. The addition algorithm, the repetitive steps used to add numbers, specifies that we count by "ones" in the right hand column, by "tens" in the next column to the left, by "hundreds" in the next column to the left and so on. When the sum in a column exceeds nine, the amount over 10 is retained and the rest is carried to the next column to the left. To add 67 and 97 we add each column. Adding each column gives us 14 "ones" and 15 "tens." Using the addition algorithm, the 14 "ones" are equal to 1 "ten" plus 4 "ones" and so we carry one ten to the column to the left. This then gives us 16 "tens" and 4 "ones." The additional algorithm dictates that 10 "tens" are equal to one hundred and so the number 1 is inserted into the "hundreds" column and the remainder of 6 left in the "tens" column. The algorithm thus yields the correct answer of 164 ($67 + 97 = 164$).

The addition algorithm does not work for other systems of notation, such as Roman numerals.

Various methods are used to teach the essentials of the addition algorithm, so the description above may not be exactly what you remember from early elementary school. Regardless, whatever the words you use to describe the operation of addition—and the other operations of subtraction, multiplication, and division—those terms describe an algorithm in action.

ARCHEOLOGY

Archeologists (scientists who study past civilizations) collect as much information as possible as they explore a site or find. A small grave or ancient trash dump may ultimately result in thousands of measurements of pieces of bone, pottery, or other fragments of the past. Other scientists who study the past, including archaeoastronomers (scientists who study and make calculations about what past civilizations may have observed in the skies with regard to the movement of the Sun, planets, and stars), also compile thousands of observations to yield clues about ancient humankind. Algorithms are used to analyze those mountains of data to yield clues regarding the building of ancient temples, monuments, and tombs or upon the movements and cultural practices of ancient civilizations.

COMPUTER PROGRAMMING

Computers are particularly adept at utilizing algorithms, and algorithms lie at the heart of computer programming, the set of instructions that computers use to analyze data. The creation of elegant (a term used by mathematicians to describe something simple yet powerful) and thus faster algorithms has become an important consideration in the study of theoretical computer science.

Logical algorithms (rules and steps based upon patterns of mathematical logic or proof) including a class of algorithms known as "backtracking" algorithms were developed in the 1960s to explore methods of solving computational problems. Such algorithms can be designed to test possible combinations of sub-problems and such algorithms result in tree-like solutions. A particular solution can be traced back to through prior solutions that are analogous to backtracking through the increasing larger more inclusive branches of a tree that ultimately lead to the trunk. Navigating the solution tree, analogous to a squirrel climbing through the limbs of a tree, produces computational solutions that can then be described as "longest" or "shortest" path solutions. Many computer-programming languages rely on backtracking. For example, if a particular sub-problem solution (a particular branch of the solution tree) proves to be incorrect, the computational algorithm "backtracks" and tries another path to solve the problem.

CREDIT CARD FRAUD DETECTION

Every time you use a credit card, the purchase made is analyzed by computers programmed with algorithms to detect the crime of credit card fraud. Banks and financial institutions that issue credit cards program their computers to use a series of algorithms that compare each purchase to the established pattern of use. For example, if you usually use your credit card in the New York area to make purchases totaling $150 a month, the computer's algorithms should be able to quickly determine that a series of charges from a foreign country totaling thousands of dollars is "unusual." The fact that the purchase does not fit an established pattern can be determined by algorithms and is often enough to alert bank security officials that further investigation is required before authorizing a particular purchase. Upon investigation, they may discover that the user is the authorized cardholder enjoying a vacation or semester abroad. On the other hand, investigation may reveal that the credit card number has been stolen, and that the intended purchase is unauthorized. The use of algorithms to analyze purchases can thus save the bank—or credit card holder—thousands of dollars.

CRYPTOLOGY

In 1977, Ronald Rivest, Adi Shamir, and Leonard Adleman published an algorithm (known as the RSA algorithm—a name derived from the first letters of the founder's last names) that marked a major advancement in cryptology. The RSA algorithm factors very large composite numbers. As of 2004, the RSA algorithm was the most commonly used encryption and authentication algorithm in the world. The RSA algorithm was used in the development of Internet web browsers, spreadsheets, data analysis, email, and word processing programs.

DATA MINING

Association of data in a data mining process involves the use of algorithms that establish relationships or patterns in data. Such algorithms use "nested" or sub-algorithms that rely on statistics and statistical analysis to make associations between data. Usually the algorithm designer (e.g., a computer programmer) specifies desired associations or patterns to be established. Algorithms can be written, however, to perform what is termed exploratory analysis, a form of analysis where associations between data are sought without a preconception or guess as to what patterns might exist.

DIGITAL ANIMATION AND DIGITAL MODEL CREATION

Moviemakers rely on mathematical algorithms to construct digital animation and models. Such algorithms relate points on known surfaces to points on a drawing (often a computer drawing) or digital model. For example, data points for the movement of an arm or leg can be obtained by actors wearing special gloves or sensors that translate movements such as waving or walking into data

(sets of numbers) that can be analyzed by algorithms designed to fill in the gaps between data points. Such algorithms allow animation experts to subsequently draw and animate figures with increasingly realistic features and movement. Model makers can construct digital models at a fraction of the cost needed to construct and test physical models.

DNA OR GENETIC ANALYSIS

Biochemists use algorithms, more commonly referred to in the laboratory as "lab procedures" to identify DNA markers that allow scientists and physicians to determine genetic relatedness (identification of parents or family members) to settle a court case or find a suitable organ donor, determine a patient's risk of disease susceptibility risk, or to evaluate the effectiveness (efficacy) of drug treatments.

In addition to physical testing algorithms, mathematical and computer algorithms can be used to determine or predict patterns of genetic inheritance. The pundit square used in beginning biology is a simple yet powerful use of algorithms that result in the diagrammatic representation of potential gene combinations. In some cases, the pundit square allows the calculation of the odds of having a child who might develop sickle cell anemia or be a carrier of the gene that might lead to actual sickle cell disease in their children.

The task of analyzing massive amounts of data generated by DNA testing is daunting even for very powerful computers. New technologies, including so-called "bio-flip" technologies use specialized computer algorithms to detect small and differences and changes in the structure of DNA (i.e., variation in genetic structure).

ENCRYPTION AND ENCRYPTION DEVICES

Although the technology exists to allow the construction of cryptographic devices intended to protect private communications from unauthorized users while at the same time assuring that authorized government agencies (e.g., those agencies such as the FBI who might obtain a court order) can quickly decode (decrypt) and read messages as needed, such technologies remain highly controversial. So-called "clipper-chips" and "capstone chips" would allow use United States law and intelligence agencies to use specific algorithms to decode encrypted (coded) messages. Certain authorized agencies would then hold the algorithmic "keys" (the step-by step procedures and codes) to any communication using the encrypting technology.

Use of the clipper chip was first adopted and authorized in 1994 by the National Institute of Standards and Technology (NIST). The United States Department of the Treasury was initially designated to hold the keys (algorithms) to decode messages. Rules regarding access to the keys are defined in state and national security wiretap laws. The clipper chip utilizes the SKIPJACK algorithm—a symmetric cipher (code) with a fixed key length of 80 bits. A bit is shorthand for "binary digit," a unit of information (a "1" or "0" in binary notation).

A cipher uses algorithms (i.e., sets of fixed rules) to transform a legible message (clear text or plaintext) into an apparently random string of characters (ciphertext or coded text). For example, a cipher might be defined by the following rule: "For every letter of plaintext, substitute a two-digit number specifying the plaintext letter's position in the alphabet plus a constant (or key) between 1 and 73 that shall be agreed upon in advance." This would result in every letter in the alphabet being represented by a number between 17 and 99 (depending on the particular constant used). For example, if 16 is the agreed-upon constant, then the plaintext word PAPA enciphers to 32173217 as follows: $P = 16 + 16 = 32$; $A = 1 + 16 = 17$; $P = 16 + 16 = 32$; $A = 1 + 16 = 17$. Real keys would, of course be longer and more complex, but the basic idea remains the same: an algorithm-based encryption key allows messages to be locked (enciphered) or unlocked (deciphered), just as a physical key fits into a lock and allows it to be locked and unlocked. Without a key, a cipher algorithm is missing its most critical part. In fact, so important is the key that many times the algorithm itself is widely known and distributed—it is only the keys that remain secret. For this reason, the algorithm used to code messages may remain the same for months or years, but the keys change daily.

Other algorithms remain a mystery. In 1943, Alan Turing (1912–1954), Tommy Flowers (1905–1998), Harry Hinsley (1922–1998), and M. H. A. Newman at Bletchley Park, England, constructed a computational device called Colossus to crack the Nazi German encryption codes created by the top secret Enigma machine used by the Germans. The decryption algorithms used by Colossus remain secret.

In the late 1970s, the United States government set a specific cipher algorithm for standard use by all government departments. The digital encryption standard (DES) is a transposition-substitution algorithm that offers 2^{56} different possible keys (a number roughly equivalent to a 1 followed by 17 zeroes). As larger a number of different keys as that number represents, modern higher speed computers might allow hackers (who also use algorithms) to too easily crack codes with this many keys, and so a new algorithm, known as the advanced encryption standard, is replacing the old algorithm.

Security is increasing as a function of who can develop the most sophisticated algorithms to either protect data, or hack into protected algorithmic codes.

IMAGING

The digitization of images used in modern digital computers would not be possible without the use of algorithms to translate the images into numbers and back again into a viewable image. Digital cameras can be in a vacationer's backpack or be mounted in satellites in orbit hundreds of miles above the Earth. Digital cameras offer higher resolution (the ability to distinguish small objects) than cameras that use light-sensitive photographic film. Digital photo manipulation has also revolutionized photography, including commercial advertising, and offers new security challenges to uncover altered photos. Fractal image compression algorithms allow much greater compression in the storage of images.

Digital cameras capture reflected light on a chip or charged coupled device (CCD). The surface of the CCD contains light-sensitive cells (photo diodes). Each cell or diode represents a pixel and so the pixel becomes a basic unit of a digital image. Light-stimulated diodes produce a signal (often using a transistor) with a voltage that corresponds to the light intensity recorded by the diode. An algorithm in the camera's processing unit then translates that signal into binary code—1s and 0s—that can later be reconverted by the reverse algorithm back into a viewable image. For example, algorithms may assign a code sequence between 0 and 255 to color data (0 is black and 255 indicates an intense red color). These codes are then turned into eight digit binary code sequences (00000000 for black, 11111111 for the most intense red).

Digital photo manipulation involves the alteration of the binary code (i.e., the digital 1s and 0s) that represents the image. While algorithms can be used to alter photos, they can also be used to detect forgery or alteration. Algorithms can compare values of pixels in the background of an image and determine whether they are consistent. Other parts of an image can be protected by altering certain pixels to form a digital watermark that can only be removed by application of a particular algorithm to the image binary code.

INTERNET DATA TRANSMISSION

All information sent by a computer over the Internet contains the sending computer's hardware source address

(MAC address). This is similar to a return address included on a piece of physical mail (snail mail). Conversely, all the information that the computer accepts must be addressed to its unique hardware address (or often a more common "broadcast address" that is similar to a zip code used in regular mail). When an packet of data (e.g., a portion of a text) is received, the receiving computer subjects the incoming packet of data to a processing algorithm, a mathematical formula or set of procedures that determines whether the address information is correct and the message intended for that computer. If the packet of data is accepted, additional algorithms are used to decode the binary message (a series of ones and zeros such as "100010110" into text, pictures, or sound).

MAPPING

Algorithms can analyze data measurements of height, depth, and distance to construct maps. For example, bathymetric maps (maps that depict the oceans as a function of depth) help develop a model of a body of water as depth increases. Such maps are important to fishermen and similar algorithmic programs analyze data in navigational and "fish-finding" equipment aboard many commercial and sport fishing boats.

To construct a precise map of the region, whether of land or at sea, it is necessary to perform detailed measurements, a task increasingly performed by satellites (or in the case of bathymetric maps, by ships with echo sounding surveying equipment that bounce sound waves off the ocean floor). Such data is then set into an array (a particular grid or pattern) that are analogous to strips of a map. Algorithms perform calculations that link the data between various strips and allow the construction of a larger map of the area.

THE GENETIC CODE

Humans themselves are the result of an molecular algorithm that operates at the genetic and cellular level. The genetic code ultimately relates a sequence of chemicals called nitrogenous bases found in deoxyribonucleic acid (DNA) to the amino acid sequences of proteins (polypeptides). These proteins control the biochemistry of the body. The algorithm that describes this process allows scientists to understand the genetic and molecular basis of heredity and many genetic disorders.

In humans, DNA is copied to make mRNA (messenger RNA), and mRNA is used as the template to make proteins. Formation of RNA is called transcription and formation of protein is called translation. This process is the fundamental control mechanism for the development (morphogenesis), growth and regulation of the body and complex physiological processes.

The structure of DNA—and the sequences formed during transcription—can, for example be predicted from an algorithm (based upon the physical shape of the molecules themselves) that specifies that the nucleotide with the nitrogenous base adenine will pair only with a nucleotide that contains a thymine base (an A-T pair). Likewise, nucleotides with a cytosine base will pair only with a nucleotide that contains a guanine base (a C-G pair). The molecular algorithm allows the prediction of bases (e.g., the ATTATCGG sequences) that in triplet sequences (three base sequences) then form the backbone of genes.

A sequence, such as A-T-T-C-G-C-T . . . etc., might direct a cell to make one kind of protein, while another sequence, such as G-C-T-C-T-C-G . . . etc., might code for a different kind of protein.

MARKET OR SALES ANALYSIS

Algorithms are routinely used to analyze buying and selling patterns. Businesses rely on algorithms to make decisions regarding which products to sell, or to which portion of the overall market advertising can be most effectively targeted (e.g., an advertising campaign designed to sell blue jeans to teenagers). Such marketing algorithms make associations between buying patterns and established demographic data (i.e., data about the age, sex, race, income groups, etc.) of the user.

For example, market specialists use algorithms to study data developed from test groups. If a product receives a sufficiently favorable rating from a very small test group, a manufacturer may skip the costs and delays of further testing and move straight into production of a product. The algorithms might be simple (e.g., if 75% of the initial test group likes the product the manufacture knows from experience that there will be sufficient sales to make a profit) or complex (e.g., a complex relation or weighting of responses to the demographics of the test group). Less enthusiastic results may require further testing or a decision not to develop a particular product.

Algorithms can also be used to compare observed responses of a test group to anticipated responses (or responses from other test groups) to determine which products might gain a market advantage if manufactured in a certain way. For example, algorithms can be used to analyze data to find the most favored color, size, and potential name of a skateboard and intended for sale to 10–13-year-old boys.

LINGUISTICS, THE STUDY OF LANGUAGE

The study of the elements of language (e.g., English, French, ancient Native American languages, etc.) is increasingly a quantitative science that relies on mathematical analysis to yield clues about the source of a language and the placement of a language within larger families of languages. This quantitative analysis often requires sophisticated computer-processing and algorithms designed to sift through large databases to determine statistically significant relationships (more than accidental or random relationships) between words and the use of words.

SECURITY DEVICES

Closed circuit television (CCTV) is increasing a part of security measures. CCTV is widely used in the United Kingdom (CCTV is so widespread in London, for example, that it can be used to detect violations of traffic laws) and its use is growing in Europe and in the United States. Many major cities, including New York and Washington, D.C., now utilize widespread public-surveillance CCTV systems, most often operated by the local police.

As a response to terrorism, images from CCTV cameras, especially those located in airports and other transportation hubs, are analyzed by algorithms that compare biometric data (e.g., height, shape of head, etc.). Facial recognition systems use algorithms to compare observed facial features that are hard to change (e.g., head size, width/length of nose, distance between the eyes) against databases containing photographs of known terrorists and other criminals.

SPORTS STANDING AND "SEEDINGS"

Many sports tournaments such as the NCAA men's and women's basketball championships—or the pairings for the annual football bowl games that are now used as "national championship" games—rely on algorithms that are designed to take the bias (an unwarranted predisposition in favor of someone or something) and prejudice (an unwarranted predisposition against someone or something) out of the selection process.

Depending on the sport, "seeding" algorithms can be designed to use data derived from poll results, strength of schedule points, points earned through competition, individual race or game results, conference or league standings, etc. Algorithms are used to determine everything from which lane a runner or swimmer starts a particular race to more fundamental questions as to whether an athlete or team is invited to participate in a competition.

Trevecca Nazarene's Alex Renfroe drives around Lewis-Clark State's Danny Allen during the second half of a 2005 NAIA Division I tournament game. Algorithms are often used in sports to determine bids and placement of teams in match brackets. AP/WIDE WORLD PHOTOS. REPRODUCED BY PERMISSION.

For example, the NCAA algorithms were designed to replace a simple reliance upon polls of coaches or sportswriters than were often driven by publicity and favoritism toward certain teams or teams from certain areas of the country. Although the new algorithm-driven selection processes are not perfect, they are an attempt to make the selection processes more fair.

TAX RETURNS

Some tasks that seem complex and difficult can be broken down into simpler steps (i.e., an algorithm can be written to accomplish the task). For example, completing a tax return form can be a time consuming and frustrating task for many people—especially young students who may have just started working.

At first glance, tax forms often seem overly and needlessly complex. The forms, however, can be completed by using a series of algorithms that are described in the instructions for each form. The instructions themselves are keyed to the step-by-step (systematic) completion of the tax form they describe. Completing a series of smaller,

Key Terms

Algorithm: A fixed set of mathematical steps used to solve a problem.

Operation: A method of combining the members of a set so the result is also a member of the set. Addition, subtraction, multiplication, and division of real numbers are everyday examples of mathematical operations.

Program: A sequence of instructions, written in a mathematical language, that accomplish a certain task.

usually less complex steps, allows taxpayer to correctly complete the tax form.

Potential Applications

ARTIFICIAL INTELLIGENCE

Since their development, electronic computers have been programmed with algorithms to accomplish specific tasks (e.g., a numerical calculations). Current research and development seeks to develop sophisticated algorithms that when combined with more flexible rules of operation may result in what is termed "artificial intelligence." The exact differences in computers that perform intricate algorithms and those with "artificial intelligence" is often a hotly debated topic among scientists and engineers. Regardless, one element or defining characteristic of artificial intelligence that is widely agreed upon is that a computer using artificial intelligence will use flexible rules rather that rigid algorithms for seeking solutions.

Artificial intelligence programming may even allow computers to modify their own programming rules and develop their own algorithms for tackling problems.

Where to Learn More

Books

Edelstein, Herbert A. *Introduction to Data Mining and Knowledge Discovery*, 3rd ed. Potomac, MD: Two Crows Corporation, 2000.

Grahm, Alan. *Teach Yourself Basic Mathematics.* Chicago: McGraw-Hill Contemporary, 2001.

Sherman, Chris, and Gary Price. *The Invisible Web: Uncovering Information Sources Search Engines Can't See.* Medford, NJ: CyberAge Books, 2001.

Web sites

National Institute of Standards and Technology. "Advanced Encryption Standard: Questions and Answers." Computer Resource Security Center. March 5, 2001. <http://csrc.nist.gov/encryption/aes/round2/aesfact.html> (June 16, 2004).

Architectural mathematics uses mathematical formulae and algorithms for designing various architectural structures. Most of these structures are buildings such as museums, galleries, sport complexes and stadiums, theatres, churches, cathedrals, offices, houses, and so on. Architectural mathematics is also used to design open spaces in cities and towns, recreational places including gardens, parks, playgrounds, water bodies such as lakes, ponds and fountains, and a variety of physical construction and development.

Put simply, architectural math pertains to mathematical concepts that are central to architecture. These architectural math concepts are also used in several other activities that we see in our daily lives. They are extensively used in sports, technology, design, aviation, medicine, astronomy, and much more.

Understanding architectural math requires knowledge of various two-dimensional as well as three-dimensional shapes such as square, rectangle, triangle, cube, cuboids, sphere, cone, and cylinder. There are also other concepts such as symmetry and proportion that are integral to architecture math. The pyramids of Egypt, for example, are based on these principles.

Architectural Math

Fundamental Mathematical Concepts and Terms

As architectural designs are strongly inspired and implemented using various shapes and forms, basic architectural mathematics involves understanding underlying principles that derive such shapes and forms. This requires understanding geometric equations associated with its visual representations. In other words, architectural mathematics is not expressed as simple numbers but rather as graphical or visual forms. What follows are some of the most widely used architectural math concepts.

RATIO

A ratio is a comparison by division of two quantities expressed in the same unit of measure. In other words, you get a ratio by dividing two numbers of quantities. The ratio may be represented in words or in symbols. For example, if segment Line A is one inch long and Line B is two inches long, we say that the ratio of Line A to B is one to two. In terms of mathematical symbols, the ratio may be denoted in fractional form as ½, or it may be expressed as 1:2.

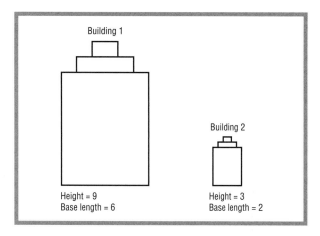

Figure 1.

Since ancient times, Greeks and Romans were known to be the most elaborate builders. They had a flair for architecture and built structures that were pleasing to the eye. They were convinced that architectural beauty was attained by the interrelation of universally valid ratios. Frequently, complicated mathematical ratios were used by architects to accomplish their goals. Take, for example, the golden ratio (or phi)—1.618. This ratio that has applications in many areas, has been extensively used in architecture—both modern and ancient.

PROPORTION

Proportion, like ratio has always been a vital component of architectural math. The ancient Greeks and Romans followed certain mathematical proportions (and ratio) to attain order, unity, and beauty in their buildings. Using simple mathematical formulae (based on proportion) they were able to establish a unique relationship among various parts of buildings. Such relationships have been used for generations.

To better understand the concept of proportion, consider an example of two buildings variable in height and base, however displaying the same proportion (see Figure 1).

In Figure 1, there are four terms that would define proportion of one building to another. These are 2, 3, 6, and 9. A proportion is an equation that states the ratios of comparison are equal. Thus, in the above example we would say that Building 1 is in proportion (or proportionate) to Building 2 if 6/9 = 2/3, or 9/6 = 3/2. This is the case, and hence the statement that the buildings are in proportion holds.

If the ratios for two objects are not equal, they would not be in proportion. Proportion can also be used to

calculate the ratio of the total magnitude (in this case size) of the two objects. For example, in our case 9/6 can also be expressed as 3/2 × 3. Thus, we can say that Building 1 is three times the size of Building 2.

SYMMETRY

In architecture, one way to attain balance and response while designing structures is by the use of symmetry. Architecture is based on principles of balance. Basically, if most architectural forms are divided into two equal parts by a line in the center, the opposites sides of the dividing line would be similar (or even identical). This concept that can also be seen in all basic geometric forms (square, rectangle, circle, triangle, and so on) is known as symmetry and is extensively used architectural structures—modern and ancient.

See Figure 2 to understand symmetry further. In this figure (2a), triangle ABC and triangle BCD are symmetric about line m. The corresponding sides and corresponding angles of the triangles are similar. In other words, triangle CBD is the reflection of triangle ABC, and m is the line of symmetry. Such type of symmetry is also known as symmetry by reflection.

Forms can also be created using translation or sliding symmetry. This type of symmetry involves two or more similar forms of the same size and facing the same direction. In figure (2b), all images are similar to each other and face the same direction. Another way to understand this is imagine that the same image has been slid on the line p (and thus the name sliding symmetry).

The third method of attaining symmetry is by rotation. If a figure, after rotating it around a central point by less than 360°, remains unchanged, then it has rotation symmetry. For example, in figure 2c, if this form is rotated from the central point B by 180°, the resulting form would be the same. Thus, the figure has rotation symmetry for a rotation of 180°.

It is interesting to note that all geometric forms (square, rectangle, triangle, pyramid, hexagon, etc.) can have reflection, translation, and rotation symmetry. It is for this reason, they are used extensively in architecture.

SCALE DRAWING

For designing any structure (a building, a house, or a city), an architect is required to convert his ideas to drawings. These drawings provide homeowners, contractors, carpenters, and others with a small diagram of the final structure. The drawings show in detail the sizes, shapes,

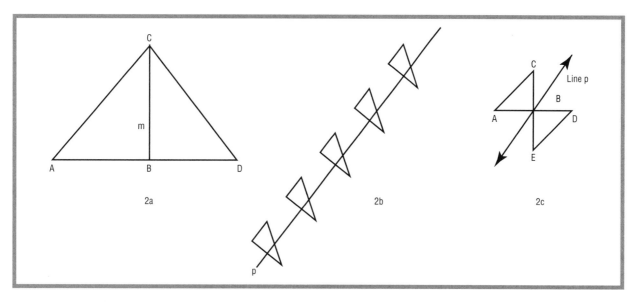

Figure 2.

arrangements of rooms, structural elements, windows, doors, closets, and other important details of construction. For example, a drawing for a house would specify the area (length, width, and height) of every room including the living area, bedroom, and bathroom at every floor. Such miniature reproductions of the structure are called scale drawings.

Scale drawings that represent parts of a structure must be in proportion to the actual structure.

To do this, architects use a specific scale corresponding to the actual size. For example, a scale such as ¼ inch = 1 foot, would suggest that a length of ¼ inch on the scale drawing is equal to one foot within the actual structure.

Scale drawings allow an architect to visualize a structure before building it.

MEASUREMENT

Measurement is important and is required while designing any building, right from the planning stage to the actual construction work. The instrument used to measure objects is the ruler, or a measuring tape. Architects, carpenters, and designers use measure-ments to come up with accurate scale drawings before starting construction work. Throughout the process of building any structure, measurement is extensively used.

Measurement can be expressed as inches, feet, and yards (English system), or centimeters, and meters (metric system).

A Brief History of Discovery and Development

There is a commonality between the seventeenth-century Round Tower of Copenhagen, the thirteenth-century Leaning Tower of Pisa, Houston's Astrodome, (the first indoor baseball stadium built in the United States), the vast dome of the Pantheon in Rome, a Chinese pagoda, and the Sydney Opera House. All these buildings were built using architectural math concepts such as scale, measurement, ratio, proportion, and symmetry.

Architectural mathematics has always been a vital part of structural design. The pyramids of Egypt used basic principles of the geometric "pyramid"—a square base and an apex tapering as the elevation of the pyramid increased. The pyramid shape provides higher stability compared to other structures as it is able to counter wind forces and natural forces, such as earthquakes, much more effectively than compared to most other shapes.

The same concept of visual geometry was used while constructing the Eiffel Tower in Paris. It has a square base with a narrowing apex as one moves toward the top of the structure. One key aspect of mathematics to be considered while building a pyramid structure is the use of ratio and proportion in designing the base and apex of the tower. Higher the ratio of base to the apex, the higher will be the stability of the structure to withstand the various forces. This has been kept in mind while designing the

Carpenters use a variety of everyday math skills. CORBIS.

above mentioned structures (and many other around the world).

The influence of mathematics on architecture and its principles can be seen since the time of the Greek mathematician Pythagoras (569 B.C.–475 B.C.). Pythagoras, and his followers, believed that all things could be represented in numbers. This concept has been used extensively in architecture, ever since.

Pythagoras, after conducting many experiments, found out that music depended considerably on mathematics. He concluded that musical scales (notes) depended on ratios of small integers. Architects adopted this principle and designed buildings based on ratios of smaller integers or units. These smaller units could be units of length, size, or dimension. For example, a simple wall consisted of smaller blocks equal in length.

Pythagoras also believed that all numbers could be represented as geometrical shapes. Furthermore, he developed an idea that geometrical symmetry based on proportion is far more appealing, visually. This is the very concept that architects used while designing buildings and

other structures. The structures that were then built in ancient Greece and Rome were based on symmetry.

Architectural math concepts were eventually used in a variety of other areas including astronomy, carpentry, jewelry design, and more. It is not known how many of these concepts were used in architecture first and then later re-used in other fields of work. However, they are interlinked and there is a high possibility that these concepts inspired other designers and engineers to use mathematics effectively to justify the form or function of other objects and devices, just the way architects used it to express their building designs.

Real-life Applications

ARCHITECTURE

It is evident by now that architecture is the most common application of architectural mathematics. These concepts are used in a number of ways by architects the world over.

Carpentry

"Measure twice and cut once" is a principle slogan of carpentry. Cutting out pieces of wood in such a way that they subsequently fit together to make a beautiful cabinet, desk, cupboard or whatever calls for correct measurements of length, width and height. The manufacture of these and other items calls for the correct measurement of slope; no one wants to try to eat off of a table that is so slanted that the soup spills onto the floor! Nailing pieces of wood together calls for a mathematical distinction between perpendicular and an angle less than 90 degrees.

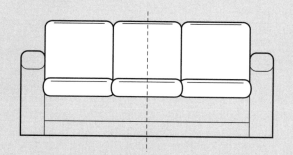

A carpenter needs to understand the size and proportion of each object depending on the person who is expected to use it. These are again based on ergonomic standards (see the section on Ergonomics). Subsequently, the designs reflect most of the basic concepts of architectural mathematics.

For example, table tops often have wood pieces cut and put together to form a symmetric design, exemplifying their beauty. Symmetries and ratios are clearly evident in the design of a couch as well, which is based on bilateral symmetry (see figure at above right).

The myriad number of carpentry processes that go into the construction of a house are rooted in math. Carpenters get involved in house construction following the laying of the foundation. Construction of a floor frame is essential. Often the wood frame sits directly on the foundation, and is not fixed or bolted to the foundation. The weight of the house will provide the force to keep the frame intact. But, for this operation to be successful, the frame must be the same length, width and shape as the foundation and have enough cross braces to provide strength. Proper measuring is crucial, as is the fitting together of the slabs of wood that make up the skeleton of the frame.

The cross braces, which are also called "joists" are attached to a center beam that runs down the center of the house. Once again, proper positioning of the beam is essential to establishing the support needed for the floors to come.

Once a floor is installed on the floor frame, walls can be built. This construction is fraught with measurements. For example, since special vertical supports need to be in the right places to accommodate the interior walls. Other side supports are usually positioned 8 to 16 inches apart and comprise the supporting studs. Doors, windows and other exterior openings must be properly located. In the case of windows, a special structure called a header needs to be built above the window opening. It will give the wall enough strength over the expanse of glass to support the roof. Typically, this phase of the construction requires detailed plans of the structure, with accurate measurements. A blueprint of the project is a necessary and prudent tool.

USE OF SPACE

Architecture is all about using physical three-dimensional space. The ways in which this space is used is one of the most important aspects of architecture. Many mathematical concepts are visible in architectural designs that include spaces. For example, a building design may leave an open space in the center, which is often geometrical in shape such as a square, rectangle, or a circle, with several other geometric designs surrounding the space. Spaces are always designed based on mathematical principles or ratio, proportion, and symmetry.

Designs based on symmetry and space can be seen in several open spaces and squares. One great example of the same is St. Peter's Square in Vatican City, Rome, where a circular symmetry of elements is focused on the central piece. The intersecting lines in the circle have a focal point of concentration, a place where people can gather around.

USE OF GRIDS

Elements in a building are often arranged forming a grid that resembles a set of parallel and perpendicular lines on a piece of paper. Again these elements are based on symmetry and patterns.

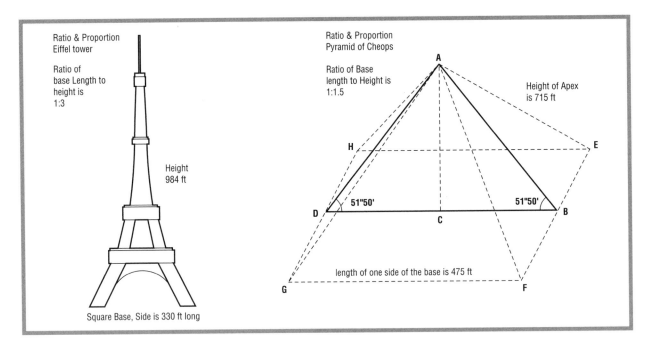

Figure 3.

USE OF RATIO AND PROPORTION

Historical monuments and modern buildings, alike, have used architectural mathematics extensively. This is reflected in several landmarks. As stated earlier, the Pyramids of Egypt are a very good example of the use of a simple form (the geometric pyramid shape), having a square base and tapering as its height increases. These are built such that their base and height have specific ratio and proportion. This is done to impart greater strength and stability. The same is the case with the Eiffel tower in Paris.

For example, the ratio of the base of the Pyramids with their height is almost 1:1.5. The ratio gives the structure higher stability. Conversely, the Eiffel tower has a ratio of 1:3 for base to height. Both these structures have different ratios. However, they both impart stability to the structure due to their shape and design. (See Figure 3.)

Architectural designs also use ratio and proportion to justify the dimensions of elements within the buildings. This includes length and width of the corridor, its proportion to doors that lie within the corridor, the height of the ceiling from the floor with respect to the type of building, and the ratio of size of the steps on a staircase with respect to the total height of the staircase—all these aspects are considered vital while designing a structure.

USE OF ARCHITECTURAL SYMMETRY IN BUILDINGS

Reflection symmetry (also known as bilateral symmetry) is the most common type of symmetry in architectural designs. In bilateral symmetry, the halves of a composition mirror each other. Such symmetry exists in the Pantheon in Rome. We find the same symmetry in the mission-style architecture of the Alamo in San Antonio, Texas.

Bilateral symmetry existed in several buildings built during the Roman or Greek periods. Modern architects also use such symmetry widely for various structures (see Figure 4).

Additionally, translation and rotation symmetry are also employed considerably in modern architectural designs.

USE OF RECTANGLE AS "GOLDEN RECTANGLE" AND "GOLDEN RATIO"

Since ancient times, architectural designs have used the golden ratio (1.618) in various ways. One of the best examples being the Parthenon, the main temple of the goddess Athena, built on the Acropolis in Athens. The front of the Parthenon is a triangular area that fits inside a rectangle whose sides are equivalent to the golden ratio (the rectangle is popularly known as the golden rectangle). The golden ratio and its related figures were incorporated into every piece and detail of the Parthenon.

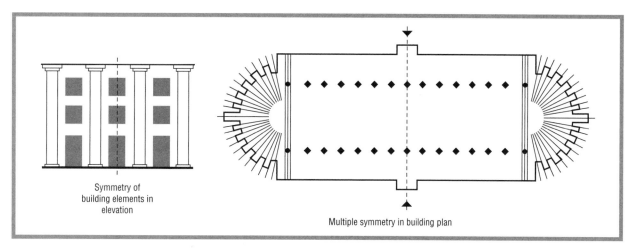

Symmetry of
building elements in
elevation

Multiple symmetry in building plan

Figure 4.

The same math principles that allowed the construction of the Arch of Constantine in Rome also allow designers to shape modern home interiors. The arch distributes load. TRAVELSITE/DAGLI ORTI. REPRODUCED BY PERMISSION.

Decorating

Numerous symmetrical shapes and forms are used while decorating furniture and home accessories ranging from a flower vase to the kitchen sink. The use of architectural shapes and concepts is clearly visible in every decorative aspect of the complete design.

Besides the arrangement of these shapes and forms (see figures below), the concept of symmetry also plays a vital role in the layout of these. For example, while decorating a room, most interior designers would ensure that the entire layout of the room (and how all elements within the room are placed) is based on symmetry principles. The main purpose is to give a decorative touch to the room to make it visually more appealing.

Home accessories, especially decorative artwork (vase, glassware, china pottery, and so on) are often made of wood or ceramics. These decorative pieces more often than not are also based on principles of symmetry. Their shape, exterior designs, and colors are amazingly symmetrical, and in many cases are based on basic geometric forms and shapes—much like designs in architecture.

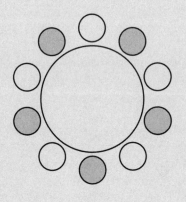
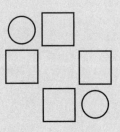
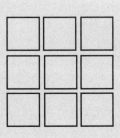

The Triumphal Arch of Constantine, and the Colosseum—an amphitheater in Rome built in around A.D. 75 (both in Rome)—are other great examples of ancient use of golden relationships in architecture. The main idea behind employing this ratio was to make the structure visually appealing and also more stable.

USE OF BASIC FORMS AND SHAPES OF GEOMETRY

Apart from mathematical concepts such as ratio, proportion, and symmetry, most architectural designs are based on basic geometric shapes and forms including triangle, rectangles, pyramids, cones, cylinders, and more. Although, when viewed as a whole these structures would have basic shapes, their interiors can always be represented by the above mentioned mathematical concepts.

The Taj Mahal in India is an example of the use of a basic shape or form—the cube. The Taj Mahal was built as a cube, where the four minarets and the center burial tomb of the queen all are contained in a perfect cube. The length, breadth, and height of all sides are equal in dimension. Additionally, the sense of ratio, proportion, and symmetry of this structure is precise and spell-binding.

A modern example of the use of basic shapes is the Pentagon, in Washington, D.C. The Pentagon's five sides are equal in length, denoting a perfect pentagon. Within the main structure, there are five concentric pentagons of corridors and offices. Again, these internal pentagons are symmetric and in proportion to each other.

SPORTS

Geometric shapes and forms, symmetry, ratio, and proportion have found a place in sports as well. Practically, every field sport uses architectural math principles. A tennis court, basketball court, football, hockey, soccer

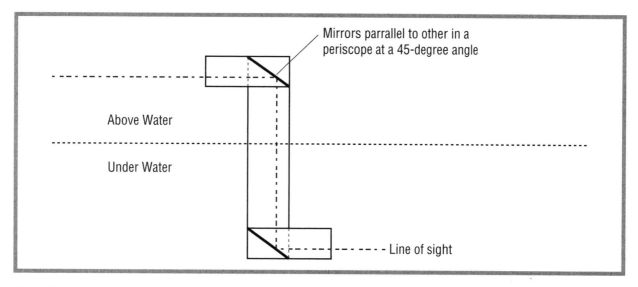

Mirrors parrallel to other in a periscope at a 45-degree angle

Above Water

Under Water

Line of sight

Figure 5.

fields—all of them are rectangles having a center line dividing each into two halves for each team or players. These are perfect examples of reflection symmetry. Besides, many of these sports have fields that have sides forming the "Golden Rectangle." The ratio of a side to its length is based on the golden ratio—a concept adopted by architects to depict buildings during the Roman and Greek periods, as discussed earlier.

In addition, other concepts of mathematics that are commonly applicable to architectural designs, such as measurements and scales also apply to field sports.

TECHNOLOGY

Technology tools and devices use architectural math concepts of symmetry and proportion to facilitate their underlying functions. Equipments such as a periscope used in submarines, guns in aircrafts, and satellite transmission use the principle of symmetry.

A periscope is commonly used in submarines. It is a device that can help view objects such as ships and other water vessels above the water surface, while still being underwater. A periscope has two mirrors placed at a 45° angle to the eye's line of sight along with another mirror placed at 45° parallel to the first one at a variable height (see Figure 5). This allows a person to view objects using the laws of reflection at different heights while maintaining his/her own position and eye level.

The principle of periscope is clearly visible in the rotational symmetry concept of many architectural building designs. The use of two mirrors can be compared with two parallel lines to reflect a design and make

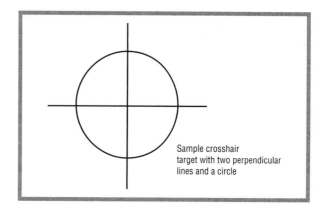

Sample crosshair target with two perpendicular lines and a circle

Figure 6.

it look symmetrical. The concept is the same, however its application and use varies drastically.

Fighter aircraft guns are often assisted by the visual cross-hair—two perpendicular lines, where the point of intersection is often pinpointing or locating a target (see Figure 6). There are several aircraft guns that have two cross-hairs, one for the aircraft gun itself, and the other for the object. Once the two cross hairs coincide with each other, or are symmetrically aligned with each other, the target object is in line with the gun point. In other words, the crosshairs are now pointing at the target. The target can be locked and shot. This entire mechanism is based on principles of symmetry.

Another such technological tool is the fan. A fan works on the principles of architectural math, mainly symmetry, and is used in several applications including aircrafts, helicopters, wind mills, and air conditioners, as well as industrial/home establishments such as kitchens (exhaust fans). A fan has typically three wings, which are similar in size, shape, ratio, and proportion. On rotation, they (wings) generate air and help run several mechanical as well as electrical devices. A simple fan is one of the most commonly used mathematical applications of symmetry and geometric shape (circle).

USING SYMMETRY IN CITY PLANNING

Since the early 1930s, most cities in the world have developed in similar ways. City planners have always focused on symmetrical models for planning a new city or even developing existing cities further. Many cities have a central area known as the central business district (CBD). This is where businesses within a city are concentrated. The areas around the CBD are mostly residential.

The manner in which residential areas have developed over the years is comparable (around the world). A city can be thought of as a group of clusters, where each cluster comprises of a number of buildings, roads, and other structures. The entire city consists of numerous such clusters arranged symmetrically. In other words, a city would consist of a central area (CBD) and several similar clusters around the CBD placed in a symmetrical pattern. This concept is based on the principle of translation symmetry (also known as a fractal or motif).

Some of the biggest cities in the world, including New York, London, Paris, Beijing, and so on are in many ways based on the architectural mathematical concept of symmetry. That said, the late 1990s and early 2000s have started witnessing newer cities that are far more decentralized. In other words, the concept of a central business area and symmetrical clusters of residential areas around it is losing popularity. Clusters within some cities have become dispersed and random rather than symmetric.

ERGONOMICS

Ergonomics is a science that studies technology and how well it suits the human body. Ergonomics involves understanding basic body parts, their functions and abilities to operate equipments, machinery, products, and other technological devices. Ergonomics is commonly used while designing cars, among other things. Ergonomic car designs are based on the principles of ratio and proportion. In other words, car designers use principles and math concepts that are used considerably in architecture to come up with designs that better fitand serve the human body.

For example, the height from the surface, inclination, and movements patterns in a car seat for drivers are all designed in proportion to the human body. The ratios are extremely critical here. The size of the seat has to be in proportion with the size of an average human driver.

Besides, you do not expect a person to have a giant wheel in front of him/her, the size of the wheel (the diameter of the wheel) has to be in proportion to the size of the hand grip, shoulder width, and distance between the wheel and person driving the car. All these elements are carefully incorporated into the design of cars.

Similarly, interior designers also use ratios to design various objects (such as beds, tables, chairs, and so on) within a house. These ratios are based on ergonomic standards. For example, a bed is designed such that it is in proportion to the human body. In Sweden, beds have a length of 7 feet (2.1 m), while beds in Japan are rarely 6 feet (1.8 m) long. This is due height differences in the populations. The average height of an individual in Japan is 5 feet 2 in (1.5 m), while the average height of a person in Sweden is about six feet (1.8 m). This also influences other design standards such as height of the bed from the floor, width of the bed, and portability of the bed.

The size of the window is also often based on the proportion of human body. A window in a house will be smaller, compared to a window in a public building. The proportion of both windows may be same implying that the ratio of their width/height is equal. However, their sizes would differ.

Architectural mathematical concepts such as ratio and proportion form an integral part of ergonomics, especially when it comes to design related issues.

JEWELRY

Ornaments made of gemstones, diamonds, gold, and silver use symmetry of arrangement extensively. A cut of a diamond often displays several shapes and forms. Gold is molded into several geometric forms to add value to an ornament.

Consider, for example, pendants that are more often than not designed using principles of symmetry. The symmetry in such pendants is visible in architectural structures as it is in nature (arrangement of flowers and fruits on trees). Ornament designs are often very intricate and require a finer view to understand their symmetry, ratio, and geometric shapes. Such symmetric designs are not limited to pendants but are also visible in rings, bracelets, and several other ornaments.

Compare these with symmetric designs in the ceilings of several domes and museum galleries and a stark resemblance is clearly evident. Mirrors, stained glass, and other shinny materials that are commonly used to signify architectural designs in building interiors are very similar to ornament designs—with respect to their visual arrangement and their underlying mathematical principles.

Cathedrals are a common example, where the architecture is inspired by arranging materials and objects in symmetry, similar to that in ornaments and jewelry.

ASTRONOMY

Fundamentals of architectural math including distance, size, and proportion are also visible in various astronomical advancements. The telescope is one such example. Telescopes are used to view stars and planets located in far away galaxies. The distance is measured in light years. The distance traveled by light in one year is known as a light year (light travels at a speed of 186,000 miles per second). This gives an indication of the distances between the Earth and some of the stars and planets.

Telescopes are used to magnify the image of these objects. This is done by using different lenses. Larger telescopes, such as the Hubble telescope, are able to magnify objects situated at a larger distance. Smaller telescopes in comparison have lower magnification implying lower visibility and clarity.

One of the basic mathematical principles of telescopes is scaling—a concept extremely common in architecture. Just like architects draw scale diagrams using ratio and proportion, telescopes use the same principles to magnify objects situated at large distances. In other words, telescopes present a scale model of an object that is not otherwise visible (or too tiny) with the naked eye.

Although, larger telescopes magnify objects that are further away, as compared to the smaller telescopes, the degree of magnification (of both types of telescopes) is always in proportion.

TEXTILE AND FABRICS

Cloth or fabrics are used for a variety of purposes. This includes bed sheets, covers, clothes, apparels, wipes, and more. Fabrics are textile products that require knitting. These are made from fibers of cotton, nylon, or other types. However, most of these fabrics do not have any value until a design is printed or woven on them. In other words, fabric prints carry considerable value to a plain piece of fabric or cloth. People would usually buy fabrics with visually appealing prints, rather than those that are plain.

Symmetry, which is used commonly used in architecture, is often reflected in fabric or cloth designs. Most fabric designs are composed of motifs. Motifs are repetitive use of a single design concept, style, or shape—Motifs signify symmetry (translation symmetry). The type of motifs could range from a leaf or a flower of same color or style repeated over the entire fabric print. The design varies depending on the final use of the fabric. Bed sheets, clothes, fashion apparels, and so on have different symmetry motifs depending the type of fabric, their manufacturing price, and quality of print.

Motifs are not limited to floral or color patterns but are often extend to lines, simple geometric shapes (squares, circles, rectangle, etc.), blocks, and much more. In some cases, once the fabric is cut or is stitched to make the final product, the symmetry may be lost. Nevertheless, the design is still based on the very principle of symmetry.

This is one of the most common applications in daily life that uses mathematical concepts of architecture in a very different way.

ARCHITECTURAL CONCEPTS IN WHEELS

Commuting has become an integral part of our daily life. We drive (on our own or in public transportation) to work, to school, to attend meetings, to go shopping or buy groceries. We require transportation to reach different places. Today, transportation is seen as a necessity.

Transportation is facilitated by public buses, railways, airplanes, and cars. All of these use wheels. A wheel, be it of rubber, magnet, or iron, is a vital component of any automobile. The wheel consists of a bar in its center known as the axle. The width of the axle is governed by the width of the carriage (weight of the automobile) required. Subsequently, the width varies in trains, buses, and cars. While designing wheels, engineers must ensure that the size of the wheel and the axle is in proportion to the total weight of the vehicle (including the people it carries) as well as the speed at which the vehicle can travel.

Ratio and proportion play a very important role in defining the diameter, width and the number of wheels that have to be attached with a vehicle. Higher the load to be carried, the more number of wheels (and even stronger wheels) will be required. Similarly, the longer the length of the vehicle, more the number of wheels required. Airplanes do not travel on wheels but require them to land and take off. However, the proportion of their wheels is much greater when compared with other vehicles as the amount of load is much higher. Besides, the size of the plane is also much larger when compared with other vehicles.

In short, wheels have to compliment the size of the vehicle and its intended purpose. Automobile design uses mathematical concepts of ratio and proportion, similar to those used in architecture. These are also based on ergonomical standards (see section on Ergonomics).

Key Terms

Proportion: An equality between two ratios.

Ratio: The ratio of a to b is a way to convey the idea of relative magnitude of two amounts. Thus if the number a is always twice the number b, we can say that the ratio of a to b is "2 to 1." This ratio is sometimes written 2:1. Today, however, it is more common to write a ratio as a fraction, in this case 2/1.

Scale: The ratio of the size of an object to the size of its representation.

Symmetry: An object that is left unchanged by an operation has a symmetry.

Where to Learn More

Books

Rossi, Corinna. *Architecture and Mathematics in Ancient Egypt.* Cambridge University Press, 2004.

Williams, Kim. *Nexus III: Architecture and Mathematics.* Pacini Editore, 2000.

Web sites

University College London, Department of Geography. "Fractals New Ways of Looking at Cities" <http://www.geog.ucl.ac.uk/casa/nature.html>(April 9, 2005).

Yale New Haven Teachers Institute. "Some Mathematical Principles of Architecture" <http://www.cis.yale.edu/ynhti/curriculum/units/1983/1/83.01.12.x.html> (April 9, 2005).

Overview

An area is a measurement of a defined surface such as a face, plane, or side. Conceptually, an object's area can be compared quantitatively to the amount of paint needed to cover the object completely. However, in contrast to measures of volume in pints, liters, or gallons, area measurements are expressed in units such as square feet, square meters, or square miles. Calculations of area are basic to science, engineering, business, buying and selling land, medicine, and building.

Fundamental Mathematical Concepts and Terms

AREA OF A RECTANGLE

Every real-world object and every geometrical figure that is not a point or a line has a surface. The amount or size of that surface is the object's or figure's area. There are many standard formulas for calculating areas, the simplest and most commonly used being the formula for the area of a rectangle. To find the area of a rectangle, first measure the lengths of its sides. If the rectangle is W centimeters (cm) wide and H cm high, then its area, A, is given by $A = W \text{ cm} \times H \text{ cm}$.

Centimeters are used here only as an example. The units used to measure length—centimeters, inches, kilometers, miles, or anything else—do not change the basic formula: area equals width times height. So, for example, a typical sheet of typing paper, which is 8.5 inches wide and 11 inches high, has area $A = 8.5 \text{ inches} \times 11 \text{ inches} = 93.5$ square inches.

UNITS OF AREA

Area has now been explained in terms of "square inches" (or centimeters). This means that on the right-hand side of the formula $A = W \text{ cm} \times H \text{ cm}$, four terms are multiplied: W, H, and cm (twice). These four terms can be reordered to give $W \times H \times \text{cm} \times \text{cm}$. It is customary in mathematics to use the square notation when a term is multiplied by itself, so $\text{cm} \times \text{cm}$ is always written cm^2, which is centimeters squared, or square centimeters. Another way of writing the rectangle area formula is, therefore, $A = WH \text{ cm}^2$. Area is therefore measured in units of square centimeters—or square inches, square feet, square kilometers, square miles, or any other length measure squared. For example, a square with edges 1 foot long has an area of 1 square foot.

When talking about physical materials such as cloth, land, sheet steel, plywood, or the like, it is important to

Area

Areas of geometric shapes

Geometric figure	Dimensions	Formula for area
rectangle	width *W*, height *H*	$A = WH$
square	side length *H*	$A = H^2$
circle	radius *R*	$A = R^2$
triangle	base *B*, height *H*	$A = 1/2\ BH$
parallelogram	base *B*, height *H*	$A = BH$
trapezoid	base *B*, top *T*, height *H*	$A = 1/2\ (B + T)H$

Figure 1.

standard formulas were worked out centuries ago for simple rounded objects like cones, spheres, and cylinders; these formulas are listed in many math books. For example, the area of a sphere of radius *R* is $A = 4\pi R^2$ (π, pronounced "pie," is a special number approximately equal to 3.1416; see the article on "Pi" in this book). The Earth, which is basically sphere-shaped, has an average radius of 6,371 kilometers (km), or about 3,956 miles. Its surface area is therefore $A = 4\pi 6{,}371^2 = 510{,}060{,}000\ \text{km}^2$, which is about 316,750,000 square miles. The Earth is 53 times the area of the United States.

give correct units for length and area. However, in mathematics it is common to not use units. The norm is to say that an imagined rectangle has a length of 4, a height of 5, and an area of $4 \times 5 = 20$.

AREAS OF OTHER COMMON SHAPES

The simplest rectangle is a square, which is a rectangle whose four sides are all of equal length. If a square has sides of length *H*, then its area is $A = H \times H = H^2$.

The standard formulas for finding the areas of other simple geometric figures are depicted in Figure 1.

Notice that in all the area formulas, two measures of length are multiplied, not added. This means that whenever an object is made larger, its area increases faster than its height or width. For example, a square that has sides of length 2 has area $A = 2^2 = 4$, but a square that is twice as tall, with sides of length 4, has area $A = 4^2 = 16$, which is four times larger. Likewise, making the square three times taller, with sides of length 6, makes it area $A = 6^2 = 36$, which is nine times larger. In general, a square's area equals its height squared; therefore its area "increases in proportion to" or "goes as" the square of the side length. Consequently, a common rule of thumb for sizes and areas is, increasing the size of a flat object or figure makes its area grow in proportion to the square of the size increase.

AREAS OF SOLID OBJECTS

Three-dimensional objects such as boxes or balls also have areas. The area of a box can be calculated by adding up the areas of the rectangles that make up its sides. For example, the formula for the area of a cube (which has squares for sides) is just the area of one of its sides, H^2 multiplied by the number of sides, which is 6: $A = 6H^2$.

Calculating the area of a rounded object like a ball is not as simple, because it has no flat sides and none of the standard formulas for simple geometric shapes can be used to find the areas of parts of its surface. Fortunately,

A Brief History of Discovery and Development

The calculation of areas was one of the earliest mathematical ideas to be developed by ancient civilizations, preceded only by counting and length measurement. The ability to calculate areas was originally needed in the buying and selling of land. Four thousand years ago the Egyptian and Babylonian civilizations also knew how to calculate the area of a circle, having worked out approximate values for the number π. The ability to calculate areas was also useful in construction projects. The pyramids of Egypt, for example, could only have been constructed with the help of sophisticated geometric knowledge, including formulas for the areas of basic shapes. Calculation of the areas of spheres and other solid objects also dates back to the ancient Egyptian and Babylonian civilizations. Similar knowledge was discovered independently by Chinese mathematicians at about the same time.

In the seventeenth century, the calculation of the areas of shapes with smoothly curving boundaries was an important goal of the inventors of the branch of mathematics known as calculus, especially the English physicist Isaac Newton (1642–1727) and the German mathematician Gottfried Wilhelm von Leibniz (1646–1716). One of the two basic operations of calculus, integration, describes the area under a curve. (To understand what is meant by the area under a curve, one must imagine looking at the flat end of a building with an arch-shaped roof. The area of the wall at the end of the building is the area under the curve marked by the roofline.) The area under a curve may stand for a real physical area—if, for example, the curve describes the edge of a piece of metal or a plot of land—or, it may stand for some other quantity, such as money earned, hours lived, fluid pumped, fuel consumed, energy generated. The extension of the area concept through calculus over the last three centuries has made modern technology possible.

About 70% of the surface area of Earth is covered with water. U.S. NATIONAL AERONAUTICS AND SPACE ADMINISTRATION (NASA).

Real-life Applications

DRUG DOSING

The amount of a drug that a person should take depends, in general, on their physical size. This is because the effect of a drug in the body is determined by how concentrated the drug is in the blood, not by the total amount of drug in the body. Children and small adults are therefore given smaller doses of drugs than are large adults. The size of a patient is most often determined by how much the patient weighs. However, in giving drugs for human immunodeficiency virus (HIV, the virus that causes AIDS), hepatitis B, cancer, and some other diseases, doctors do not use the patient's weight but instead use the patient's body surface area (BSA). They do so because BSA is a better guide to how quickly the kidneys will clear the drug out of the body.

Doctors can measure skin area of patients directly using molds, but this is practical only for special research studies. Rather than measuring a patient's skin area, doctors use formulas that give an approximate value for BSA based on the patient's weight and height. These are similar in principle to the standard geometric formulas that give the area of a sphere or cone based on its dimensions, but less exact (because people are all shaped differently). Several formulas are in use. In the West, an equation called the DuBois formula is most often used; in Japan, the Fujimoto formula is standard. The DuBois formula estimates BSA in units of square meters based on the patient's weight in kilograms, Wt, and height in centimeters, Ht: BSA $= .007184 Wt^{.425} Ht^{.725}$

In recent years, doctors have debated whether setting drug doses according to BSA really is the best method. Some research shows that BSA is useful for calculating doses of drugs such as lamivudine, given to treat the hepatitis B virus, which is transmitted by blood, dirty needles, and unprotected sex. (Teenagers are a high-risk group for this virus.) Other research shows that drug dosing based on BSA does not work as well in some kinds of cancer therapy.

BUYING BY AREA

Besides addition and subtraction to keep track of money, perhaps no other mathematical operation is performed so often by so many ordinary people as the calculation of areas. This is because the price of so many common materials depends on area: carpeting, floor tile, construction materials such as sheetrock, plywood, exterior siding, wallpaper, and paint, whole cloth, land, and much more. In deciding how much paint it takes to paint a room, for example, a painter measures the dimensions of the walls, windows, floor, and doors. The walls (and ceiling or floor, if either of those is to be painted) are basically rectangles, so the area of each is calculated by multiplying its height by its width. Window and door areas are calculated the same way. The amount of area that is to be painted is, then, the sum of the wall areas (plus ceiling or floor) minus the areas of the windows and doors. For each kind of paint or stain, manufacturers specify how much area each gallon will cover, the spread rate. This usually ranges from 200 to 600 square feet per gallon, depending on the product and on the smoothness of the surface being painted. (Rough surfaces have greater actual surface

area, just as the lid of an egg carton has more surface area than a flat piece of cardboard of the same width and length.) Dividing the area to be painted by the spread rate gives the number of gallons of paint needed.

FILTERING

Surface area is important in chemistry and filtering because chemical reactions take place only when substances can make contact with each other, and this only happens on the surfaces of objects: the outside of a marble can be touched, but not the center of it (unless the marble is cut in half, in which case the center is now exposed on a new surface). Therefore a basic way to take a lump of material, like a crystal of sugar, and make it react more quickly with other chemicals is to break it into smaller pieces. The amount of material stays the same, but the surface area increases.

But don't larger cubes or spheres have more surface area than small ones? Of course they do, but a group of small objects has much more surface area than a single large object of the same total volume. Imagine a cube having sides of length L. Its area is $L = 6L^2$. If the cube is cut in half by a knife, there are now two rectangular bricks. All the outside surfaces of the original cube are still there, but now there are two additional surfaces—the ones that have appeared where the knife blade cut. Each of these surfaces is the same size as any of the cube's original faces, so by cutting the cube in half there has added $2L^2$ to the total area of the material. Further cuts will increase the total surface area even more.

Increasing reaction area by breaking solid material down into smaller pieces, or by filling it full of holes like a sponge, is used throughout industrial chemistry to make reactions happen faster. It is also used in filtering, especially with activated charcoal. Charcoal is solid carbon; activated charcoal is solid carbon that has been treated to fill it with billions of tiny holes, making it spongelike. When water is passed through activated charcoal, chemicals in the water stick to the carbon. A single teaspoonful of activated charcoal can contain about 10,000 square feet of surface area (930 square meters, the size of an American football field). About a fourth of the expensive bottled water sold in stores is actually city tap water that has been passed through activated charcoal filters.

CLOUD AND ICE AREA
AND GLOBAL WARMING

Climate change is a good example of the importance of area measurements in earth science. For almost 200 years, human beings, especially those in Europe, the

United States, and other industrialized countries, have been burning massive quantities of fossil fuels such as coal, natural gas, and oil (from which gasoline is made). The carbon in these fuels combines with oxygen in the air to form carbon dioxide, which is a greenhouse gas. A greenhouse gas allows energy from the Sun get to the surface of the Earth, but keeps heat from escaping (like the glass panels of a greenhouse). This can melt glaciers and ice caps, thus raising sea levels and flooding low-lying lands, and can change weather patterns, possibly making fertile areas dry and causing violent weather disasters to happen more often. Scientists are constantly trying to make better predictions of how the world's climate will change as a result of the greenhouse effect.

Among other data that scientists collect to study global warming, they measure areas. In particular, they measure the areas of clouds and ice-covered areas. Clouds are important because they can either speed or slow global climate change: high, wispy clouds act as greenhouse filters, warming Earth, while low, puffy clouds act to reflect sunlight back into space, cooling Earth. If global warming produces more low clouds, it may slow climate change; if it produces more high wispy clouds, it may speed climate change. Cloud areas are measured by having computers count bright areas in satellite photographs.

Cloud areas help predict how fast the world will get warmer; tracking ice area helps to verify how fast the world has already been getting warmer. Most glaciers around the world have been melting much faster over the last century—but scientists need to know exactly how much faster. To find out, they first take a satellite photo of a glacier. Then they measure its outline, from which they can calculate its area. If the area is shrinking, then the glacier is melting; this is itself an important piece of knowledge. Scientists also measure the area of the glacier's accumulation zone, which is the high-altitude part of the glacier where snow is adding to its mass. Knowing the total area of the glacier and the area of the accumulation zone, scientists can calculate the accumulation area ratio, which is the area of the glacier's accumulation zone divided by its total area. The mass balance of a glacier—whether it is growing or shrinking—can be estimated using the accumulation area ratio and other information.

CAR RADIATORS

Chemical reactions are not the only things that happen at surfaces; heat is also gained or lost at an object's surface. To cool an object faster, therefore, surface area needs to be increased. This is why elephants have big ears: they have a large volume for their body surface area, and their large, flat ears help them radiate extra heat. It is also

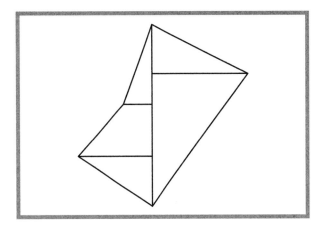

Figure 2.

why we hug ourselves with our arms and curl up when we are cold: we are trying to decrease our surface area. And it is how cars engines are kept cool. A car engine is supposed to turn the energy in fuel into mechanical motion, but about half of it is actually turned into heat. Some of this heat can be useful, as in cold weather, but most of it must simply be expelled. This is done by passing a liquid (consisting mostly of water) through channels in the engine and then pumping the hot liquid from the engine through a radiator. A radiator is full of holes, which increase its surface area. The more surface area a radiator has, the more cool air it can touch and the more quickly the metal (heated by the flowing liquid inside) can get rid of heat. When the liquid has given up heat to the outside world through the large surface area of the radiator, the liquid is cooler and is pumped back through the engine to pick up more waste heat. Car designers must size radiator surface area to engine heat output in order to produce cars that do not overheat.

SURVEYING

If a parcel of land is rectangular, calculating its area is simple: length $\times$ width. But, how do surveyors find the area of an irregularly shaped piece of land—one that has crooked boundaries, or maybe even a winding river along one side?

If the piece of land is very large or its boundaries very curvy, the surveyor can plot it out on a map marked with grid squares and count how many squares fit in the parcel. If an exact area measurement is needed and the parcel's boundary is made up of straight line segments, which is usually the case, the surveyor can divide a drawing of the piece of land into rectangles, trapezoids, triangles. The area of each of these can be calculated separately using a standard formula, and the total area found as the sum of the parts. Figure 2 depicts an irregular piece of property that has been divided into four triangles and one trapezoid.

Today, it is also possible to take global positioning system readings of locations around the boundary of a piece of property and have a computer estimate the inside area automatically. This is still not as accurate as an area estimate based on a true survey, because global positioning systems are as yet only accurate to within a meter or so at best. Error in measuring the boundary leads to error in calculating the area.

SOLAR PANELS

Solar panels are flat electronic devices that turn part of the energy of sunlight that falls on them—anywhere from 1% or 2% to almost 40%—into electricity. Solar panels, which are getting cheaper every year, can be installed on the roofs of houses to produce electricity to run refrigerators, computers, TVs, lights, and other machines. The amount of electricity produced by a collection of solar panels depends on their area: the more area, the more electricity. Therefore, whether a system of solar panels can meet all the electricity demands of a household depends on three things: (1) how much electricity the household uses, (2) how efficient the solar panels are (that is, how much of the sun energy that falls on them is turned into electricity), and (3) how much area is available on the roof of the house.

The average U.S. household uses about 9,000 kWh of electricity per year. A kWh, or kilowatt-hour, is the amount of electricity used by a 100-watt light bulb burning for 10 hours. That's equal to 1,040 watts of around-the-clock use, which is the amount of electricity used by ten 100-watt bulbs burning constantly. A typical square meter of land in the United States receives from the Sun about 150 watts of power per square meter (W/m^2), averaged around the clock, so using solar panels with an efficiency of 20% we could harvest about 30 watts per square meter of panel (on average, around the clock). To get 1,040 watts, therefore, we need 1,040 W / 30 W/m^2 = 34 m^2 of solar panels. At a more realistic 10% panel efficiency, we would need twice as much panel area, about 68 m^2. This would be a square 8.2 meters on a side (27 feet). Many household rooftops in the United States could accommodate a solar system of this size, but it would be a tight fit. In Europe and Japan, where the average household uses about half as much electricity as the average U.S. household, it would be easier to meet all of a household's electricity demands using a solar panel system. Of course, it might still a good idea to meet some of a household's electricity needs using solar panels, even where it is not practical to meet them completely that way.

Where to Learn More

Web sites

Math.com. "Area Formulas." 2005. <http://www.math.com/tables/geometry/areas.htm> (March 9, 2005).

Math.com. "Area of Polygons and Circles." 2005. <http://www.math.com/school/subject3/lessons/S3U2L4GL.html> (March 9, 2005).

O'Connor, J.J., E.F. Robertson. "An Overview of Egyptian Mathematics." December 2000. <http://www-groups.dcs.st-and.ac.uk/~history/HistTopics/Egyptian_mathematics.html> (March 9, 2005).

O'Neill, Dennis. "Adapting to Climate Extremes." <http://anthro.palomar.edu/adapt/adapt_2.htm> (March 9, 2005).

Overview

An average is a number that expresses the central tendency of a group of numbers. Another word for average, one that is used more often in science and math, is "mean." Averages are often used when people need to understand groups of numbers. Whenever groups of measurements are collected in biology, physics, engineering, astronomy or any other science, averages are calculated. Averages also appear in grading, sports, business, politics, insurance, and other aspects of daily life. An average or mean can be calculated for any list of two or more numbers by adding up the list and dividing by how many numbers are on it.

Fundamental Mathematical Concepts and Terms

ARITHMETIC MEAN

There are several ways to get at the "average" value of a set of numbers. The most common is to calculate the arithmetic mean, usually referred to simply as "the mean." Imagine any group of numbers—say, 140, 141, 156, 169, and 170. These might stand for the heights in centimeters (cm) of five students. To find their mean, add them up and divide by the number of numbers in the list, in this case, 5:

$$\text{Mean} = \frac{140 + 141 + 156 + 169 + 170}{5}$$

$$= \frac{776}{5}$$

$$= 155.2$$

Figure 1: Calculation of an average or mean.

The average or mean height of the students is therefore 155.2 centimeters (about 5 ft 1 in). Mentioning the mean is a quicker, easier way of describing about how tall the students in the group are than listing all five individual heights.

This is convenient, but to pay for this convenience, information must be left out. The mean is a single number formed by blending all the numbers on the original list together, and can only tell us so much. From the mean, we cannot tell how tall the tallest person or shortest person in the group is, or how close people in the group tend to be to the mean, or even how big the group is—all things that we might want to know. These details are often given by listing other numbers as well as the mean, such as the minimum (smallest number), maximum (largest number), and standard deviation (a measure of how spread out the list is).

More than one list of numbers might have the same mean. For example, the mean of the three numbers 155, 155.2, and 155.4 is also 155.2.

GEOMETRIC MEAN

The kind of average found by adding up a list of numbers and dividing by how many there are is called the "arithmetic" mean to distinguish it from the "geometric" mean. When numbers on a list are multiplied by each other, they yield a product; the geometric mean of the list is the number that, when multiplied by itself as many times as there are numbers on the list, gives the same product. Take, for example, the list 2, 6, 12. The product of these three numbers is $2 \times 6 \times 12 = 144$. The geometric mean of 2, 6, and 12 is therefore 5.24148 because $5.24148 \times 5.24148 \times 5.24148$ also equals 144.

The geometric mean is not found by adding up the numbers on the list and dividing by how many there are, but by multiplying the numbers together and finding the nth root of the product, where n stands for how many numbers there are on the list. So, for instance, the geometric mean of 2, 6, and 12 is the third (or "cube") root of $2 \times 6 \times 12$:

$$\text{Geometric mean} = \sqrt[3]{2 \times 6 \times 12}$$

$$= \sqrt[3]{144}$$

$$= 5.24148$$

The geometric mean is used much less often than the arithmetic mean. The word "mean" is always taken as referring to the arithmetic mean unless stated otherwise.

THE MEDIAN

Another number that expresses the "average" of a group of numbers is the median. If a group of numbers is listed in numerical order, that is, from smallest to largest, then the median is the number in the middle of the list. For the list 140, 141, 156, 169, 170, the median is 156.

The mean and the median are similar in that they both give a number "in the middle." The difference is that the mean is the "middle" of where the listed numbers are on the number line, whereas the median is just the number that happens to be in the middle of the list. Consider the list 1, 1, 1, 1, 100. The mean is found by adding them up and dividing by how many there are:

$$\frac{1 + 1 + 1 + 1 + 100}{5} = \frac{104}{5} = 20.8$$

The median, on the other hand—the number in the middle of the list—is simply 1. For this particular list, therefore, the mean and median are quite different. Yet for the list of heights discussed earlier (140, 141, 156, 169, 170), the mean is 155.2 and the median is 156, which are similar. What makes the two lists different is that on the list 1, 1, 1, 1, 100, the number 100 is much larger the others: it makes the mean larger without changing the median. (If it were 1 or 10 instead of 100, the median would still be 1—but the average would be smaller.) A number that is much smaller or larger than most of the others on a list is called an "outlier." The rule for finding the median ignores outliers, but the rule for finding the mean does not.

If a list contains an odd number of numbers, as does the five-number list 1, 1, 1, 1, 100, one of the numbers is in the middle: that number is the median. If a list contains an even number of numbers, then the median is the number that lies halfway between the two numbers nearest the middle of the list: so, for the four-number list 1, 1, 2, 100 the median is 1.5 (halfway between 1 and 2).

WHAT THE MEAN MEANS

The mean is not a physical entity. It is a mathematical tool for making sense of a group of numbers. In a group of students with heights 140, 141, 156, 169, 170 cm and average height 155.2 cm, no single person is actually 155.2 cm tall. It does not usually mean much, therefore, when we are told that somebody or something is above or below average. In this group of students, everybody is above or below average.

Further, averages only make sense for groups of numbers that have a gist or central tendency, that are fairly evenly scattered around some central value. Averages do not make sense for groups of numbers that cluster around two or more values. If a room contains a mouse weighing 50 grams and an elephant weighing 1,000,000 grams, you could truly say that the room contains a population of animals weighing, on average, $(50 + 1,000,0000) / 2 = 500,025$ grams, half as much as a full-grown elephant, but this would be somewhat ridiculous. It is more reasonable to say simply that the room contains a 50-gram mouse and a 1,000,000-gram elephant and forget about averaging altogether in this case. If the room contains a thousand mice and a thousand elephants, it might be useful to talk about the mean weight of the mice and the mean weight of the elephants, but it would still probably not make sense to average the mice and the elephants together. The weights of the mice and elephants belong on different lists because mice and elephants are such different creatures. These two lists will have different means. In general, the average or arithmetic mean of a list of numbers is meaningful only if all the numbers belong on that list.

A Brief History of Discovery and Development

The concept of the average or mean first appeared in ancient times in problems of estimation. When making an estimate, we seek an approximate figure for some number of objects that cannot be counted directly: the number of leaves on a tree, soldiers in an attacking army, galaxies in the universe, jellybeans in a jar. A realistic way to get such a figure—sometimes the only realistic way—is to pick a typical part of the larger whole, then count how many leaves, soldiers, galaxies, or jellybeans appear in that fragment, then multiply this figure by the number of times that the part fits into the whole. This gives an estimate for the total number. If there are 100 leaves on a typical branch, for instance, then we can estimate that on a tree with 1,000 branches there will be 100,000 leaves. By a "typical" branch, we really mean a branch with a number of leaves on it equal to the average or mean number of leaves per branch. The idea of the average is therefore embedded in the idea of estimation from typical parts. The ancient king Rituparna, as described in Hindu texts at least 3,000 years old, estimated the number of leaves on a tree in just this way. This shows that an intuitive grasp of averages existed at least that long ago.

By 2,500 years ago, the Greeks, too, understood estimation using averages. They had also discovered the idea of the arithmetic mean, possibly to help in spreading out losses when a ship full of goods sank. By 300 B.C., the Greeks had discovered not only the arithmetic mean but the geometric mean, the median, and at least nine other forms of average value. Yet they understood these averages only for cases involving two numbers. For example, the philosopher Aristotle (384–322 B.C.) understood that the arithmetic mean of 2 and 10 was 6 (because 2 plus 10 divided by 2 equals 6), but could not have calculated the

average height of the five students in the example used earlier. It was not until the 1500s that mathematicians realized that the arithmetic mean could be calculated for lists of three or more numbers. This important fact was discovered by astronomers who realized that they could make several measurements of a star's position, with each individual measurement suffering from some unknown, ever-changing error, and then average the measurements to make the errors cancel out. From the late 1500s on, averaging to reduce measurement error spread to other fields of study from astronomy. By the nineteenth century averaging was being used widely in business, insurance, and finance. Today it is still used for all these purposes and more, including the calculation of grade-point averages in schools.

Real-life Applications

BATTING AVERAGES

A batting average is a three-digit number that tells how often a baseball player has managed to hit the ball during a game, season, or career. A player's batting average is calculated by dividing the number of hits the player gets by the number of times they have been at bat (although this is not the number of times they have stepped up to the plate to hit because there are also special rules as to what constitutes a legal "at bat" to be used in calculating a player's batting average). Say a player goes to bat 3 times and gets 0 hits the first time, 1 the second, and 0 the third (this is actually pretty good). Their batting average is then $(0 + 1 + 0) / 3 = .333$. (A batting average is always rounded off to three decimal places.) A batting average cannot be higher than 1, because a player's turn at bat is over once they get a hit: if a player went up three times and got three hits, their batting average would $(1 + 1 + 1) / 3 = 1.000$.

But this would be superhumanly high. Not even the greatest hitters in the Baseball Hall of Fame got a hit every time they went to bat—or even half the time they went to bat. Ty Cobb, for instance, got 4,191 hits in 11,429 turns at bat for a batting average of .367, the highest career batting average ever. The highest batting average for a single season, .485, was achieved by Tip O'Neill in 1887.

In cricket, popular in much of the world outside the United States, a batsman's batting average is determined by the number of runs they have scored divided by the number of times they have been out. A "bowling average" is calculated for bowlers (the cricket equivalent of pitchers) as the number of runs scored against the bowler divided by the number of wickets they have taken. The

A motorcyclist soars high during motocross freestyle practice at the 2000 X Games in San Francisco. Riders and coaches make calculations of average "hang time" and length of jumps at various speeds so that they know what tricks are safe to land. AP/WIDE WORLD PHOTOS. REPRODUCED BY PERMISSION.

higher a cricket player's batting average, the better; the lower a player's bowling average, the better.

GRADES

In school, averages are an everyday fact of life: an English or algebra grade for the marking period is calculated as an average of all the students' test scores. For example, if you do four assignments in the course of the marking period for a certain class and get the scores 95, 87, 82, and 91, then your grade for the marking period is

$$\frac{95 + 87 + 82 + 91}{4} = 88.75$$

In many schools that assign letter grades, all grades between 80 and 90 are considered Bs. In such a school, your grade for the marking period in this case would be a B.

WEIGHTED AVERAGES IN GRADING

What if some of the assignments in a course are more important than the others? It would not be fair to count them all the same when averaging scores to calculate your grade from the marking period, would it? To make score-averaging meaningful when not all scores stand for equally important work, teachers use the weighted-average method. Calculation of a weighted average assigns a weight or multiplying factor to each grade. For example, quizzes might be assigned a weight of 1 and tests a weight of 2 to signify that they are twice as important (in this particular class). The weighted average is then calculated as the sum of the grades—each grade multiplied by its weight—divided by the sum of the weights. So if during a marking period you take two quizzes (grades 82 and 87) and two tests (grades 95 and 91), your grade for the marking period will be

$$\frac{82 + 87 + (2 \times 95) + (2 \times 91)}{1 + 1 + 2 + 2} = \frac{541}{6} = 90.2$$

Because you did better on the tests than on the quizzes, and the tests are weighted more heavily than the quizzes, your grade is higher than if all the scores had been worth the same.

In most colleges and some high schools, weighted averaging is also used to assign a single number to academic performance, the famous (or perhaps infamous) grade point average, or GPA. Like individual tests, some classes require more work and must be given a heavier weight when calculating the GPA.

WEIGHTED AVERAGES IN BUSINESS

Weighted averages are also used in business. If in the course of a month a store sells different amounts of five kinds of cheese, some more expensive than others, the owner can use weighted averaging to calculate the average income per pound of cheese sold. Here the "weight" assigned to the sales figure for each kind of cheese is the price per pound of that cheese: more expensive cheeses are weighted more heavily. Weighted averaging is also used to calculate how expensive it is to borrow capital (money for doing business) from various lenders that all charge different interest rates: a higher interest rate means that the borrower has to pay more for each dollar borrowed, so money from a higher-interest-rate source costs more. When a business wants to know what an average dollar of capital costs, it calculates a weighted average of borrowing costs. This commonly calculated figure is known in business as the weighted average cost of capital. Spreadsheet software packages sold to businesses for calculating profit and loss routinely include a weighted-averaging option.

AVERAGING FOR ACCURACY

How long does it take a rat to get sick after eating a gram of Chemical X? Exactly how bright is Star Y? Each rat and each photograph of a star is a little different from every other, so there is no final answer to either of these questions, or to any other question of measurement in science. But by performing experiments on more than one rat (or taking more than one picture of a star, or taking any other measurement more than once) and averaging the results, scientists can get a better answer than if they look at just one measurement. This is done constantly in all kinds of science. In medical research, for instance, nobody performs an experiment or gathers data on just one patient. An observation is performed as many times as is practical, and the measurements are averaged to get a more accurate result. It is also standard practice to look at how much the measurements tend to spread out around the average value—the "standard deviation."

How does averaging increase accuracy? Imagine weighing a restless cat. You weigh the cat four times, but because it won't hold still you get a scale reading each time that is a little too high or a little too low: 5.103 lb, 5.093 lb, 5.101 lb, 5.099 lb. In this case, the cat's real weight is 5.1 lb. The error in the first reading, therefore, is .003 lb, because 5.1 + .003 = 5.103. Likewise, the other three errors are −.003, .001, and −.001 lb. The average of these errors is 0:

$$\frac{.003 + (-.003) + .001 + (-.001)}{4} = \frac{0}{4} = 0$$

The average of the four weights is therefore the true weight of the cat:

$$\frac{5.103 + 5.093 + 5.101 + 5.099}{4} = \frac{20.4}{4} = 5.1$$

Although in real life the errors rarely cancel out to exactly zero, the average error is usually much smaller than any of the individual errors. Whenever measurement errors are equally likely to be positive and negative, averaging improves accuracy.

In astronomy, this principle has been used for the star pictures taken by the International Ultraviolet Explorer satellite, which took pictures of stars from 1978 to 1996. To make final images for a standard star atlas (a collection of images of the whole sky), two or three images for each star were combined by averaging. In fact, a weighted average was calculated, with each image being weighted by its exposure: short-exposure images were dimmer, and were given a heavier weight to compensate. The resulting star atlas is more accurate than it would have been without averaging.

HOW MANY GALAXIES?

As scientists discovered in the early twentieth century, the Universe does not go on forever. It is finite in size, like a very large room (only without walls, and other strange properties). There cannot, therefore, be an infinite number of galaxies because there is not an infinite space.

Scientists use averages to estimate such large numbers. Galaxies, like leaves on a large tree, are hard to count. Many galaxies are so faint and far away that even the powerful Hubble Space Telescope must gaze for days a small patch of sky to see them. It would take many years to examine the whole sky this way, so instead the Hubble takes a picture of just one part of the sky—an area about as big as a dime 75 ft (22.86 m) away. Scientists assume that the number of galaxies in this small area of the sky is about the same as in any other area of the same size. That is, they assume that the number of galaxies in the observed area is equal to the average for all areas of the same size. By counting the number of galaxies in that small area and multiplying to account for the size of the whole sky, they can estimate the number of galaxies in the Universe.

In 2004, the Hubble took a picture called the Ultra Deep Field, gazing for 300 straight hours at one six-millionth of the sky. The Ultra Deep Field found over 10,000 galaxies in that tiny area. If this is a fair average for any equal-sized part of the sky, then there are at least twenty billion galaxies in the universe. Most galaxies contain several hundred billion stars.

THE "AVERAGE" FAMILY

Any list of numbers has an average, but an average that has been calculated for a list of numbers that does not cluster around a central value can be meaningless or misleading. In such a case, the "distribution" of the numbers—how they are clumped or spread out on the number line—can be important. This knowledge is lost when the numbers are squashed down into a single number, the average.

In politics, numbers about income, taxes, spending, and debt are often named. It is sometimes necessary to talk about averages when talking about these numbers, but some averages are misleading. Sometimes politicians, financial experts, and columnists quote averages in a way that creates a false impression.

For example, public figures often talk about what a proposed law will give to or take away from an "average" family. If the subject is income, then most listeners probably assume that an "average" family is a family with an income near the median of the income range. For instance, if 99 families in a certain neighborhood make $30,000 a year and one family makes $3,000,000, the median income will be $30,000 but the average income—the total income of the neighborhood divided by the number of families living there—will be $59,700, twice as much as all but one of the families actually make. To say that the "average" family makes almost $60,000 in this neighborhood would be mathematically correct but misleading to a typical listener. It would make it sound like a wealthier neighborhood than it really is.

This problem is that there is an unusually large value in the list of incomes, namely, the single $3,000,000 income—an outlier. This makes the arithmetic average inappropriate. A similar problem often arises in real life when political claims are being made about tax cuts. A tax cut that gives a great deal of money to the richest one percent of families, and a great deal less money to all the rest, might give an "average" of, say, $2,500.00 each year. "My tax cut will put $2,500 back in the pocket of the average American family!" a politician might say, meaning that the sum of all tax cuts divided by the number of all families receiving cuts equals $2,500.00. Yet only a small number of wealthier families might actually see cuts of $2,500 or larger. Middle-class and poorer families, to whom the number "$2500.00" sounds more important because it a bigger percentage of their income—the great majority of voters hearing the politician's promise—might actually have no chance of receiving as much as $2,500. An average figure can misused to convey a false idea while still being mathematically true.

SPACE SHUTTLE SAFETY

Many of the machines on which lives depend—jet planes, medical devices, spacecraft, and others—contain thousands or millions of parts. No single part is perfectly reliable, but in designing complex machines we would like to guarantee that the chances of a do-or-die part failing during use is very small. But how do we put a number on a part's chances for failing?

For commonplace parts, one way is to hook up a large number of them and watch to see how many fail, on average, in a given period of time. But for a complex system like a space shuttle, designers cannot afford to wait and they cannot afford to fail. They therefore resort to a method known as "probabilistic risk assessment." Probabilistic risk assessment tries to guess the chances of the complex system failing based on the reliability of all its separate parts. Reliability is sometimes expressed as an average number, the "mean time between failures" (MBTF). If the MBTF for a computer hard drive is five years, for example, then after each failure you will have to wait—on average—five years until another failure occurs. The MBTF is not a minimum, but an average: the next failure might happen the next day, or not for a decade.

MBTF is not an average from real data, but a guess about the average value of numbers that one does not know yet. MBTF estimates can, therefore, be wrong. In the 1980s, in the early days of the space shuttle program, NASA calculated an estimated MBTF for the space shuttle. Its estimate was that the shuttle would suffer a catastrophic accident, on average, during 1 in every 100,000 launches. That is, the official MBTF for the shuttle was 100,000 launches.

But it was at the 25th shuttle launch, that of the space shuttle *Challenger*, that a fatal failure occurred. Seventy-six seconds after liftoff, *Challenger* exploded. This did not prove absolutely that the MBTF was wrong, because the MBTF is an average, not a minimum—yet the chances were small that an accident would have happened so soon if the MBTF were really 100,000 launches. NASA therefore revised its MBTF estimate down to 265 launches. But in 2003, only 88 flights after the *Challenger* disaster, *Columbia* disintegrated during re-entry into the atmosphere. Again, this did not prove that NASA's MBTF was wrong, but if it were right then such a quick failure was very unlikely.

STUDENT LOAN CONSOLIDATION

Millions of students end up owing tens of thousands of dollars in student loans by the time they finish college. Usually this money is borrowed in the form of several different loans having different interest rates. After graduation, many people "consolidate" these loans. That is, several loans are combined into one loan with a new interest rate, and this new, single loan is owed to a different institution (usually one that specializes in consolidated loans). There are several advantages to consolidation. The new interest rate is fixed, that is, it cannot go up over time. Also, monthly payments are usually lower, and there is only one payment to make, rather than several.

The interest rate on a consolidated student loan is calculated by averaging the interest rates for all the old loans that are being consolidated. Say you are paying off two (rather small) student loans. You still owe $100 on one loan at 7% interest and $200 on another at 8% interest. When the loans are consolidated you will owe

$100 + $200 = $300, and the interest rate will be the weighted average of the two interest rates:

$$\text{New interest rate} = \frac{100 \times .07 + 200 \times .08}{300} = .07667$$

The weights in the weighted average are the amounts of money still owed on each loan: the interest rate of the bigger loan counts for more in calculating the new interest rate, which is 7.667%. In practice, the rate is rounded up to the nearest one eighth of a percent, so your real rate would be 7.75%.

AVERAGE LIFESPAN

We often read that the average human lifespan is increasing. Strictly speaking, this is true. In the mid nineteenth century, the average lifespan for a person in the rich countries was about 40 years; today, thanks to medical science and public health advances such as clean drinking water, it is about 75 years. Here the word "average" means the arithmetic mean, that is, the sum of all individual lifespans in a certain historical period divided by the number of people born in that period.

Some have argued that because average lifespan has been increasing, it must keep on increasing without limit, making us immortal. For example, computer scientist Ray Kurzweil said in "the eighteenth century, we added a few days to the human life expectancy every year. In the nineteenth century, we added a few weeks every year. Now we're adding over a hundred days per year to human life expectancy . . . Many observers, including myself, believe that within ten years we will be adding more than a year—every year—to human life expectancy. So as you go forward a year, human life expectancy will move away from us." (Kurzweil, R. "The Ascendence of Science and Technology [a panel discussion]." *Partisan Review*. Sept 2, 2002.)

The problem with this argument is that it mixes up average lifespan with maximum lifespan. The average lifespan is not increasing because people are living to be older than anyone ever could in the past: they are not. A few people have always lived to be 90, 100, or 110 years old. The reason average lifespan is higher now than in the past is that fewer people are dying in childhood and youth. Today, at least in the industrialized countries, most people do not die until old age. However, the ultimate limit on how old a person can get has not increased, and the average lifespan cannot be increased beyond that limit by advances that keep people from dying until they reach it. Perhaps in the future, medical science will increase the maximum possible age, but that is only a possibility. It has nothing to do with past increases in average lifespan.

INSURANCE

In the industrial world, virtually everyone, from their late teens on up, has some kind of insurance. For example, all European Union states and most U.S. states require that all drivers buy liability insurance—that is, insurance to pay for medical care for anyone that the driver may injure in an accident that is their fault. Insurance is basic to business, health care, and personal life—and it is founded on averages.

Insurance companies charge their customers a certain amount every month, a "premium," in return for a commitment that the insurance company will pay the customer a much greater amount of money if a problem should happen—sickness, car accident, death in the family, house fire, or other (depending on the kind of insurance policy). This premium is based on averages. The insurance company groups people (on paper) by age, gender, health, and other factors. It then calculates what the average rate of car wrecks, house fires, or other problems for the people in each group, and how much these problems cost on average. This tells it how much it has to charge each customer in order to pay for the money that the company will have to pay out—again, on average. To this amount is added the insurance company's cost of doing business and a profit margin (if the insurance company is for-profit, which not all are).

Insurance costs are higher for some groups than for others because they have higher average rates for some problems. For example, young drivers pay more for car insurance because they have more accidents. The average crash rate per mile driven for 16-year-olds is three times higher than for 18- and 19-year olds; the rate for drivers 16–19 years old, considered as a single group, is four times higher than for all older drivers. What's more, young male drivers 16–25, who on average drive more miles, drink more alcohol, and take more driving risks, have more accidents than female drivers in this age group: two thirds of all teenagers killed in car crashes (the leading cause of death for both genders in the 18–25 age group) are male.

More crashes, injuries, and deaths mean more payout by the insurance company, which makes it reasonable, unfortunately, for the company to charge higher rates to drivers in this group. Some companies offer reduced-rate deals to young drivers who avoid traffic tickets.

EVOLUTION IN ACTION

Averaging makes it possible to see trends in nature that can't be seen by looking at individual animals. Averages have been especially useful in studying evolution, which happens to slowly to see by looking at individual animals and their offspring. The most famous example of observed evolutionary changes is the research done by the biologists Peter and Rosemary Grant on the Galapagos Islands off the west coast of South America. Fourteen or 15 closely related species of finches live in the Galapagos. The Grants have been watching these finches carefully for decades, taking exact measurements of their beaks. They average these measurements together because they are interested in how each finch population as a whole is evolving, rather than in how the individual birds differ from each other. The individual differences, like random measurement errors, tend to cancel each other out when the beak measurements are averaged. When a list of data is averaged like this, the resulting mean is called a "sample mean."

The Grants' measurements show that the average beak for each finch species changes shape depending on what kind of food the finches can get. When mostly large, tough seeds are available, birds with large, seed-cracking beaks get more food and leave more offspring. The next generation of birds has, on average, larger, tougher beaks. This is exactly what the Darwinian theory of evolution predicts: slight, inherited differences between individual animals enable them to take advantage of changing conditions, like food supply. Those birds whose beaks just happen to be better suited to the food supply leave more offspring, and future generations become more like those successful birds.

Key Terms

Mean: Any measure of the central tendency of a group of numbers.

Median: When arranging numbers in order of ascending size, the median is the value in the middle of the list.

Where to Learn More

Books

Tanur, Judith M., et al. *Statistics: A Guide to the Unknown.* Belmont, CA: Wadsworth Publishing Co., 1989.

Wheater, C. Philip, and Penny A. Cook. *Using Statistics to Understand the Environment.* New York: Routledge, 2000.

Web sites

Insurance Institute for Highway Safety. "Q7&A: Teenagers: General." March 9, 2004. <http://www.iihs.org/safety_facts/qanda/teens.htm#2> (February 15, 2005).

Mathworld. "Arithmetic mean." Wolfram Research. 1999. <http://mathworld.wolfram.com/ArithmeticMean.html> (February 15, 2005).

Wikelsky, Martin. "Natural Selection and Darwin's Finches." Pearson Education. 2003. <http://wps.prenhall.com/esm _freeman_evol_3/0,8018,849374-,00.html> (February 15, 2005).

Overview

In everyday life, a base is something that provides support. A house would crumble if not for the support of its base. So it is too with math. Various bases are the foundation of the various ways we humans have devised to count things. Counting things (enumeration) is an essential part of our everyday life. Enumeration would be impossible if not for based valued numbers.

Fundamental Mathematical Concepts and Terms

In numbering systems, the base is the positive integer that is equal to the value of 1 in the second highest counting place or column. For example, in base 10, the value of a 1 in the "tens" column or place is 10.

A Brief History of Discovery and Development

The various base numbering systems that have arisen since before recorded history have been vital to our existence and have been one of the keys that drove the formation of societies. Without the ability to quantify information, much of our everyday world would simply be unmanageable. Base numbering systems are indeed an important facet of real life math.

The concept of the base has been part of mathematics since primitive humans began counting. For example, animal bones that are about 37,000 years old have been found in Africa. That is not the remarkable thing. The remarkable thing is that the bones have human-made notches on them. Scientists argue that each notch represented a night when the moon was visible. This base 1 (1, 2, 3, 4, 5, . . .) system allowed the cave dwellers to chart the moon's appearance. So, the bones were a sort of calendar or record of the how frequent the nights were moonlit. This knowledge may have been important in determining when the best was to hunt (sneaking up on game under a full moon is less successful than when there is no moon).

Another base system that is rooted in the deep past is base 5. Most of us are familiar with base 5 when we chart numbers on paper, a whiteboard or even in the dirt, by making four vertical marks and then a diagonal line across these. The base 5-tally system likely arose because of the construction of our hands. Typically, a hand has four fingers and a thumb. It is our own carry-around base 5 counting system.

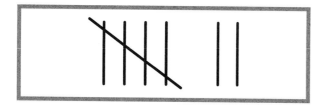

Figure 1: Counting to seven in a base 5 tally system.

In base 5 tallying, the number 7 would be represented as depicted in Figure 1.

Of course, since typically we have two hands and a total of ten digits, we can also count in multiples of 10. So, most of us also naturally carry around with us a convenient base 10 (or decimal) counting system.

Counting in multiples of 5 and 10 has been common for thousands of years. Examples can be found in the hieroglyphics that adorn the walls of structures built by Egyptians before the time of Christ. In their system, the powers of 10 (ones, tens, hundreds, thousands, and so on) were represented by different symbols. One thousand might be a frog, one hundred a line, ten a flower and one a circle. So, the number 5,473 would be a hieroglyphic that, from left to right, would be a pattern of five frogs, four lines, seven flowers and three circles.

There are many other base systems. Base 2 or binary (which we will talk about in more detail in the next section) is at the heart of modern computer languages and applications. Numbering in terms of groups of 8 is a base-8 (octal) system. Base 8 is also very important in computer languages and programming. Others include base 12 (duodecimal), base 16 (hexidecimal), base 20 (vigesimal) and base 60 (sexagesimal).

The latter system is also very old, evidence shows its presence in ancient Babylon. Whether the Babylonians created this numbering system outright, or modified it from earlier civilizations is not clear. As well, it is unclear why a base 60 system ever came about. It seems like a cumbersome system, as compared with the base 5 and 10 systems that could literally rely on the fingers and some scratches in the dirt to keep track of really big numbers. Even a base 20 system could be done manually, using both fingers and toes.

Scholars have tried to unravel the mystery of base 60's origin. Theories include a relationship between numbers and geometry, astronomical events and the system of weights and measures that was used at the time. The real explanation is likely lost in the mists of time.

BASE 2 AND COMPUTERS

Base 2 is a two digit numbering system. The two digits are 0 and 1. Each of these is used alternately as numbers grow from ones to tens to hundreds to thousands and upwards. Put another way, the base 2 pattern looks like this: 0, 1, 10, 11, 100, 101, 110, 111, 1000, . . . (0, 1, 2, 3, 4, 5, 6, 7, 8, . . .).

The roots of base 2 are thought to go back to ancient China but base 2 is as also fresh and relevant because it is perfect for the expression of information in computer languages. This is because, for all their sophistication, computer language is pretty rudimentary. Being driven by electricity, language is either happening as electricity flows (on) or it is not (off). In the binary world of a computer, on is represented by 1 and off is represented by 0.

As an example, consider the sequence depicted in Figure 2.

off-off-on-off-on-on-on-off-on-on

Figure 2: Information series.

In the base 2 world, this sequence would be written as depicted in Figure 3.

0010111011

Figure 3: Information series translated to Base 2.

View the fundamental code for a computer program and you will see line upon line of 0s and 1s. Base 2 in action!

Each 0 or 1 is known as a bit of information. An arrangement of four bits is called a nibble and an arrangement of 8 bits is called a byte (more on this arrangement below, in the section on base 8).

A base 2 numbering system can also involve digits other than 0 and 1, with the arrangement of the numbers being the important facet. In this arrangement, each number is double the preceding number. This base 2 pattern looks like this: 1, 2, 4, 8, 16, 32, 64, 128, 256, . . . It is also evident that in this series, from one number to the next, the numbers of the power also double. For example, compare the numbers 64 and 128. In the larger number, 12 is the double of 6 and 8 is the double of 4.

Base 8

In the base 8 number system, each digit occupies a place value (ones, eights, sixteens, etc.). When the number 7 is reached, the digit in that place switches back to 0 and 1 is added to the next place. The pattern looks like this: 0, 1, 2, 3, 4, 5, 6, 7, 10, 11, 12, 13, 14, 15, 16, 17, 20, 21, 22, ...

Each increasing place value is 8 times as big as the preceding place value. This is similar to the pattern shown above for base 2, only now the numbers get a lot bigger more quickly. The pattern looks like this: 1, 8, 64, 512, 4096, 32768, . . .

As mentioned in the preceding section, the base 2 digits can be arranged in groups of 8. In the computer world, this arrangement is called a byte. Often, computer software programs are spoken of in terms of how many bytes of information they consist of. So, the use of the base 8 numbering system is vital to the operation of computers.

Base 10

The base 10, or decimal, numbering system is another ancient system. Historians think that base 10 originated in India some 5,000 years ago.

The digits used in the base 10 system are 0 through 9. When the latter is reached, the value goes to 0 and 1 is added to the next place. The pattern look like this: 0, 1, 2, 3, 4, 5, 6, 7, 8, 9, 10, 11, 12, 13, 14, . . .

Each successive place value is 10 times greater than the preceding value, which results in the familiar ones, tens, hundreds, thousands, etc. columns with which we usually do addition, subtraction, multiplication and division.

Where to Learn More

Books

Devlin, K.J. *The Math Gene: How Mathematical Thinking Evolved & Why Numbers are like Gossip*. New York: Basic Books, 2001.

Gibilisco, S. *Everyday Math Demystified*. New York: McGraw-Hill Professional, 2004.

Web sites

Loy, J. "Base 2 (Binary)." <http://www.jimloy.com/math/base2.htm> (October 31, 2004).

Poseidon Software and Invention. "Base Valued Numbers." <http://www.psinvention.com/zoetic/basenumb.htm> (October 31, 2004).

Smith, J. "Base Arithmetic." <http://www.jegsworks.com/Lessons/reference/basearith.htm> (October 30, 2004).

Business Math

Money is the difference between leisure activity and business. While enjoying leisure activity one can expect to pay to have a good time by purchasing a ticket, supplies or paying a fee to gain access to whatever they wish to do. Business activity in any form spends money to earn money. In both cases, numbers are the alphabet of money and math is its universal language.

Computing systems have displaced manual information gathering, recordkeeping, and accounting at an ever-increasing rate within the business world. Advancing computer technology has made this possible and, to some extent, decreasing math skills among the general populations of all nations have made it necessary. One of the initial motivating factors that have led more and more stores to investing large amounts of money to install and operate code-scanning checkout systems is the increasing difficulty in finding an adequate number of people with the necessary math skills to consistently and reliably make change at checkout counters. The introduction of these systems has improved merchants' ability to keep accurate records of what they sell, what they need to order, and to recognize what their customers want so that they may maintain a ready supply. However, for all of the advances business computing has made in generating real-time management reports, none of it is of any value without people who can interpret what it means and, to do that, one must understand the math used by the computing system. Simply because a computer prints out a report does not ensure that it is accurate or useful.

It is worth stating that those people with good math skills will have the best opportunities to excel in many ways in jobs and careers within the business world. Math is not just an exercise for the classroom, but is a critical skill if one is to succeed now and in the future. All money is being monitored and managed by someone. One's personal future depends on how well they manage their money. The future of any employer, and the local, state, and national governments in which one lives, depends on how well they manage money. Money attracts attention. If a person or the business and the governmental institutions they depend on do not use the math skills necessary to wisely manage the money in their respective care, someone else will and they are not likely have the best interest of others in mind. Math skills are one of the most essential means for one to look after their own best interest as an individual, employee, investor, or business owner.

Fundamental Mathematical Concepts and Terms

Business math is a very broad subject, but the most fundamental areas include budgets, accounting, payroll, profits and earnings, and interest.

BUDGETS

All successful businesses of any size, from single individuals to world-class corporations, manage everything according to a budget. A budget is a plan that considers the amount of money to be spent over a specific time schedule, what it is to be spent on, how that money is to be obtained, and what it is expected to deliver in return. Though this sounds simple, it is a very complicated concept.

Businesses and governments rise and fall on their ability to perform reliably according to their budgets. Budgets include detailed estimates of money and all related activities in a format that enables the state of progress toward established goals and objectives to be monitored on a regular basis through various business reports. The reports provide the information necessary for management to identify opportunity and areas of concern or changing conditions so that proper adjustments may be made and put into action in timely fashion to improve the likelihood of success or warn of impending failure to meet expectations. In a budget, all actions, events, activities, and project outcomes are quantified in terms of money.

The basic components of any budget are capital investments, operating expense and revenue generation. Capital investments include building offices, plants and factories, and purchasing land or equipment and the related goods and services for new projects, including the cost of acquiring the money to invest in these projects. Expense outlays include personnel wages, personnel benefits, operating goods and services, advertising, rents, royalties, and taxes.

Budgets are prepared by identifying and quantifying the cost and contributions from all ongoing projects, as well as new projects being put in place and potential new projects and opportunities expected to be begun during the planning cycle. Typically, budgets cover both the immediate year and a longer view of the next three to five years. Historical trends are derived by taking an after-look at the actual results of prior period budgets compared to their respective plan projections. Quite often the numerical data is converted to graphs and charts to aid in spotting trends and changes over time. A simple budget is represented by Figure 1.

The math involved in this simplistic example budget is addition, subtraction, and multiplication, where Revenue from shoe and sandal sales = Number of pairs of sold multiplied by the price received; Personnel Expense = Number of people employed each month multiplied by individual monthly wages; Federal Taxes = The applicable published tax rate multiplied times Income Before Tax.

As the year progresses, a second report would be prepared to compare the projections above with the actual performance. If seasonal shoe sales fall below plan, then the company knows that they need to improve the product or find out why it is not selling as expected. If shoe sales are better than expected, they may need to consider building another factory to meet increasing demand or acquire additional shoes elsewhere.

This somewhat boring exercise is essential to the A.Z. Neuman Shoe Factory to know if it is making or losing money and if it is a healthy company or not. This information also helps potential investors decide if the company is worth investing money in to help grow, to possibly buy the company itself, or to sell if they own any part of it. As a single year look at the company, A.Z. Neuman seems to be doing fine. To really know how well the company is doing, one would have to look at similar combined reports over the past history of the company, its outstanding debts, and similar information on its competitors.

ACCOUNTING

Accounting is a method of recordkeeping, commonly referred to as bookkeeping, that maintains a financial record of the business transactions and prepares various statements and reports concerning the assets, liabilities, and operating performance of a business. In the case of the A.Z. Neuman Shoe Factory, transactions include the sale of shoes and sandals, the purchase of supplies, machines, and the building of a new store as shown in the budget. Other transactions not shown in detail in the budget might include the sale of stocks and bonds or loans taken to raise the necessary money to buy the machines or build the new store if the company did not have the money on hand from prior years' profits to do so.

People who perform the work of accounting are called accountants. Their job is to collect the numbers related to every aspect of the business and put them in proper order so that management can review how the company is performing and make necessary adjustments. Accountants usually write narratives or stories that serve to explain the numbers. Computing systems help gather and sort the numbers and information, and it is very important that the accountant understand where the

A. Z. Neuman Shoe Factory - Projected Annual Budget – Figures rounded to $MM (millions)

Months:	J	F	M	A	M	J	J	A	S	O	N	D	Total
Revenue													
Shoe sales	3	4	4	16	14	2	2	3	18	4	3	1	74
Sandal sales	0	1	1	3	4	3	3	2	1	1	0	0	19
Total	3	5	5	19	18	5	5	5	19	5	3	1	93
Operating Expense													
Personnel	1	2	2	2	1	1	1	1	1	1	1	1	15
Supplies	2	2	2	2	2	2	2	2	1	1	1	1	20
Electricity	1	1	1	1	1	0	1	0	1	0	1	0	8
Local Taxes	0	0	0	1	0	0	1	0	0	0	0	1	3
Total	4	5	5	6	4	3	5	3	3	2	3	3	46
Net Contribution (Revenue – OpExp.)													
	−1	0	0	13	14	2	0	2	16	3	0	−2	47
Capital Investments													
Machines	1	3	3	4	0	0	0	0	0	0	0	0	11
New Store	0	0	0	0	5	0	0	0	0	0	0	0	5
Total	1	3	3	4	5	0	0	0	0	0	0	0	16
Income Before Tax (IBT = Net Contribution – Capital)													
	−2	−3	−3	9	9	2	0	2	16	3	0	−2	31
State & Federal Tax (Minus = credit)													
	−1	−1	−2	3	3	1	0	0	5	1	0	−1	8
Income After Tax (IAT = IBT – S&FT)													
	−1	−2	−1	6	6	1	0	2	11	2	0	−1	23

Figure 1: A simple budget.

computing system got its information and what mathematical functions were performed to produce the tables, charts, and figures in order to verify that the information is true and correct. Management must understand the accounting and everything involved in it before it can fully understand how well the company is doing.

When this level of understanding is not achieved for any reason, the performance of the company is not likely

Troy McConnell, founder of Batanga.com at his office. The center broadcasts alternative Hispanic music on dedicated Internet channels to consumers between 12 and 33 years of age. In addition, studies at the center include all aspects of business math. AP/WIDE WORLD PHOTOS. REPRODUCED BY PERMISSION.

to be as expected. It would be like trying to ride a bicycle with blinders on: one hopes to make to the corner without crashing, but odds are they will not. Recent and historical news articles are full of stories of successful companies that achieved positive outcomes because they were aware of what they were doing and managed it well. However, there are almost as many stories of companies that did not do well because they did not understand what they were truly doing and mismanaged themselves or misrepresented their performance to investors and legal authorities. If they only mismanage themselves, companies go out of business and jobs are lost and past investments possibly wasted. If a company misrepresents itself either because it did not keep its records properly, did not do its accounting accurately, or altered the facts and calculations in any untruthful way, people can go to jail. The truth begins with honest mathematics and numbers.

PAYROLL

Payroll is the accounting process of paying employees for the work performed and gathering the information for

budget preparation and monitoring. An employee sees how much money is received at the end of a pay period, while the employer sees how much it is spending each pay period and the two perspectives do not see the same number. Why? A.Z. Neuman wants to attract quality employees so it pays competitive wages and provides certain benefits. Tom Smith operates a high-tech machine that is critical to the shoe factory on a regular 40-hour-per-week schedule, has been with the company a few years, and has three dependents to care for. How much money does Tom take home and what does it cost A.Z. Neuman each month? Figure 2 lays out the details.

This is just an example. Not all companies offer such benefits, and the relative split in shared cost may vary considerably if the cost is shared at all. If Tom is a member of a labor union, dues would also be withheld. As is shown in Figure 2, the company has to spend approximately $2 for every $1 Tom takes home as disposable income to live on. Correspondingly, Tom will take home only about half of any raise or bonus he receives from the company. At the end of each tax year, Tom then has to file both State and Federal income tax and may discover that

	Tom	A.Z. Neuman	Government
Gross Pay ($25/hour, 40 hours/week, 4 weeks per month)	$4,000	-$4,000	
Withholdings: (Required by law)			
Social Security 12.4% split 6.2% each	-$248	-$248	$ 496
Medicare 2.9% split 1.45% each	-$58	-$58	$ 116
State & Federal Unemployment Insurance		-$120	$ 120
Federal Income Tax	-$800		$ 800
State Income Tax	-$200		$ 200
Savings Plan (Tom can put up to 4%, company matches)	-$160	-$160	
Insurance (Cost split between Tom and company)			
Life	-$62	-$62	
Medical	-$72	-$72	
Net Pay	$2,400	-$4,720	$1,732

Figure 2: Sample of payroll accounting.

he is either due a refund or owes even more depending on his individual situation. Tax withholdings are required by State and Federal law at least in part to fund the operation of governmental functions throughout the year. In some regions of the country, there are other local and city taxes not shown in this example. If people had to send their tax payments in every month in place of having them automatically withheld they would be more mindful of the burden of taxation. In theory, Tom will get his contributions to Social Security and Medicare back in the future in his old age. Tom's contribution to the savings plan is his own attempt to ensure his future.

PROFITS

Unless Mr. A.Z. Neuman just really enjoyed making shoes, he founded the business to make a profit. A profit is realized when the income received is greater than the sum of all expenditures. As shown in the example budget in Figure 1, the company does not make a profit every month and is very dependent on a few really good months when shoe sales are in season to yield a profit for the year. Most businesses operate in this up and down environment. Some business segments have even longer profit and loss cycles, such that they may lose money for several years before experiencing a strong year and hopefully making enough profit to sustain them through the next down cycle. If they fail to make a profit long enough, companies go out of business and this occurs to a large percentage of all companies every year. Without the effective application of good math skills in accounting and

business evaluations as well as the ability to understand their meaning to direct future decisions, companies have no idea if they are in fact growing or dying, but they can be sure they are doing one or the other.

EARNINGS

Fundamentally, profits and earnings are defined in very similar terms. However, earnings are often thought of as the return on capital investments as distinguished from expense, as shown in the budget example in Figure 1. One has to know how much capital has been invested throughout the life of the company to fairly calculate the earnings or return on capital employed. The example budget is limited to only one year and suggests that A.Z. Neuman is expected to earn $23 million while investing $16 million in that year. The budget shows that the company had product to sell before investing in new equipment and a new store; thus, the year shown is benefiting from prior year investments of some unknown magnitude. In the developing period of any company, annual earnings are negative (losses) until the initial investments have generated earnings of equal amount to reach what is called payout. Once past payout, companies can begin generating a positive annual return on capital employed. Some industries require continued annual capital investment to expand or replace their asset base, and this will continue to hold down their annual rate of return until such time as there are no more attractive investment opportunities and they are in the later stage, but high earnings generating phase, of their business life. Typically, businesses that can

A presentation in a "business" environment. PREMIUM STOCK/CORBIS.

generate a 15% rate of return on capital employed over a period of several years have done very well. Most companies struggle to deliver less than half that level of earnings.

INTEREST

Interest is money earned on money loaned or money paid on money borrowed. Interest rates vary based on a variety of factors determined in financial markets and by governmental regulations. Low interest rates are good for a borrower or anyone dependent on others' ability to borrow money to buy goods and services. High interest rates are good for those saving or lending. When the A.Z. Neuman Shoe Factory wants to buy additional equipment or build new factories or stores, it has to determine where the money will come from to do so. If interest rates are low, it may elect to borrow instead of spending its own cash. If interest rates are high, it will have to consider other courses of raising the money needed to fund investments if it has a cash reserve and wishes to hold on to it for protection or other investments. The two primary ways businesses raise capital, other than borrowing, are to sell stocks and bonds in the company.

A share of stock represents a fractional share of ownership in the company for the price paid. The owner of stock shares in the future performance of the company. If the company does well, the stock goes up and the investor does well, and can do very well under the right circumstances. If the company does poorly, the investor does poorly and can lose the entire amount invested. Stock ownership has a definite share of risk while it has a definite attraction of significant growth potential. Companies will pay a return, or dividend, that might be thought of as interest to stockholders when it can afford to do so as incentive for them to continue to own the stock.

Bonds are generally less risky than stocks, but only those ensured by cash reserves or the assets of sound national governments are secure. A company issues a bond, or guaranty, to investors willing to buy them that over a specified period of time interest will be paid on the amount invested and that the original investment will be returned to the buyer when the bond matures. However, the security of a bond is only as good as the company issuing it. It is in the best interest of a company to meet its bond obligations or it may never sell another bond.

Key Terms

Balance: An amount left over, such as the portion of a credit card bill that remains unpaid and is carried over until the following billing period.

Bankruptcy: A legal declaration that one's debts are larger than one's assets; in common language,

when one is unable to pay his bills and seeks relief from the legal system.

Interest: Money paid for a loan, or for the privilege of using another's money.

The advantage of a bond to the company is that ownership is not being shared among the buyers, the upside potential of the company remains owned by the company, and the interest rate paid out is usually less than the interest rate that would have to be paid by the company on a loan. The benefit to the buyer is that bonds are not as risky as stock and, while the return is limited by the established interest rate, the initial investment is not at as great a risk of loss. Bonds are safer investments than stocks in that they tend to have guaranteed earnings, even if considerably lower than the growth potential of stock without the downside risk of loss.

Companies pay the interest on loans, the interest on bonds, and any dividends to stockholders out of their earnings; thus, the rate of return as mentioned earlier is an important indicator to potential investors of all types. The assessment of business risks and opportunity can only be performed through extensive mathematical evaluation,

and the individuals performing these evaluations and using them to consider investments must possess a high degree of math skills. In the end, the primary difference between evaluating a business and balancing one's own personal checkbook is the magnitude of the numbers.

Where to Learn More

Books

Boyer, Carl B. *A History of Mathematics.* New York: Wiley and Sons, 1991.

Bybee, L. *Math Formulas for Everyday Living.* Uptime Publications, 2002.

Devlin, Keith. *Life by the Numbers.* New York: Wiley and Sons, 1998.

Westbrook, P. *Math Smart for Business: Essentials of Managerial Finance.* Princeton Review, 1997.

Overview

A calculator is a tool that performs mathematical operations on numbers. Some of the simplest calculators can only perform addition, subtraction, multiplication, and division. More sophisticated calculators can find roots, perform exponential and logarithmic operations, and evaluate trigonometric functions in a fraction of a second. Some calculators perform all of these operations using repeated processes of addition.

Basic calculators come in sizes from as small as a credit card to as large as a coffee table. Some specialized calculators involve groups of computing machines that can take up an entire room. A wide variety of calculators around the world perform tasks ranging from adding up bills at retail stores to figuring out the best route when launching satellites into orbit. Calculators, in some form or another, have been important tools for mankind throughout history. Throughout the ages, calculators have progressed from pebbles in sand used for solving basic counting problems to modern digital calculators that come in handy when solving a homework problem or balancing a checkbook.

People regularly use calculators to aid in everyday calculations. Some common types of modern digital calculators include basic calculators (capable of addition, subtraction, multiplication, and division), scientific calculators (for dealing with more advanced mathematics), and graphing calculators. Scientific calculators have more buttons than more basic calculators because they can perform many more types of tasks. Graphing calculators generally have more buttons and larger screens allowing them to display graphs of information provided by the user. In addition to providing a convenient means for working out mathematical problems, calculators also offer one of the best ways to verify work performed by hand.

Fundamental Mathematical Concepts and Terms

Modern calculators generally include buttons, an internal computing mechanism, and a screen. The internal computing mechanism (usually a single chip made of silicon and wires, called a microprocessor, central processing unit, or CPU) provides the brains of the calculator. The microprocessor takes the numbers entered using the buttons, translates them into its own language, computes the answer to the problem, translates the answer back into our numbering system, and displays the answer on the screen. What is even more impressive is that it usually does all of this in a fraction of a second.

A student works on his Texas Instruments graphing calculator. American students have been using graphing calculators for over a decade, and Texas Instruments accounts for more than 80% of those sales, according to an industry research firm. Texas Instruments faces what may turn out to be a more serious challenge: software that turns handheld computers into graphing calculators. AP/WIDE WORLD PHOTOS. REPRODUCED BY PERMISSION.

The easiest way to understand the language of a calculator is to compare it to our numbering system, which is a base ten system. This is due to the fact that we have ten fingers and ten toes. For example, consider how humans count to 34 using fingers. You basically keep track of how many times you count to ten until you get to three, and then count four more fingers. This idea is represented in our numbering system. There is a three in the tens column and a four in the ones column. The tens column represents how many times we have to go through a set of ten fingers, and the ones column represents the rest of the fingers required. A calculator counts in a similar way, but its numbering system is based on the number two instead of ten. This is known as a binary numbering system, meaning that it is based on the number two.

Our ten-based numbering system is known as the decimal numbering system. Much in the same way that each column of a decimal number represents one of the ten numbers between zero and nine, a number in binary

form is represented by a series of zeros and ones. Though binary numbers may seem unintuitive and confusing, they are simpler than decimal numbers in many ways, allowing complex calculations to be carried out on tiny microprocessor chips.

The columns (places) in the decimal numbering system each represent multiples of ten: ones, tens, hundreds, thousands, and so forth. After the value of a column reaches nine, the next column is increased. Similarly, the columns in binary numbers represent multiples of two: ones, twos, fours, eights, and so on. Counting from zero, binary numbers go 0, 1, 10, 11, 100, 101, 110, 111, 1000, etc. 110 represents six because it has a one in the fours column, a one in the twos column, and a zero in the ones column. Because binary notation only involves two values in different columns, it is common to think of each column either being on or off. If a column has a 1 in it, then the value represented by the column (1, 2, 4, 8, 16, 32, and so on) is included in the number. So a 1 can be seen to mean that the column is on, and a 0 can be seen to mean that the column is off. This is the essence of the binary numbering system that a calculator uses to perform mathematical operations.

As an example, add the numbers 6 and 7 together. Using fingers to count in decimal numbers, count 6 fingers and then count 7 more fingers. When all ten fingers are used, make a mental tally in the tens column, and then count the last three fingers to get a single tally in the tens column and three in the ones column. This represents one ten and three ones, or 13. When you input 6 plus 7 into a calculator, the calculator firsts translates the two numbers into binary notation. In binary notation, 6 is represented by 110 (a one in the fours column, a one in the twos column, and a zero in the ones column) and 7 is represented by 111 (a one in the fours column, a one in the twos column, and a one in the ones column). Next, the two numbers are added together by adding the columns together. First, adding up the values in the ones column (0 and 1) results in a one in the ones column. Next, adding the values in the twos column results in a 2 so the twos column of the sum get a 0 and the next column over, the fours column, is increased by one (just like the next column in the decimal numbering system is increased when a column goes beyond nine). Adding this to the other values in the fours column results in a 3 in the fours column (because the two numbers being added together each have a 1 in the fours column), so the eights column now has a 1 in it, and a 1 is still left in the fours column. Now listing the columns together reveals the answer in the binary form: 1101. Finally, the calculator translates this answer back into decimal form and displays it on the screen: $8 + 4 + 0 + 1 = 13$. As illustrated by this example, the columns in the binary numbering system cause each other to increase much quicker than the columns in the decimal number system. Many calculators use this form of addition as the basis for the most complicated of operations.

Most calculators allow combinations of operations, but paying attention to the order of operations is essential. For example, a calculator can find the value of four plus six and then divide by two to arrive at five, or it can find the value of four plus the value of six divided by two to arrive at seven. If the numerical and operational numbers (e.g., addition and division) are pressed in the wrong order, the (correct) answer to the wrong question will be found. For example, adding two numbers, dividing by two, and then adding another number usually results in a different value than adding three numbers and then dividing by two.

The ability to store numbers is a valuable function of a calculator. For example, if it takes a long series of operations to find a number that will be used in future calculations, the number can be stored in the calculator (usually by pressing a button labeled STO) and then recalled when needed (usually by pressing a button labeled RCL). Some universally important numbers have been permanently stored in most calculators. Most scientific calculators, for example, have a button for recalling a reasonable approximation of the value of pi, the number that defines how a circle's radius is related to its circumference and area. The exact value of pi cannot be represented on a typical scientific calculator, and repeatedly typing in the numbers involved in the approximation of pi would be tedious to say the least. The ability to quickly provide important numbers is one of most significant benefits of electronic calculators.

Calculators that are capable of more than basic addition, subtraction, multiplication, and division usually have the ability to work in three different modes: degrees, radians, and gradians. These modes pertain to different units for measuring angles. Degrees are used for most of the basic operations. A right angle is 90 degrees and a circle encompasses 360 degrees. Radians measure angles in terms of pi, where pi represents the same angle as 180 degrees (a straight line or half way around a circle). Most calculators indicate that they are working in the degree mode by displaying DEG in the screen. A right angle is half of pi and a circle is represented by pi multiplied by two. When a calculator is working in terms of radians, RAD usually appears in the screen. In gradians, a circle is represented by 400; so a right angle is 100 gradians and a straight line is 200 gradians. This mode is usually indicated by GRAD displayed in the screen.

A Brief History of Discovery and Development

As previously mentioned, the decimal numbering system is based on the number ten because the earliest calculating devices were the ten fingers found on the human body. As human intelligence developed, calculators evolved to incorporate pebbles and sticks. In fact, the word calculator comes from a form of the late fourteenth century word calculus, which originally referred to stones used for counting. Long before the inception of the word, many different ancient civilizations used piles of stones (as well as twigs and other small plentiful things) to count and perform basic addition. However, counting out large piles of stones had limitations (imagine counting 343 stones and then adding 421 stones to find the sum). As civilizations progressed, needs for more efficient calculators increased. For example, more and more merchants were selling their goods in the growing towns, and keeping track of sales transactions became a common need.

Around 300 B.C., the Babylonians used the first counting board, called the Salamis Tablet, which consisted of a marble tablet with parallel lines carved into it. Stones were set on each line to indicate how many of each multiple of five were needed to represent the number. Counting boards similar to the Salamis Tablet eventually appeared in the outdoor markets of many different civilizations. These counting boards were usually made of large slabs of stone and intended to remain stationary, but people with more money could afford more portable boards made of wood.

The abacus took the counting board methods to another level by allowing beads to be slid up and down small rods held together by a frame. The word abacus stems from the Greek word abax, meaning table, which was a common name for the counting boards that became obsolete with the popularization of the abacus. Historians believe that the first abacus was invented by the Aztecs between A.D. 900 and 1000. The Chinese version of the abacus, which is still the calculator of choice in many parts of Asia, first appeared around A.D. 1200. In A.D. 1600, a Russian form of abacus was invented. A Japanese style of Abacus was invented in 1930 and is still widely used in that country. The rods of most abaci are divided into two sections (called decks) by a bar, with the beads above the bar representing multiples of five. A top bead in the ones column represents five, a top bead in the tens column represents 50, and so on. Some abaci have more than two decks. In 1958, the Lee abacus was invented by Lee-Kai-chen. This abacus is still used in some areas. It can be thought of as two abaci (the plural

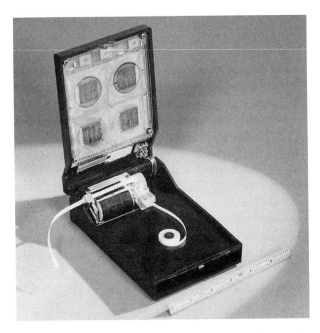

View of the inside of the first miniature calculator, invented at Texas Instruments in 1967. CORBIS/SYGMA.

of abacus) stacked on top of each other, and is supposed to facilitate multiplication, division, and other more complicated operations.

Mathematical tables and slide rules were two of the most common computational aids before small electronic calculators became reasonably affordable in the 1970s. Mathematical tables were used for thousands of years as a convenient way to find values of certain types of mathematical problems. For example, finding the value of 23 multiplied by 78 on a multiplication table only requires finding the row next to the number 23 and then following that row until reaching the column labeled 78; no computation is necessary, and finding the value takes little time.

The first slide rule was created in 1622. A typical slide rule consists of a two or more rulers marked with numeric scales. At least one of the rulers slides so that two or more of the scales move along each other. Different types of slide rules can be used to reduce various complex operations to simple addition and subtraction. By aligning the scales in the proper positions and observing the positions of other marks on the rulers, a trained user can make quick computations by reducing multiplication and more complex operations to simple addition. Slide rules, along with mathematical tables, remained two of the most useful mathematical tools until they were made obsolete in most areas of computation by the invention of electronic calculators.

The invention of the slide rule was dependent on the discovery of logarithms about a decade earlier because the scales on a slide rule involve logarithms. John Napier was the first to publish writings describing the concept of logarithms, though historians also point out that the idea was most likely conceived a few years earlier by Joost Bürgi, a Swiss clockmaker. The math behind the discovery and development of logarithms is beyond the scope of this text, but their main contribution to science and mathematics lies in their ability to reduce multiplication to addition, division to subtraction. Furthermore, exponents can be found using only multiplication; and finding roots only involves division. For example, when using a table of logarithmic values to multiply two large numbers, one only needs to find the logarithmic values for both of the numbers and add them together. The invention of the slide rule made it possible to work with logarithms without searching through large tables for values.

Many mechanical calculators were invented before the electronic technology used in modern calculators came about. One such mechanical calculator, the Pascaline, was invented in 1642 by 19-year-old French mathematician Blaise Pascal. The Pascaline was based on a gear with only one tooth attached to another gear that had ten teeth. Every time the gear with one tooth completed a turn it would cause the other gear to move a tenth of the way around, so the gear with ten teeth completed one turn for every ten turns of the gear with one tooth. Using multiple gears in this way, the Pascaline mechanically counted in way similar to a person counting on their fingers or using an abacus. The concepts first explored in the Pascaline mechanical calculator are still used in things like the odometer that keeps track of how far an automobile has gone, and the water meter that keeps track of how much water is used in a household.

Compact electronic calculators were made readily available in the early 1970s and changed mathematics forever. Not only were these calculators small and easily portable, they substituted for both slide rules and mathematical tables with their ability to store important and commonly used numbers and to use them in complex operations. With clearly labeled buttons and a screen that shows the answer, these calculators were easier to use and required less practice to master. Like slide rules, many modern electronic calculators use logarithms to reduce mathematical operations to repeated operations of addition.

Personal computers are powered by the same type of technology as handheld calculators. Most computers include a software program that simulates the look and feel of a handheld calculator, with buttons that can be clicked with the mouse. The main difference between computers and calculators is that computers are capable of handling complex logical expressions involving unknown values. This basically means that computers are capable of processing more types of information and performing a wider variety of tasks. Making the jump from calculators to computers is an important technological milestone. Just as people a thousand years ago could not have imagined a small battery-operated mathematical tool, it is difficult to imagine a technology that will replace electronic calculators and computers.

Real-life Applications

FINANCIAL TRANSACTIONS

When it comes to personal finances, electronic calculating devices have gone far beyond helping people balance checkbooks. Cash registers and automatic teller machines (ATMs) have shaped how people trade money for products and services.

Cash Registers A cash register can be thought of as a large calculator with a secured drawer that holds money. The cash register was originally invented in 1879 to prevent employee theft. The drawer on most cash registers can only be opened after a sales transaction has taken place so that employees can not purposely fail to record a transaction and pocket the money. Manually opening the drawer requires either a secret code or a key that is kept safe by the store manager or owner. The buttons on a cash register are different from the buttons on calculators intended for personal use. The basic buttons of a calculator that are applicable to money (e.g., the numbers and the decimal point) are present on a cash register; but the remaining buttons can usually be customized to fit the needs of the organization that uses it. For example, a restaurant can program a group of buttons to store the prices of their various menu items; or cash registers in certain geographic locations might have buttons for computing the regional sales tax. The screen can usually be turned so that the merchant and the customer can both see the prices, taxes (if any), and total. Like many calculators, a cash register has a roll of paper and a printing device used for creating printed records of calculations (called receipts in the case of monetary transactions). The inside of a cash register works (and always has worked) almost exactly like a calculator. Modern cash registers include electronic microprocessors similar to those found in handheld calculators; but when calculators were powered by the turning of mechanical gears, cash registers were also powered by similar gear mechanisms.

ATM Machines Automatic teller machines (ATMs) were first used in 1960 when a few machines were placed in bank lobbies to allow customers to quickly pay bills without talking to a bank teller. Later in the decade, the first cash dispensing ATMs were introduced, followed by ATMs that could accept and read bank cards. The fact that ATMs are unmanned requires that they possess greater security. To ensure the safety of the bank's money, the materials that make up the ATM and connect it to a building are precisely constructed and physically strong. To thwart attempts to pose as another person in order to take that person's money out of an ATM, transactions require two forms of identification: physical possession of a bank card and knowledge of a personal identification number (PIN). While the inner workings of an ATM are more complicated than that of a cash register, the technology and concepts of the electronic calculator provide the basis for computing the values of every transaction.

The introduction of check cards has combined the technological benefits of cash registers and ATMs to further facilitate the storage and expenditure of money. A check card can be used to make purchases using money that is stored in a checking account at a bank in another location. Other advancements in technology (e.g., scanners that quickly scan barcodes on items, self-checkout stations that allow customers to scan their own items, and secure Internet transactions that use calculators operating on a computer thousands of miles away from the computer being used by the customer) continue to revolutionize how humans buy and sell products and services. However, none of these accomplishments would have been possible without tools that automatically perform the mathematical operations that take place in every monetary transaction.

NAUTICAL NAVIGATION

For hundreds of years, sailors used celestial navigation: navigating sea vessels by keeping track of the relative positions of stars in the sky. Through the ages, a wide variety of tools have been created to help a navigators navigate boats and ships from one point to another in a safe and timely manner. Different colored buoys warn of shallow waters or fishing nets, and ensure that ships do not collide when nearing docks and harbors. A compass is an essential tool for determining and maintaining directional bearings. Tables of tides and detailed nautical maps help to determine the quickest and safest route and foresee potential obstacles and dangers. For centuries,

Beat the Abacus

Contests throughout the world have pitted individuals equipped with an abacus against individuals equipped with a handheld digital calculator. In most cases, the person with the abacus wins, no matter how complicated the mathematical operations involved. This, of course, does not mean that even the most skilled person with an abacus can make calculations faster than a calculator; the time that it takes to press the buttons accounts for most of the time that it takes to use a calculator to solve a problem. Nonetheless, even in operations as complicated as multiplying and dividing 100 pairs of numbers with up to 12 digits (trillions), a proficient abacus user beats a skilled calculator user almost every time.

navigation of the seas required an in-depth understanding of trigonometry (relationships between lengths and angles) and intensive calculations performed by hand; and, as many navigators have discovered the hard way, small directional errors can result in devastating miscalculations over a trip of thousands of miles. Handheld electronic calculators have proven to be an essential navigational aid since they became reasonably affordable. They are often used aboard sea vessels as either the primary tool for calculating directions and distances on the water or the secondary tool for double-checking calculations carried out by hand.

For every type of navigational problem that can be solved with the help of a handheld electronic calculator, there is also a specialized calculator for solving the specific problem. Often found either on a sea vessel or on the Internet, several calculators have been programmed to take a few pertinent values and find a specific answer. One example of a specialized nautical calculator is a speed-distance-time calculator for finding the time that it will take to get from one point to another if traveling at a certain speed. Most of these calculators require two of the three values (speed, distance, and time) in order to calculate the third value. The time that it takes to get from one point to another is the product of the distance between the two points and the speed at which the ship is traveling (time is equal to distance multiplied by speed). Similarly, to figure out how fast the ship needs to travel in order to get from one point to another in a

specified amount of time requires dividing the distance by the desired time of travel (speed is equal to distance divided by time). Finally, to figure out far a ship will go if traveling at a given speed for a specified amount of time, the speed and time must be multiplied together (distance is equal to speed multiplied by time). Due to the fact that all of these operations involve only multiplication and division, this type of calculator only needs to be capable of multiplication and division. More sophisticated navigation calculators exist to quickly determine values that help a ship's navigator make crucial decisions. These decisions range from determining the fuel necessary for completing a trip and planning appropriate stops for refueling, and finding the true direction in which to steer the ship in order to maintain a desired heading (direction) while taking into account forces such as wind and the current of the water. Specialized calculators are also often used to ensure that a ship is built properly. One such calculator measures a ship's resistance to capsizing (turning upside-down in the water) based on the width of the widest part of the ship and the weight of the ship.

Although modern global positioning system (GPS) technology allows precise and accurate position measurements, calculators (whether external or internal) are used determine vectors (directions and distance) to execute course changes or to determine the best path.

COMPOUND INTEREST

Banking can be a highly profitable business. For example, a bank can use the money in a savings account for other investments as long as the money is stored at the bank; so the more money present in the bank's various accounts at any given time, the more money the bank can earn on its own investments. As an incentive for banking customers to store their money with a bank, savings accounts earn compound interest. That is, the bank pays a savings account holder a relatively small amount of money based on the amount of money in the savings account. The basic idea that drives this investment chain is that the bank makes more money in its own investments than it pays out to its account holders.

The amount of money that is earned on a savings account containing a given amount of money is determined by a compound interest formula. Compound interest is an example of exponential growth: the larger the number becomes, the faster the number grows. The term compound refers to the idea that the growth depends both on how much money is deposited into the account as well as the amount of interest already earned in past growth periods. These growth periods are referred to as compounding periods. Interest is typically compounded annually or monthly, but may also be compounded weekly, or even daily. More frequent compounding benefits the account holder and may be offered to attract more account holders in order to increase the bank's profits.

Determining the amount of interest earned and predicting future account values requires calculations of inverses (1 divided by a number) and exponents (one number raised to the power of another number), both of which are usually rather messy operations, especially when performed by hand. A handheld scientific calculator allows account holders to calculate these values quickly and accurately in order to compare banks and track earnings with ease.

MEASUREMENT CALCULATIONS

How calculators function to solve an array of measurement and conversion problems is perhaps best illustrated by example. Imagine a local high school is hosting a regional basketball tournament. On the day of the tournament, the athletic director discovers that the supply closet has been vandalized and all of the basketballs have been damaged. As the athletic director begins to make the announcement that the tournament will have to be delayed due to the lack of basketballs in the building, a student in the crowd reveals that she has a basketball in her backpack and throws it down to the court. Before the tournament can resume, the officials must determine whether or not the ball is regulation size. All of the writing, including the size of the ball, has been worn off by years of use. Fortunately, one of the referees knows that the diameter of a full-sized basketball (the distance from one side of the ball to the other measured through the center of the ball) is about 9.4 inches. The high school home economics teacher, who happens to be in the crowd, quickly produces a tape measure from her purse, hoping to be of assistance. However, an accurate measurement of the diameter of the basketball cannot be determined with a tape measure. The referee measures the circumference of the ball (the longest distance around the surface of the ball) and finds that it is 29.5 inches. Not knowing the circumference of a regulation-size basketball, the referee asks if anyone in the crowd might know how to solve this problem.

A student speaks up, stating that he has been studying circles and spheres in his math class. He was able to recall an important fact that would help to determine the diameter of the basketball: the circumference of a sphere (such as a basketball) is equal to the diameter of the sphere multiplied by pi. So the diameter of the basketball

is 29.5 divided by pi. The student cannot remember a good approximation of the value of pi, but his scientific calculator has a button for recalling the value of pi (approximated to the ten digits that his calculator can display). He enters 29.5, presses the / button (for division), recalls the value of pi (which displays 3.141592654), and presses the = button (the equal sign). The calculator displays the answer as 9.390141642. This value rounds to 9.4, which is the value that the referee indicated as the diameter of a regulation basketball. The ball is accepted by the officials and the tournament continues.

RANDOM NUMBER GENERATOR

When conducting scientific experiments, it is often necessary to generate a random number (or a set of multiple random numbers) in order to simulate real-life situations. For example, a group of scientists attempting to model the way that fire spreads in a forest need to account for the fact that a burning tree may or may not ignite a nearby tree. Unpredictable factors like shifting winds and seasonal levels of moisture make incorporating the probability of fire spreading in a certain direction into models next to impossible because the nature of wildfires is seemingly random. However, this randomness can be loosely accounted for in scientific wildfire models by strategically inserting random numbers into the mathematical formulas that are used to describe the nature of the fire. These models are often run repeatedly in order to evaluate how well they fit real-world observations. Each time the formula is used, different random numbers are generated and inserted into the formula.

Cryptography Another important area of study that benefits from the generation of random numbers is cryptography, in which messages are encrypted (scrambled) so that they cannot be understood if they are intercepted by an unauthorized party. A message is encrypted according to mathematical formulas. Most of these encryption formulas incorporate random numbers in order to create keys that must be used to decrypt (unscramble) the message. The decryption key is available only to the message sender and the intended message reader.

Random number generators are important tools in many other scientific endeavors, from population modeling to sports predictions. Fortunately, most scientific calculators and graphing calculators include buttons for generating random numbers. Some calculators have a single button (often labeled RAND or RND) for generating a random three-digit number, between 000 and 999. Each time the button is pressed, a new random number is generated. Other calculators also allow the user to adjust the number of digits and the placement of the decimal point in the random generated numbers. Other calculators allow the user to define upper and lower bounds for random numbers; and some can generate multiple random numbers at once. On such a calculator, inputting a set of three numbers that looks something like (1, 52, 9) will cause the calculator to display nine random numbers with values between one and 52. These values can then be used to represent the drawing of nine cards from a deck of 52 playing cards, where each card is assigned a number between one and 52.

Random number generators included in calculators (and various computer software programs) not only make it easy to generate the random numbers needed to simulate real-life situations; using random number generators also ensures that the numbers are truly random. The idea of structured randomness may seem strange; but in order to fully simulate the true randomness found in real-life situations, random number generators use mathematical formulas to generate the numbers according to certain guidelines. One such guideline ensures that the numbers are distributed in certain ways (e.g., to ensure that the numbers are not all close together or equally spaced). Different methods for achieving randomness are used to generate random numbers, and choosing a method is an important consideration in scientific modeling scenarios. Random numbers generated according to mathematical formulas are referred to as pseudorandom numbers.

With the hordes of numbers and unknowns required to model real-life situations, it can be easy to lose sight of the essential ideas behind the data. Using random numbers in mathematical models makes it possible to imitate experiments and focus on the underlying patterns and ideas.

BRIDGE CONSTRUCTION

The construction of large suspension bridges requires almost unfathomable amounts of calculations to ensure that the structures can withstand the multitude of forces acting on a bridge at any given time. Although a suspension bridge looks solid, it is a complex structure that is constantly swaying and twisting; if it were rigid, it would snap under heavy winds and other forces. The weight of the roadway alone would cause the bridge to crumble if swooping cables attached to strong towers were not accurately designed and built. Some forces, including gravity and the weight of the materials that make up the bridge, are constant (unchanging). Other factors are constantly changing: the weight of the automobiles, the strength of the wind, the strength of the water current pushing on the supporting structure, varying temperatures, earthquakes

and other disastrous activity. Bridge engineers must ensure that a bridge can withstand the worst possible situations. For example, a worst-case calculation might examine the stability of the bridge supporting the maximum number of automobiles while under the pressure of high winds and strong water currents during a reasonably large earthquake. When building a large bridge, the slightest miscalculation has the potential of endangering hundreds of human lives.

Before the invention of electronic calculators, the colossal calculations involved in building a safe and long-lasting bridge were performed (and rechecked many times) by hand with the assistance of slide rules and enormous mathematical tables. When the Golden Gate Bridge was built in San Francisco, California, it was the longest suspension bridge in the world. Most experts believed the distance that needed to be spanned in order to build a bridge across the Golden Gate Straight was too large. Furthermore, the many other regional complications—including characteristically high winds, strong tidal currents, the weight of water formed by dense fog, and frequent earthquake activity—made most bridge engineers skeptical to say the least. However, Joseph B. Strauss, who worked on hundreds of bridges in his life, successfully planned and headed the construction of the Golden Gate Bridge. Strauss and his team of engineers worked for months using circular slide rules and making (and rechecking) calculations involving more than 30 unknowns (e.g., the height of the towers, the lengths and arcs of the cables, the thickness of the roadway, the speed of the wind, the strength of water currents, and the weight of automobiles). The bridge took over four years to build, spanned 4,200 feet (1,280 m), and cost over 30 million dollars. To someone accustomed to using a handheld electronic calculator, even the task of approximating the cost of the bridge—taking into account the amounts of materials, the number of people required for construction, and the predicted amount of time needed—seems daunting.

The invention of electronic calculating devices greatly reduced the amount of time needed to perform and repeat immense calculations. For example, the stability of a suspension bridge depends heavily on the lengths of the cables, the heights of the towers to which the cables are connected, and the angles between them. A typical scientific calculator has buttons labeled SIN, COS, and TAN. These buttons are related to trigonometry, the study of triangles that defines the relationship between lengths and angles, and greatly reduce the time needed to calculate and confirm crucial measurements for the parts of a bridge.

Bridge engineering continued to advance as calculating devices evolved into computing technology that could quickly and accurately simulate diverse situations involving numerous adjustable factors. The record for longest suspension bridge has been broken many times since the completion of the Golden Gate Bridge. In 1998, the Akashi Kaikyo Bridge was constructed across the Akashi Strait between Kobe and Awaji-shima in Japan. This massive steel bridge was the longest (and tallest) suspension bridge as of 2005, spanning a total of 12,828 feet (3,910 m). It took over ten years and 4.3 billion dollars to build.

COMBINATORICS

Combinatorics is the mathematical field relating to the possible combinations of a given number of items. A common example investigates the number of possible arrangements (called permutations) for a standard deck of 52 playing cards. It can be shown that 52 cards can be arranged in a surprisingly large number of ways, which is essential in making cards games unpredictable and interesting.

To grasp the idea, start with the order in which cards are usually organized when the pack is first opened: increasing from two to king (with the ace on one end or the other) and separated by suit (hearts, spades, diamonds, clubs, not necessarily in that order). That's one combination (permutation). Next, take the card on the top of the deck and move it down one position. That's two combinations. Continue to move that card down one position until it reaches the bottom of the deck. That is 52 different combinations obtained by moving a single card. Next, take the next card in the deck (the card that is now on the top of the stack) and move it through all of the possible positions. Keep in mind that the first combination was already accounted for when the first card was in its final position on the bottom of the stack; so that is another 51 combinations of cards. The next card will provide another 50 combinations, and so on. It turns out that the total number of combinations is the product of the numbers between 52 and one: 52 multiplied by 51 multiplied by 50, and so on down to one. This type of value (the product of every whole number between a given number and one) appears often in combinatorics and has a standard notation. The number of combinations for 52 cards, for example, is written as 52! and pronounced fifty-two factorial. This type of multiplication is difficult and time-consuming to work out by hand. Fortunately, typical scientific calculators include a button for performing factorial operations (usually labeled n! and pronounced n factorial). Entering 52 and then pressing n! returns a number larger than eight multiplied by ten to the 67th power. That means that number of possible combinations for a deck of 52 cards is more than 8 followed by 67

zeros! A million has only six zeros; a billion only nine. Factorial operations tend to yield large numbers and are difficult to calculate by hand; but calculators make it easy to find the values of factorials of reasonably large numbers, and even perform operations on those values.

UNDERSTANDING WEATHER

The practice of predicting the weather has been a growing art for centuries; but no advancement has influenced the field meteorology (the scientific study of Earth's climate and weather) more than the development of calculating and computing devices. As with most scientific fields, the common availability of electronic calculators affected meteorological studies by greatly reducing the amount of time required for making calculations needed to predict the weather, and updating these calculations based on frequent changes in observed weather data.

American meterologist Joanne Simpson, the first woman to earn a doctorate in meteorology, developed the first model of cloud activity in Earth's atmosphere, helped to explain the forces that power hurricanes, and discovered the cause of the air currents in tropic regions. The calculations that led to her theories were originally performed in the 1940s and 50s without the assistance of an electronic calculator. Simpson's theories were met with criticism and disbelief, but she would later stand as a shining example of electronic calculating devices verifying human calculations. Using calculators and, eventually, computers, she was able to improve her models, revolutionizing meteorological research and prediction. After years of tireless research and teaching positions at multiple universities, she went on to work at the National Aeronautics and Space Administration (NASA) for over 24 years.

While working at NASA, Simpson was integral in the advancement of meteorological studies using images and information gathered by satellites orbiting Earth. Starting in 1986, Simpson headed NASA's Tropical Rainfall Measuring Mission (TRMM). This mission involved the launch and utilization of the first satellite capable of measuring the rainfall in Earth's tropical and subtropical regions from space. This mission has been regarded as one of the most important advancements in the field of meteorological research, deepening the understanding of meteorological phenomena ranging from the affects of dust and smoke on rain clouds to the origins of hurricanes.

The scientific accomplishments of Simpson—from hand calculations leading to ground-breaking theories, to cutting-edge technological research—provide an excellent illustration of the power of a brilliant mind teamed up with technology. Having already revolutionized her field long before the availability of electronic calculating devices, she first used calculators to verify her results, and then continued to stretch the limitations of meteorology by employing the ever-growing capabilities of computers.

Potential Applications

SUPERCOMPUTERS

The earliest and simplest personal computers differed from calculators in that they could interpret instructions involving unknown values, called variables. As computers evolve, they retain the ability to perform the mathematical calculations for which calculators exist. Most personal computers have a calculator program. On the computer screen, the calculator program resembles a handheld calculator. The buttons on the screen can be clicked with the mouse, or the computer keyboard can be used to input the numbers and commands. Even the most advanced computers are based on the concepts that enable calculators to perform mathematical operations.

Supercomputers are computing systems that possess the most power and are capable of the highest level of computation in any given time period. Since the early days of digital computation, supercomputers have been employed to perform large amounts of complex calculations that go beyond the capabilities of the common computing machines of their era. Supercomputers (also known as high performance computing systems) become outdated as technology advances. For example, the most powerful personal computers in the new millennium possess power more than supercomputers from past technological eras.

Long before handheld electronic calculators were made available to the general public, electronic supercomputers were used by the United States government to break enemy codes during World War II. Military codes are usually implemented using mathematical formulas that define how information is transformed, and in turn, how to transform it back to its original, readable form. Cracking these codes requires massive amounts of calculations to determine the values of the mathematical formulas.

In the 1990s, large oil companies began deploying supercomputers developed to analyze enormous amounts of seismic data (information about vibrations in the ground). This data can be used to create images that represent the underlying contents of the terrain, helping the oil companies locate oil in the ground and leading to a dramatic increase in the accuracy of their searches.

Other uses for supercomputers include the creation of detailed three-dimensional (3-D) models of information that is difficult to grasp in its raw form. Weather patterns can be visualized, making them easier to interpret, hopefully improving future predictions. The chemical compositions of a virus can be seen in a new light that may help medical experts expose a plan of attack. The surface of planets can be recreated based on information gathered by a spacecraft. Supercomputers make it possible to understand the invisible building blocks of our world and to access the far reaches of the universe for close inspection. Like all major scientific contributions, supercomputers allow the men and women of science to achieve goals that were previously nothing more than dreams. Considering the abundance of new applications leading to breakthrough scientific discoveries each year, supercomputers seem to possess boundless potential for enhancing our understanding of science.

Where to Learn More

Books

Tao, W., and R. Adler *Cloud Systems, Hurricanes, and the Tropical Rainfall Measuring Mission (TRMM)-A Tribute to Joanne Simpson.* Boston: American Meteorological Society, 2003.

Periodicals

Petroski, H. "Keeping Swaying Bridges from Falling Down." *Science* no. 295 (2002): 2374–2375.

Web sites

Boats.com. "Navigating With a Calculator: Solving navigation problems with the touch of a button." <http://www.boats.com/boat-articles/Seamanship-148/Navigating+With+a+Calculator/2932.html> (March 18, 2005).

University of Mannheim. "Top 500 Supercomputer Sites." <http://www.top500.org/> (March 16, 2005).

Calculus

Calculus is a set of mathematical tools for solving certain problems. It assumes the methods of algebra and geometry, and in turn is used as a starting point for higher forms of mathematics. Calculus is applied to a wide range of problems in medicine, the social sciences, economics, biological growth and decay, physics, and engineering. It is a flexible language for describing the physical world: objects in motion, chemical reactions, complex surfaces and volumes, heating and cooling, the hard-to-imagine behaviors of space and time, and many other events. Not only do all students of medicine, engineering, physics, biology, and economics study calculus in college, so do many students of psychology, literature, art, business, and history.

Calculus, like geometry, is built on a foundation of simple elements that can be drawn on paper or seen in the mind: curves, slopes, and areas. And if ever a branch of mathematics was created to deal with real-life problems, this is it. English physicist Isaac Newton (1642–1727) and German mathematician Gottfried Wilhelm von Leibniz (1646–1716) invented calculus in the 1600s because it was needed to solve the cutting-edge science and math problems of their time, including how to calculate the lengths of curves, the areas bounded by curves, and the motion of objects that are accelerating (gaining speed).

More than three hundred years later, the language of calculus is common to almost every scientific or technical field. It is essential to the design of engines, computers, and all other complex machines; to economics, agriculture, and physics; in fact, to the study of pretty much everything from subatomic particles to the shape of the Universe. Almost every scientist and engineer in the world understands at least basic calculus. And even for those who will never as adults perform any mathematical operation more complex than bidding on eBay, the *concepts* of calculus, like the concepts of basic physics, can clarify our thinking about the world, make it more effective.

Geometry tells us how to deal with curves and shapes that are described by simple rules: squares, triangles, circles, ellipses, and so on. Calculus tells us how to deal with curves and shapes that are described by more complex rules, or by no rules. Calculus is more suited to the real world, where irregular curves and shapes—including graphs of changing speeds, forces, and other things that can be measured—are the rule.

Fundamental Mathematical Concepts and Terms

FUNCTIONS

To talk about calculus, we need to be able to talk about functions. A "function" is a rule that relates one group of numbers to another. For example, we can write a rule that relates each positive number, x, to some negative number, $f = -x$. This rule or function tells us that if x equals 10, f equals -10. Likewise, if x equals 2.5 then f equals -2.5, and so on. We say that "f is a function of x" when f is on the left-hand side of the equals sign, as it is here.

When f is a function of x, it is common practice to write f as $f(x)$, which we read aloud as "f of x." So the function $f = -x$ can also be written as $f(x) = -x$. If we do not want to write the whole rule out every time (or if we do not know what that rule is), we can simply write $f(x)$.

By the way, other letters besides x and f can be used to write functions. In fact, if we are talking about more than one function at a time, we have to use other letters to keep from getting confused.

It is often useful to make a picture of a function. This is done by picking values for x, applying the rule of the function, and finding out what values of f result. In this way any number x can be paired with a number f. These pairs can be graphed as dots by hand or computer. If enough of these dots are graphed, they can be joined by a smooth line. In this discussion, such graphs will be used to show what various functions look like. For example, a graph of the function $f(x) = 2x$ is shown in Figure 1.

THE DERIVATIVE

Consider the simple function shown in Figure 2, $f(x)$. The exact rule that relates f to x is not important right now; what matters to us is the shape of its graph.

The function depicted is often used in real-life math problems. It approximates one of the curves used to describe the spread of epidemics and other processes that spread geometrically in a finite medium (discussed in the article on Exponents). It also approximates part of the spectrum of tidal wetland growth (discussed further below as a real-life application of derivatives). Such a curve is shown in many introductory calculus textbooks because it offers a clear visual basis for explaining the concepts of calculus. In particular, tangent lines with positive slope can be laid against the curve without confusingly overlapping over other parts of it; also, the integral of (area under) this curve is easy to understand and to graph. Moreover, the derivative of this curve is a nontrivial bell shape that can be used to explore min-max problems.

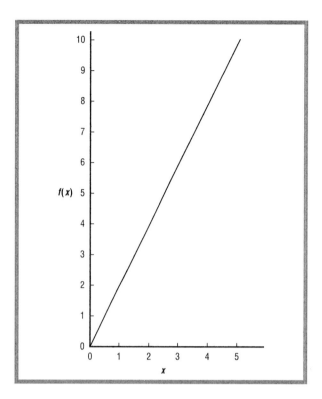

Figure 1: Graph of the function $f(x) = 2x$. The line could go on forever, but only a part can be shown.

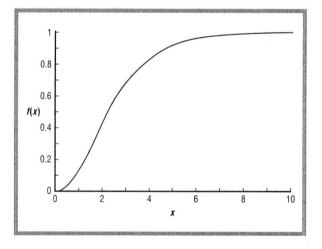

Figure 2: A simple function.

We will use this function to visualize the first ideas of calculus, but the questions we are about to ask about this particular $f(x)$ could just as well be asked about many other curves.

The two basic ideas of calculus arise from asking two questions about curves like this one. The first is: How steep is the curve at any one point? A more exact way of

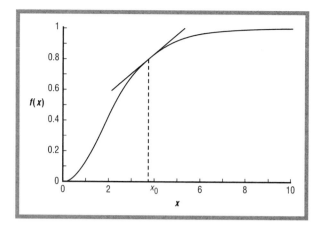

Figure 3: A line segment tangent to f(x).

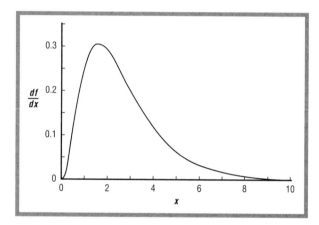

Figure 4: Derivative of the function in Figure 2.

asking this is: If we lay a straight line against the curve so that it touches it at just one point, how steep is that line? In other words, what is its slope? Figure 3 shows a line segment just touching $f(x)$ at the point directly above x_o. (The subscript "0" is just a label to distinguish x_o from other values of x.) This line segment is said to be "tangent" to the curve.

The slope of a line tangent to a curve is called the *derivative* of the curve at the point where the line and the curve touch. The derivative tells us how steep the curve is where the line touches.

A single tangent line shows the derivative only at one point, but the derivative can be found in the same way for every single point along the curve. These numbers can be graphed as a curve in their own right. From Figure 3 you can see that the slope of $f(x)$ starts out small (at zero, actually), gets bigger as $f(x)$ increases, and shrinks as $f(x)$ starts to level off. The derivative of $f(x)$ is shown in Figure 4.

The derivative of $f(x)$ is often written

$$\frac{df}{dx}$$

because it corresponds to the slope or rate of change of $f(x)$ at a single point x. The numerator df stands for a very small vertical change in $f(x)$, and the denominator dx stands for a very small horizontal change in x, so

$$\frac{df}{dx}$$

echoes the definition of a slope from elementary geometry,

$$\frac{\text{rise}}{\text{run}}.$$

Another way of looking at the derivative or slope of a function at a given point is that it tells you the *rate of change* of the function at that point.

A function $f(x)$ might be defined either by a series of measurements of some real-world quantity, or by an equation. In the case of the curve in Figures 2 and 3, the equation for $f(x)$ happens to be $f(x) = (1 - 2^{-x})^3$. There is a set of standard rules (which students in introductory calculus courses learn by heart) that says exactly how to write down what df/dx is, starting with an equation for $f(x)$. Applying these rules to $f(x)$ is called "differentiating" $f(x)$ or "taking the derivative of" $f(x)$. Some functions do not cooperate with these rules and so their derivatives cannot be written down explicitly, in which case computers must be used to find their derivatives.

Taking derivatives is one of the two fundamental operations of calculus. But what use are derivatives? Why bother with them?

Derivatives can help pilot remote vehicles (e.g., to help robot rovers navigate on other planets, such as Mars). If, for example, an engineer is piloting a robot rover on Mars by remote control and plans to drive it up a hill, he might rely on orbital photography to provide data for a graph of altitude versus distance along the rover's proposed line of travel. This graph might look something like the function in Figure 2. But if the Mars rover isn't strong enough to climb at any angle steeper than, say, 30 degrees, the pilot would want to look at the *derivative* of the altitude curve to make sure that the steepness of the proposed route never exceeds 30 degrees at any point. The peak value of the curve in Figure 4 would tell you exactly what maximum steepness your rover was going to encounter. If that value was too high, you'd have to try another route.

One more word about derivatives. The derivative of a function is just another function, and so you can take its derivative too. This function is called the "second derivative of $f(x)$" and is written

$$\frac{d^2 f}{dx^2}$$

The second derivative of the $f(x)$ graphed in Figure 2 is shown in Figure 5.

If you take the derivative of the second derivative, you have the third derivative of $f(x)$. You could find the fourth, fifth, or ten-thousandth derivative of $f(x)$, too, but the first and second derivatives are by far the most useful. For instance, the first derivative of a function that describes an object's position describes the object's speed, and the second derivative describes the object's acceleration. Any derivative beyond the first is called a "higher-order" derivative.

THE INTEGRAL

Now to ask a simple but important question about the function in Figure 2: What is the area under any given part of the curve? In Figure 6, the area under the curve between $x = 0$ and $x = x_0$ has been shaded in.

This area is called the definite integral of $f(x)$ between 0 and x_0. The definite integral, like the derivative at a single point, is simply a number. In our example, the definite integral says how many square inches the shaded area in Figure 6 is.

The definite integral can tell us the actual physical area of an object with curving edges. It can have other physical meanings, as well. For example, the integral of an equation that describes an object's velocity tells us how far the object has traveled. Consider an object moving at a steady speed or velocity, v. Velocity might change over time, so we will write v as a function of time, $v(t)$. If the object's velocity happens to be an unchanging 100 miles per hour, we can write $v(t) = 100$ miles per hour. This function is shown in Figure 7.

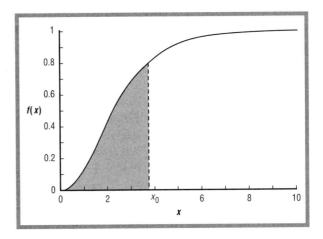

Figure 6: Shaded part is the area under $f(x)$ between 0 and x_0.

The definite integral of the velocity, $v(t)$, from time 0 to time t_0 is the area under the curve from 0 to time t_0. This area is shaded in Figure 8.

The length of the rectangle in Figure 8 is t_0 and its height is 100. Therefore, its area—the value of the definite integral—is $100 \times t_0$. The definite integral of a velocity function is useful because it gives the distance traveled in that time. In 1 hour, the object in our example will have traveled $100 \times 1 = 100$ miles; in 2.5 hours, it will have traveled $100 \times 2.5 = 250$ miles.

Here the area calculation is simple because the velocity is unchanging, so we can use the formula for the area of a rectangle. In real life, objects such as cars, bullets, spacecraft, and runners change their velocities over time. In this case the velocity curve is not a flat line (as in Figure 7) but a more complicated curve, perhaps like that in Figure 6. The more complex the curve, the more complex the mathematics needed to find its integral.

Just as with the derivative, it is possible to find a series of definite integrals and to graph them as a function in their own right. This is called *integrating* $f(x)$, and the resulting curve or function is called the *indefinite integral* (or simply the *integral*) of $f(x)$. Also, instead of graphing the integral point by point by evaluating definite integrals, it is often possible to find an exact expression for the integral. The indefinite integral of a function $f(x)$ is written as follows:

$$\int f(x)dx$$

The symbol at the far left that looks like a stretched "s", $\int$, actually *is* a stretched "s." Centuries ago it stood for "summa," which is Latin for "sum," a reference to summing up the area under the curve. This symbol is called the *integral sign*. The expression

$$\int f(x)dx$$

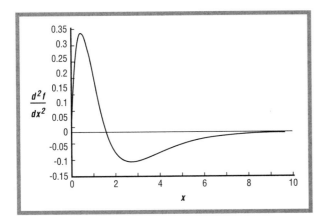

Figure 5: The second derivative of the $f(x)$ shown in Figure 2.

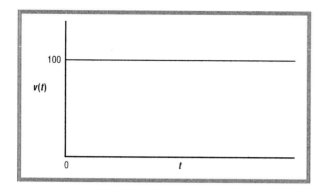

Figure 7: The velocity, $v(t)$, of an object moving at 100 miles per hour.

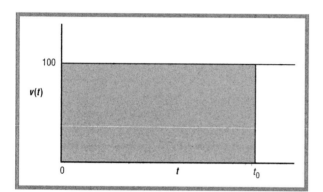

Figure 8: The area of the shaded rectangle is the definite integral of the velocity $v(t)$ from time 0 to time t_0.

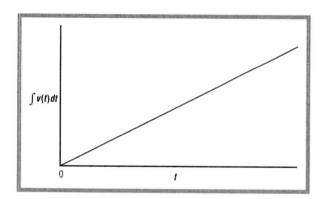

Figure 9: The indefinite integral of $v(t)$.

is read aloud as "the integral of $f(x)$ dee x." (We can use letters other than f and x whenever we like; they are just labels or names.)

In our example of an object moving at a constant 100 miles per hour, the integral of $v(t) = 100$ is easy to write down as an exact mathematical expression:

$$\int v(t)\,dt = 100t + C$$

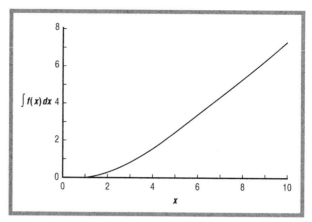

Figure 10: The indefinite integral of the function first seen in Figure 2, $\int kt\,dt = \dfrac{k}{2}\,t^2$.

The C at far right stands for "constant of integration." C is "arbitrary," meaning that it can be set equal to any number and the equation will still be true. The indefinite integral of $v(t) = 100$, namely

$$\int v(t)\,dt = 100t + C$$

is plotted in Figure 9 with C set equal to 0.

An indefinite integral is more informative than a definite integral because it can tell us the value of any definite integral. To find the value of a definite integral over a certain interval (for example, from time 0 to time t_0), we subtract the value of the indefinite integral at the left-hand end of the interval from its value at the right-hand end. In the case of the object moving at 100 miles per hour, the value of the indefinite integral at time t_0 is $100 \times t_0 + C$. At time 0 it is $100 \times 0 + C$. Subtracting, we have

$$\begin{array}{r} 100 \times t_0 + C \\ -\ 100 \times 0 + C \\ \hline 100 \times t_0 \end{array}$$

which is exactly what we found by calculating this definite integral as the area of a rectangle.

We have already seen, in Figures 4 and 5, the first and second derivatives of the curve first shown in Figure 2. The integral of that curve is shown in Figure 10.

As mentioned earlier, in a real-life application, the "area" under a curve may not correspond to a literal, physical area like 10 square miles of parking lot. In calculus, depending on what real-world quantity you're measuring, an "area" may be the number of miles driven, or the profit made by a business, or the amount of oil leaked from a beached tanker, or the probability that a rocket will explode before reaching orbit, or many other things. Derivatives are flexible in the same way. It's one of the reasons calculus is so useful.

THE FUNDAMENTAL THEOREM OF CALCULUS

The fundamental theorem of calculus is simply this: Taking the derivative is the reverse of taking the integral, and taking the integral is the reverse of taking the derivative. We have seen this sort of thing before: add 10 to a number and then subtract 10, and you are back where you started (addition and subtraction can reverse each other). Or multiply a number by 10 and then divide it by 10, and you are back where you started (multiplication and division can reverse each other).

Similarly, take the derivative of a function and then take the integral of that derivative, and you are back where you started. (Except for a constant of integration, and that can always be set to whatever we choose.) Integration and differentiation reverse each other.

Writing this out in symbols, we have

$$\frac{d}{dx} \int f(x)\,dx = f(x)$$

(taking the derivative undoes integration) and

$$\int \frac{df}{dx}\,dx = f(x)$$

(integration undoes taking the derivative). (For simplicity, the constants of integration on the right-hand sides of these two integrals, and of the rest of the indefinite integrals in this chapter, are omitted, as is often done in practical math.)

The fundamental theorem of calculus is fundamental because it tells us that a function, the derivative of that function, and the integral of that function all contain the same information in different forms. Knowing any one form, we can produce the others.

And that's it, that's calculus. Or the heart of it, at least: derivatives, integrals, and the fundamental theorem that ties them together. Related topics such as summations, limits, exponential functions, and analytic geometry are often lumped together under the term "calculus"

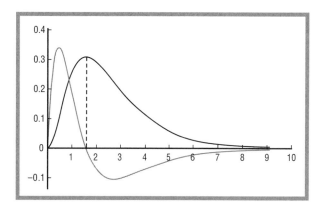

Figure 11: A function (black) and its derivative (gray) plotted together, showing that at the x value where the function levels off at a maximum, its derivative equals 0.

for convenience, but differentiation, integration, and the fundamental theorem are the big three, the core.

MAXIMA AND MINIMA

The curve in Figure 4 and its derivative, the curve in Figure 5, are plotted together in Figure 11. Notice that the place where the curve from Figure 4 (the black line) crosses the horizontal axis, around $x = 1.6$, the curve from Figure 3 (the gray line) hits a peak or maximum.

This gives us a very useful general principle: By finding out where the derivative of a function equals zero, we can determine the maximum (and minimum) points of that function. Why? Because maxima and minima are peaks and valleys, places where a function levels off briefly. And wherever a curve is level, its slope (derivative) equals zero. Being able to find maxima and minima is useful because there lots of things in life that we want to maximize or minimize—profit, cost, risk, time, distance, and more.

Second-order differentiation—finding the derivative of the derivative—is important for finding maxima and minima. The fact that the derivative equals zero at some point really only guarantees that the curve is level at that point; it doesn't say whether it is the top of a peak, or the bottom of a valley, or a ledge halfway up a slope. The second derivative, however, helps distinguishes between these possibilities. Look at the point in Figure 11 where the lighter curve crosses the horizontal axis. The curve is decreasing there, so it has a negative slope—that is, its derivative, which is the second derivative of the darker curve, is negative. Now look up at the darker curve at that point: it is at a maximum. A curve's second derivative, then, tells us which way it is bending. If the second derivative is negative, the curve is bending downward; if the second derivative is negative, the curve is bending upward. Curves bend downward at peaks (maxima) and upward at valleys (minima). This leads to the *second derivative test*: Where a function's first derivative is zero and its second derivative is negative, the function is at a maximum. Where its first derivative is zero and its second derivative is positive, the function is a maximum. If both the first and second derivatives are zero, the test fails and we have to investigate further, perhaps by investigating even higher-order derivatives.

Plotting the original curve and just looking at it is a perfectly good way to distinguish between maxima and minima when the function is simple, as in our examples, but in practice one may be looking at a function in two, three, or more dimensions, where visualization is difficult or impossible.

A Brief History of Discovery and Development

Calculus—or *the* calculus, as it is sometimes called—was invented in the 1600s independently and more or less simultaneously by the English physicist Newton and the German mathematician Leibniz. Actually, it is a slight exaggeration to say that either Leibniz or Newton invented calculus; both applied fresh insight to a mass of mathematical questions and tools that had been building up for centuries.

Mathematicians had been worrying about rates of change (what we now call derivatives) and the calculation of areas ever since the Greek mathematicians, such as Aristotle (384–322 B.C.).

But something new did come into being when the collection of techniques we now call "calculus" came together in the mid- to late 1600s. For the first time, mathematicians had a systematic way of finding derivatives and integrals, that is, rates of change on a curve and areas under a curve. Leibniz realized that to make calculus really useful, an easily understood system of notation would be needed—a way of writing down calculus that would do some of the work automatically. He achieved this by coming up with the integral sign, $\int$, and the notation for the derivative that we most commonly use today,

$$\frac{df}{dx}$$

In 1675, over 325 years ago, he was already writing calculus in his notebook using exactly the same notation we use today, such as

$$\int x^2\,dx = \frac{1}{3}\,x^3$$

This "Leibniz notation," as it is called, makes it possible for high-school and college students of normal mathematical ability to solve problems that baffled great mathematicians for centuries.

In the twentieth and twenty-first centuries, one of the most important practical developments in calculus—as in much of science and mathematics—has been the advent of the electronic digital computer, which was invented during World War II. Digital computers allow us to deal with equations that cannot be solved in "closed form," that is, reduced to a neat, algebraic expressions with the unknown variable on one side of the equals sign and all the known (or knowable) variables on the other. Equations that cannot be solved in closed form can arise even in simple problems. But using the computerized number-crunching techniques collectively referred to as "numerical methods," engineers, scientists, and others can today solve virtually any problem that can be stated using calculus, whether a closed-form solution can be found or not.

Interestingly, Leibniz and Newton still haunt the age of computerized calculus. One of the most commonly used numerical methods for solving equations was developed by Newton and is called "Newton's method," and Leibniz built one of the first mechanical calculating machines, a direct ancestor of the modern computer.

Real-life Applications

APPLICATIONS OF DERIVATIVES

Many direct applications of derivatives in real life are to situations of the sort called *min-max* problems or *extremum* problems, that is, situations where the goal is to find the minimum or maximum (the "extreme" values) of some physical, financial, or other quantity. Solving these problems requires derivatives. Derivatives are also used indirectly, as one of the many mathematical tools that engineers, scientists, and other math-using professionals need to solve their complex and many-layered problems. The following applications of calculus are simplified versions of real max-min problems, that is, they are direct applications of the concept of the derivative.

Maximizing Profits Making a profit is essential to the health and longevity of a business. And for any industrial

Credit for Calculus

Question: What do you get when you cross history with calculus? Answer: two famous mathematicians and nations arguing over who was first in its discovery.

Isaac Newton (1643–1727), staunch English Puritan and the England's champion of math and physics, developed the fundamental concepts of calculus in 1665 and 1666. He organized his ideas into a manuscript in late 1666 and showed it to a few other English mathematicians, but did not publish it. In 1672 to 1676, a German mathematician named Gottfried Leibniz (1646–1716), who started college at 15 and graduated at 17, worked privately on the same problems and came up with similar answers. Leibniz had not heard of Newton's work, and he developed notation and methods that were different from Newton's, but his ideas were essentially the same.

Leibniz first published his results in 1684 and 1686; Newton, in 1687. The math debate arose in the late 1690s, when followers of Newton began to accuse Leibniz of having stolen his calculus ideas from Newton. The fact that Leibniz had published first and Newton second might have made this impossible, but Newton and Leibniz had exchanged letters in 1676 and Leibniz had visited London in both 1673 and 1676, so it was not impossible that Leibniz had stolen Newton's ideas— merely untrue.

Newton and Leibniz actually invented calculus independently, not an uncommon event in science and mathematics. But sharing the accomplishment was not on anyone's agenda, especially in a question of national pride. Newton became so angry that he deleted all references to Leibniz's work from his scientific books (except insults). Newton and his followers publicly accused Leibniz of stealing. Leibniz asked the Royal Society of London, the major English scientific club or society of its day, to investigate this damning charge. Newton secretly stage-managed the society's investigation and Leibniz was found guilty.

Newton was buried in a cathedral with royal honors and thousands of mourners; Leibniz's funeral was attended only by his secretary. Leibniz's ultimate revenge, however, is that *his* calculus notation, not Newton's, is used today.

---

enterprise more complicated than a lemonade stand, it often calls for calculus.

Suppose that you are manufacturing x bottles of hair gel per day. It would be convenient if you could always make more profit just by cranking out more gel, but this does not work. You have to sell to make a profit, and you can only sell units as fast as your customers will buy them; making too many will result in unsold bottles and reduced profits. However, making too few units is no good either. Assuming that there are no factors affecting your profits other than how many bottles per day you manufacture—and that's a big assumption—how many bottles per day should you make?

Let's assume that you have discovered an equation that describes your daily profit, $p(x)$, as a function of how many bottles you make per day, x: $p(x) = -.03x^2 + 4x - 200$. Recall that you can find the minimum or maximum points of a function by finding where its derivative equals zero. The first step, then, is to take the derivative of the profit function. In this case, the derivative of $p(x)$ is $dp/dx = -.06x + 4$. To find where dp/dx equals zero, we solve the algebraic equation $-.06x + 4 = 0$ and find that the optimal number of bottles to manufacture per day is a whopping

$$\frac{4}{.06} = 66\frac{2}{3}$$

or, since it is presumably impossible to sell two thirds of a bottle of gel, 66 bottles.

No real-world business is this simple, but the principle is sound. Calculus, along with other branches of math (such as probability theory), is indeed used by financial analysts.

Storing Data on a Computer Disk Calculus has been used literally thousands of times in the design of every one of the electronic toys we take for granted including MP3 players, TV screens, computers, cell phones, etc. A simple example can be found inside the nearest computer.

For long-term information storage, every computer contains a "hard drive," the primary storage device that, if it crashes and you haven't made backups, will result in a loss of data. A hard drive contains a stack of thin discs

A compact disk is deteriorating along the edges and will no longer play properly. The larger the disk, the more susceptible it is to "disk rot." This and other factors limit disk size. Calculus is used to maximize the number of bits on the disk but the larger the disk the more susceptible it is to "disk rot." AP/WIDE WORLD PHOTOS. REPRODUCED BY PERMISSION.

coated with magnetic particles. A computer stores information in the form of binary digits ("bits" for short, 1s and 0s) on the surface of each disc by impressing or "writing" on it billions of tiny magnetic fields that point one way to signify "1", another way to signify "0." The bits are arranged in circular tracks, as shown in Figure 12.

To read information off the spinning disk, sensors glide back and forth between the edge of the disc and its center to place themselves over selected tracks. The track spins under the sensor, the bits are read off one by one at high speed, and within a few seconds, your favorite video game pops up. But that isn't the whole story. In designing a data-storage disc, if you should ever get the urge to do so, you will want to *optimize* the amount of data stored on the disk-that is, to store the most bits possible. How?

You might think that the way to do this would be to completely cover the disk's surface with tracks. Logical, but wrong, because there's one more real-life wrinkle, namely, that for the sake of keeping the read-write

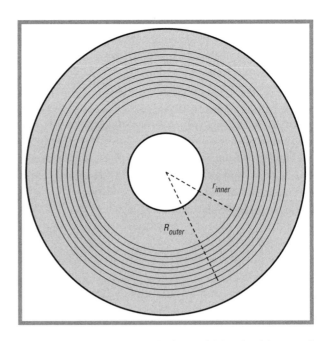

Figure 12: The shaded area is the readable-writeable part of the disc. Each circle represents a track. The radius of the innermost track is r_{inner}, that of outermost track is R_{outer}.

mechanism simpler (and therefore cheaper), every circular track has to have the same number of bits. So, by the formula for the circumference of a circle of radius r, $C = 2\pi r$, the bits are packed less densely on the outer tracks than on the inner tracks: same number of bits per track, more C to string them out on.

Say that the most bits you can pack onto each inch of track is b_i. Then, the most bits you can fit on a track of radius r is the length of the track times b_i, $2\pi r b_i$. So the smaller you make the radius of the innermost track, which we'll call r_{inner}, the fewer bits you can fit on it. But all the tracks on the disc must, as specified above, have the same number of bits, so if you make the innermost track too small, it will hold only a few bits, and so will all the other tracks, and you'll end up with an inefficient disc. On the other hand, if you make the innermost track too big, there won't be much room for additional tracks between the innermost track and the outer edge of the disc, and again your design will be inefficient. What to do?

Calculus to the rescue. Say that the most tracks that you can pack on the disk per radial inch (that is, going from the center toward the edge) is t_i. And let the radius of the outermost track on the disk be made as large as possible, the radius of the disk itself. Call this outer limit R. The number of tracks on the disk, then, is the radial distance between the innermost and outermost tracks times the number of tracks per inch that we can fit into that distance: total number of tracks equals $(R - r_{inner})t_i$. The total number of bits on the disk is the number of tracks times the number of bits per track, and the number of bits per track is limited by how many we can pack onto the smallest, innermost track, which from the previous paragraph we know to be $2\pi r_{inner}b_i$. So, writing the number of bits on the disc B as a function of r_{inner}, we have

$$B(r_{inner}) = \overbrace{2\pi r_{inner}b_i}^{\text{bits per track}}\overbrace{(R - r_{inner})t_i}^{\text{tracks per disk}} = 2\pi b_i t_i (R r_{inner} - r_{inner}^2)$$

We want to maximize this function, $B(r_{inner})$, the number of bits on the disk. Taking the derivative using the rules found in standard calculus textbooks, we get

$$B'(r_{inner}) = 2\pi b_i t_i (R - 2r_{inner}).$$

To find where a function has maxima (or minima), we look for places where the derivative equals zero. So, setting $B(r_{inner})$ equal to zero and solving for r_{inner} using elementary algebra, we find that the value of r_{inner} that maximizes the number of bits on the disk is $r_{inner} = R/2$. Disk packed, case closed.

Lenses and Rainbows Light always takes the quickest possible path through whatever transparent materials it must travel through, a fact known as Fermat's principle. A physicist can write an equation that expresses the time taken by light to pass through an optical system (say, a series of lenses and reflectors). Taking the derivative of this function to find where its minimum point is shows what path that light will take through the system. Light-path minimization based on Fermat's principle is used in some computer programs for designing optical systems.

Designing for Strength In structures that must withstand strong forces, such as bridges and rockets, the force is never distributed evenly throughout the body of the object. Certain places, depending on the object's shape, will experience more than others. Engineers need to know where these points are and how big the maximum forces are that they experience; if the steel, stone, or other material from which the object is made isn't strong enough to withstand that maximum force, the bridge, rocket, or other object will fail. To find points of maximum force, designers describe force as a function and look for points where its first derivative equals zero. Some of these points may be minima, not maxima, but there are several ways—including the second derivative test—to tell which is which. Today, design of bridges, rockets, jet or car engines, and other complex structures is often done by making a mathematical image or model of the structure that can be stored in a computer. This model predicts force throughout the object as a function of space and time; wherever this function is at a maximum (first derivative zero, second derivative negative), if the predicted force is greater than the strength of the materials being used, the design must be changed.

Failure Prediction Derivatives can be used to guess when structures will break. Before breaking, many materials develop cracks. This causes them to emit brief noises that may not be audible to the human ear, but can be recorded by machines. One method of predicting structural breakdown is to use record the noises made by an object (e.g., a large, rotating metal shaft in a generator). The noises are counted by a computer, and their frequency (how fast they are happening) is recorded as a function of time. Software then calculates the first and second derivatives of this function and uses them to decide whether the noise frequency is increasing in a way that may indicate that a breakdown is going to happen soon. If the software detects a possible impending breakdown, it warns its human operators. Some scientists have proposed using the same method for predicting earthquakes, since an

A large section of the concrete roadway in the center span of the Tacoma Narrows bridge crashes into the Puget Sound in Tacoma, WA, on November 7, 1940. High winds caused the bridge to sway, undulate, and finally collapse under the strain. Engineers use calculus derivatives to estimate the forces bridges must withstand. AP/WIDE WORLD PHOTOS. REPRODUCED BY PERMISSION.

earthquake is essentially a sudden break in a large mass of material (the crust of the earth) that is preceded by a long period of strain and cracking.

Seeing Spectra Light waves vibrate at different rates or frequencies. One frequency or vibration-rate of light affects our eyes as the color blue; a slower vibration-rate (that is, a lower frequency) affects our eyes as the color red. There are also frequencies both too low and too high for our eyes to see. Most light is a mixture of many colors or frequencies, some of which are usually brighter than others. If we graph the brightness of the different colors in a beam of mixed light as a function of frequency, the resulting curve is called a spectrum. Often, a spectrum looks like a crowded row of tall, narrow peaks and valleys,

all of different heights and depths. Spectra are used through science and engineering, for they contain information about the chemical composition of the objects giving off the light. Astronomers look at spectra to know what distant stars and planets are made of; chemists look at spectra in the laboratory to find out what chemicals their samples contain; and biologists and geologists look at spectra of the Earth's surface, as photographed by satellites, in order to map the composition and health of the earth's surface. Changes to the spectra of artificially pure beams of light (lasers) that have passed through the atmosphere are used to monitor pollution. First, second, and even higher-order derivatives of spectra are all used. In fact, in chemistry an entire sub-field, derivative spectroscopy, is devoted to the use of derivatives of spectra.

Derivatives make spectra more useful partly because they tend to *sharpen* the ups and downs in a function. If you compare a function to its first and second derivatives (as in Figures 2 and 11, for example), you'll see that where the original function is a gently curving slope, the first derivative has a peak and the second derivative has two sharper peaks. This sharpening effect can be used to clarify differences between the peaks and valleys in a spectrum, which makes it easier to tell what substances have reflected the light.

Sometimes derivatives of spectra are used even more directly. For example, the spectrum of light reflected from plants in tidal wetlands (as photographed by satellite) closely resembles, in part, the curve in Figure 2. The center of the rising part of the spectrum, where its derivative peaks (as shown in Figure 4), is called the "red edge" because here the spectrum drops to lower values for redder (lower-frequency) light. The exact location of the red edge indicates the amount of chlorophyll (a red-light-absorbing chemical) in the leaves reflecting the light, and so is an indicator of the health of the plants.

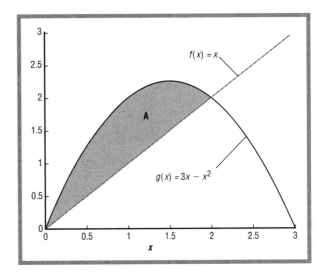

Figure 13: Difference of two functions. Integration can give an exact number value for the shaded area, A, between functions $g(x)$ and $f(x)$.

REAL-LIFE APPLICATIONS OF INTEGRALS

The Area Between Two Curves The integral of a function of one variable, as discussed earlier, is essentially the area under that function. What about the area between two functions, such as the shaded area in Figure 13?

With integration, determining the exact area of this oddly-shaped region is easy. The rule is this: To find the area between two functions $g(x)$ and $f(x)$, integrate the difference between them, $g(x) - f(x)$. In this case, $g(x) - f(x) = (3x - x^2) - x = 2x - x^2$. To find the area A between 0 and 2 in this example, we integrate this difference function between the two points, that is, we find the definite integral of $2x - x^2$ from 0 to 2. This can be written

$$A = \int_0^2 2x - x^2 \, dx$$

This expression can be evaluated using the rules of elementary integration given in all calculus textbooks. We find that A = 4/3.

Inertial Guidance Like all scientific knowledge, calculus can be applied not only to creation but to destruction. For example, the calculus-based concept of *inertial guidance* has been developed by missile-makers to a fine art.

The first ballistic missiles used in war, the V-2 rockets produced by Nazi Germany near the end of World War II (1939–1945), were fired at London from mainland

Europe. They were intended as terror or "vengeance" weapons, and so only needed to explode somewhere over the city, not over particular military objectives; yet to hit a large city such as London at such a distance, a V-2 missile needed a guidance system, a way of knowing where it was at every moment so that it could steer toward its target. It was not practical to steer by the stars or Sun, because these are hard to observe from a missile in supersonic flight and would require complex calculations. Nor was it practical to steer by sending radio signals to the missiles, for without advanced radar (not yet available) controllers on the ground would be just as ignorant of the missile's location as the missile itself. Besides, the enemy might learn to fake or jam control signals, that is, drown them out with radio noise.

The solution was inertial guidance, which exploits the calculus fact that (a) the time derivative of position is velocity and (b) the time derivative of velocity is acceleration. By the fundamental theorem of calculus, which says that integration and derivative are opposites, we know that we can follow the trail backwards: the integral of acceleration is velocity, and the integral of velocity is position.

What designers need a missile to know is its position. But position is hard to measure directly. You have to look out the window, identify landmarks (if any happen to be visible), and do some fast geometry, likewise with velocity. But acceleration is easy to measure, because every part of an object accelerated by a force experiences that force. We've all felt the seat pushing against our backs in an

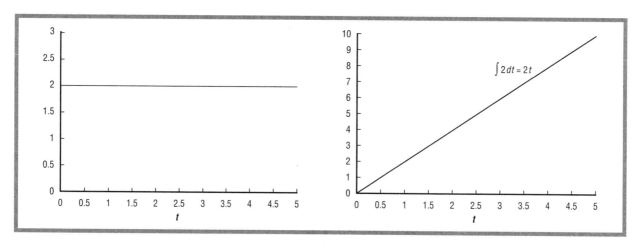

Figure 14: Left, a constant (the number 2) plotted as a function of t. Right, the definite integral of the number 2 from 0 to t. This shows that the integral of a constant is a linear function.

accelerating car or plane. In addition, unlike velocity or position, a force can be measured directly and locally, that is, without making observations of the outside world. Therefore, the V-2's engineers installed gyroscopes (spinning masses of metal) in their missile and used these to measure its accelerations. Lasers, semiconductors, and other gadgets have also been used since that time. Some are more expensive and accurate than others, but all do the same job: they measure accelerations. Any device that measures accelerations is called an accelerometer.

Thanks to its accelerometers, an inertial guidance system knows its own acceleration as a function of time. What does it do with this knowledge? Acceleration can be written as a function of time, $a(t)$. This function is known by direct measurement by accelerometers. The integral of $a(t)$ gives *velocity* as a function of time: $\int a(t)\,dt = v(t)$. And the integral of $v(t)$ gives *distance* as a function of time, which reveals one's position at any given moment: $\int v(t)\,dt = x(t)$. The real-world math of inertial guidance is of course more complex, but the principle is the same.

The bottom line for inertial guidance is that, given an accurate knowledge of its initial location and velocity, an inertial guidance system is completely independent of the outside world. It knows where it is, no matter where it goes, without ever having to make an observation.

The V-2 inertial guidance system was crude, but since World War II inertial guidance systems have become more accurate. In the early 1960s they were placed in the first intercontinental ballistic missiles (ICBMs), large missiles designed by the Soviet Union and the United States to fly to the far side of the planet in a few minutes and strike specific targets with nuclear warheads. They

were also used in the Apollo moon rocket program and in nuclear submarines, which stay underwater for weeks or months without being able to make observations of the outside world. Inertial guidance systems are today not only in missiles but in tanks, some oceangoing ships, military helicopters, the Space Shuttle and other spacecraft, and commercial airliners making transoceanic journeys.

Calculus makes inertial guidance possible, but also, in a sense, limits its accuracy. The problem is called integration drift. Integration drift is a pesky result of the fact that small "biases" are, for various technical reasons, almost certain to creep into acceleration measurements. (A bias is a small, unknown number added to all your measurements, like .000000001 m/s^2.) Now, the integral of a constant (any ordinary number, like 2.0 or .000000001) is a linear function. That is, for any constant k, $\int k\,dt = kt$.

Figure 14 shows why this works. Plotted as a function of t, a constant k (e.g., the number 2) is a flat line: it never changes, it's always just k. But the area under that line grows steadily as one takes in more of the t axis, starting from any given point, such as 0.

What's more, the integral of a linear function is a quadratic function, that is, a function containing t^2 as its highest power of t:

$$\int kt\,dt = \frac{k}{2}\,t^2$$

This function is plotted in Figure 15, for $k = 2$.

Inertial guidance depends on measuring a physical variable (acceleration) and then, in essence, integrating twice. Any bias in these acceleration measurements, any unwanted, constant number that adds itself to all the measurements, will result in a position error that increases in proportion to the square of time, just like the function in Figure 15. Notice how quickly the numbers can grow. After 1 second, a constant acceleration-measurement error of 2 m/s^2 produces a position error of 1 meter; after 3 seconds, an error of 9 meters; after 5 seconds, an error of 25 meters. These particular numbers are unrealistically large, but any degree of real-life quadratic error will eventually grow to unacceptable values. As a result, no inertial guidance system can go forever without taking an observation of the outside world to see where it really is. Increasingly, inertial guidance systems are designed to update themselves automatically by checking the global positioning system (GPS), a network of satellites that blanket the whole Earth with radio signals that can be used to determine a receiver's position accurately.

Today, inertial guidance systems have reached a fantastic degree of accuracy. A ballistic missile launched from a nuclear submarine, which may begin with a knowledge of its initial position and velocity that is several weeks old, can be launched from a still-submerged submarine by compressed air, burst through the surface of the water, ignite its rocket, fly blind to the far side of the planet, and explode its nuclear warhead within a few yards of its target.

Threading the Cosmic Needle On June 28, 2004, the Cassini space probe, a robot craft about the size of a small bus built jointly by the United States and Europe, arrived at Saturn after a seven-year journey. The plan was for it to be captured by Saturn's gravity and so become a permanent satellite, observing Saturn and its rings and moons for years to come. But to make the journey, Cassini had reached a speed of 53,000 miles per hour (85,295 km/h)—(many times faster than a rifle bullet), too fast to be captured by Saturn's gravity. At that speed it would swoop past Saturn and head out into deep space. Therefore, it was programmed to hit the brakes, to fire a rocket against its direction of travel as it approached its destination.

For objects moving in straight lines, changes in velocity can be calculated using basic algebra. Calculus is not needed. But Cassini was not moving in a straight line; it was falling through space on a curving path toward Saturn, being pulled more strongly by Saturn's gravity with every passing minute. To figure out when to start Cassini's rocket and how long to run it, the probe's human controllers on Earth had to use calculus. The effects of the important forces acting on Cassini—in particular, its own rocket motor and Saturn's gravity—had to be integrated over time. And the calculation—carried out using computers,

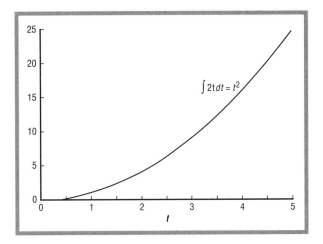

Figure 15: Showing that integrating a linear error (in this case, 2t) produces a quadratic error.

not, for the most part, on paper—had to be extremely exact, or Cassini would be destroyed. For to get deep enough into Saturn's gravitational field, Cassini would have to steer right through a relatively narrow gap in Saturn's rings called the Cassini division (named after the same Italian astronomer as the probe itself), a navigational feat comparable to threading a cosmic needle. If it missed the gap, the Cassini craft would have been destroyed by collision with the rings.

Not all of NASA's navigational calculations have been correct: in 1998, a space probe crashed into Mars because of a math mistake. But in Cassini's case, the calculations were correct. Cassini passed the rings safely, was captured by Saturn's gravity, and began its orbits of Saturn.

Energy Payback Flip the switch and the light comes on. This simple act connects us directly to a vast system of electricity production worth many billions of dollars, with transmission wires marching across the countryside to immense yet delicately adjusted generating plants where the electricity is produced. Flipping a light-switch adds to the total demand for electricity that this system must meet.

The engineers who run our electricity production system are supposed to meet not only today's demand but the demand 10 and 20 years from now, so they try to forecast what that demand will look like. One mathematical prediction or "model" that has often been used to predict growth in demand for electricity is the exponential model. (See article on Exponents in this book.) This guesses that total demand for electricity will grow by a fixed percentage every year (e.g., 2%). That is, if electric demand is 100 units in the year 2025, it will be 102 units in 2026, and so forth, if the model is correct. Demand might grow because the population is growing and there

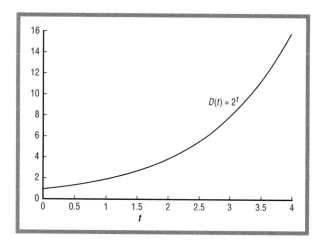

Figure 16: Exponentially growing demand for electricity modeled by $D(t) = 2^t$.

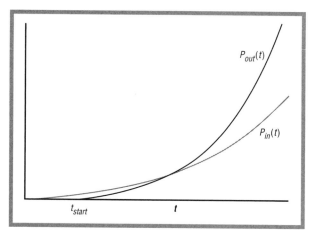

Figure 17: Total power output and power output of an exponential program for building power plants.

are more people using electricity, or because we are using more electricity to make things, or because we are wasting more electricity (by leaving our computers or lights on), or for some mixture of these reasons.

If we consider demand as a function of time, $D(t)$, then exponentially increasing demand can be written as $D(t) = D_0 k^t$, where k is some constant (fixed number) and D_0, also a constant, is the value of $D(t)$ at $t = 0$. Any number to the 0th power is 1, so at time equals 0 we have $D(0) = D_0 k^0 = D_0 \times 1 = D_0$. Exponential growth is depicted in Figure 16, with $D_0 = 1$ and $k = 2$.

Now, to meet new electricity demand, new electricity-production plants must be built. And to meet exponentially increasing demand, plants must be built at an exponentially increasing rate. So far, so clear—so why not build them? Because, even apart from the fact that on a finite planet no pattern of exponential growth can go on forever (it will eventually eat up the whole planet), we are confronted by a paradox: A program that builds power plants at an exponentially increasing rate will produce no energy for some time even after its first plants start generating electricity.

Why? Because it takes energy to *build* a power plant. That is, energy must be loaned from some other source. Before the plant can be a true energy producer, it has to pay back this energy debt. Furthermore, building power plants at an exponentially increasing rate requires an exponentially increasing amount of power. (Power is the rate of flow of energy, that is, its time derivative.)

Let's take a closer look at energy in and energy out. Call the power going into unfinished plants $P_{in}(t)$ and the power coming out of finished plants $P_{out}(t)$. We'll assume that the building program begins at time 0. At some later

time, call it t_{start}, the first plant starts delivering electricity. For a large dam, coal-fired plant, or nuclear power plant, t_{start} might be 10 years. We'll also say that each finished plant produces more power than each unfinished plant uses.

Let the power output of each finished plant be p_{out} and the power input to each unfinished plant be p_{in}. Assume for that $p_{out} > p_{in}$. As plants are finished and start putting out power, the program as a whole will produce more and more power, but it will also consume more and more power, as construction of new plants grows exponentially too.

In real life, the total amount of power required by the building program increases by little jumps of size p_{in} (as construction of each new plant is begun), and the total amount of power produced increases by little jumps of size p_{out} (as each finished plant comes online). But here we'll pretend that both curves can be treated as smooth functions of time. We'll call the total power-investment curve $P_{in}(t)$ and the total power-output curve $P_{out}(t)$. Both curves climb exponentially, but the power-investment curve $P_{in}(t)$ starts to climb as soon as construction begins, namely at $t = 0$, and the power-output curve doesn't start to climb until t_{start}, when the first finished plant kicks in.

If $p_{out} > p_{in}$, then the power-output curve climbs more steeply than the power-investment curve. This situation is shown in Figure 17.

Notice in Figure 17 that although power output gets a late start, it soon catches up with power investment (where the curves cross) and surpasses it. But the program only becomes a net power producer when its summed power output exceeds its total power debt. When does this happen?

Before we look at the actual calculus, let's first emphasize that power is the time derivative of energy. That is, if a

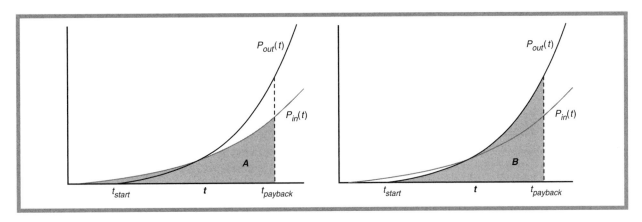

Figure 18: Area *A*, at left, and Area *B*, at right, correspond to total energy consumed and produced (respectively) in the time intervals shown. We want to find $t_{payback}$ such that $A = B$ to know when the building program pays off its energy debt.

curve records power (rate of energy transformation or "flow") as a function of time, then the area under that curve between any two times gives the energy supplied in that time. So to find out when our hypothetical program has first produced as much energy as it has consumed, we have to find that time (call it $t_{payback}$) when the area under the $P_{out}(t)$ curve equals the area under the $P_{in}(t)$ curve. That is, we have to find $t_{payback}$ such that area A equals area B in Figure 18.

Once again, calculus to the rescue. Area *A* (energy invested up to $t_{payback}$) can be written as the definite integral

$$A = \int_0^{t_{payback}} P_{in}(t)\, dt$$

Area *B* (energy produced up to $t_{payback}$) can be written

$$B = \int_{t_{start}}^{t_{payback}} P_{out}(t)\, dt$$

If A = B, then we just set these integrals equal to each other:

$$\int_0^{t_{payback}} P_{in}(t)\, dt = \int_{t_{start}}^{t_{payback}} P_{out}(t)\, dt$$

If we assign the simplest possible exponential forms to $P_{in}(t)$ and $P_{out}(t)$, this gives

$$\int_0^{t_{payback}} \left(e^{k_{in} t} - 1\right) dt = \int_{t_{start}}^{t_{payback}} \left(e^{k_{out}(t - t_{start})} - 1\right) dt$$

(Here the letter "e" stands for the constant 2.7182818 . . . , which is usually used in calculus for exponential functions.) Using the standard rules of integration found in calculus textbooks, this equation evaluates to

$$\frac{e^{k_{out}(t_{payback} - t_{start})}}{k_{out}} - \frac{1}{k_{out}} + t_{start} =$$

$$\frac{e^{k_{in} t_{payback}}}{k_{in}} - \frac{1}{k_{in}}$$

The only unknown in this equation is $t_{payback}$, so if we had particular numbers for the other variables, we could solve for $t_{payback}$ using a computer.

This general approach can be used to evaluate the realism of any proposed program to grow the electricity supply quickly. Simple-minded plans to rapidly build any kind of generating capacity—windmills, nuclear plants, coal plants, or other—in order to meet a projected energy shortage that is, say, 20 years away, may be worse than useless if $t_{payback}$ for the proposed program is 40 years!

Exponential functions have actually been used by government and industry analysts to predict growth in energy demand. A net-energy analysis such as that outlined above can reveal the long-term strengths or weaknesses of such predictions, and any energy solutions proposed to cope with such increases in demand. Exponential growth models can be relatively accurate over short periods of time and limited areas; fortunately, predictions of exponential growth in overall electricity demand have rarely turned out to be correct in the long term. This is mostly due to increases in user efficiency. For example, a typical refrigerator today uses about half as much electricity as a 10-year-old refrigerator but cools the same amount of food. Electricity costs money, so there is an ongoing economic pressure toward more efficient use.

Key Terms

Acceleration: A change of velocity (either in magnitude or direction).

Bit: The smallest unit of information storage in computers. A bit stores a 0 or a 1.

Derivative: The limiting value of the ratio expressing a change in a particular function that corresponds to a change in its independent variable. Also, the instantaneous rate of change or the slope of the line tangent to a graph of a function at a given point.

Differentiate: To determine the derivative or differential of a particular function.

Integral: The area under a curve.

Where to Learn More

Books

Edwards Jr., C.H. *The Historical Development of the Calculus.* New York: Springer-Verlag, 1979.

Price, John H. "Dynamic Energy Analysis and Nuclear Power," *Non-Nuclear Futures*, Lovins and Price, eds. Friends of the Earth International, 1975.

Sawyer, W.W. *What is Calculus About?* New York: Mathematical Association of America, 1961.

Straffin, Philip, ed. *Applications of Calculus, Vol. 3.* Washington, DC: Mathematical Association of America, 1997.

Web sites

"The Math Forum @ Drexel: Calculus" <http://mathforum.org/calculus/> (April 23, 2004).

"Visual Calculus: A collection of modules that can be used in the studying or teaching of calculus." <http://archives.math.utk.edu/visual.calculus/> (April 23, 2004).

Overview

A feature in many kitchens is a calendar hanging on a wall. Typically, there will be all sorts of notations on the calendar, charting the various events and important points in the days and weeks. In electronic form, calendars have become a time and appointment tracking tool that serve to keep individuals organized both in their work and personal lives.

The prime function of a calendar is to organize time, over a short term and even extending far ahead into the future. This function hinges on math.

A Brief History of Discovery and Development

The need and desire to map the passage of time is something that has probably been with us ever since our prehistoric ancestors started to ponder the world around them. Gazing up at the sky would have made people aware of time. Days turn into nights and back into days. Then as now the moon waxed and waned in the night sky. With the advent of telescopes, the regular movement of some celestial bodies (like the planets in our solar system) was revealed. In more northern climates, seasonal variations in temperature and weather would have been apparent. All these things and more helped form the basis of the measurement of time.

Maintaining records of the passage of time has long been with us. Carvings and scratches in rocks, bones and sticks made by people some 20,000 years ago in present-day Europe are thought to be a form of calendar, to chart the appearance of the moon. Knowledge of when the nights would be bright with a full moon, or darker and better for sneaking up on game, would be beneficial for a hunter.

Five thousand years ago, the Sumerians who dwelled in what today is Iraq had a formal calendar. Their version divided the year into equal 30-day periods, each day into 12 equal periods (corresponding to 2 of our hours) and each of these daily periods into 30 parts (corresponding to about 4 of our minutes). Stonehenge, the jumble of stones assembled in southwest England over 4,000 years ago, was likely built at least in part to help chart universal events such as lunar eclipses and the passage of seasons.

Ancient Egyptians originally had a calendar based on the monthly cycle of the moon. However, they came to realize that a star (we know it as Sirius) appeared in the sky next to the sun every 365 days, at around the same time as the great river Nile flooded. This lead them,

Ancient Egyptians originally had a calendar based on the monthly cycle of the moon. This ancient Egyptian calendar is from Ptolemaic Alexandria, and show zodiac signs. BETTMANN/CORBIS. REPRODUCED BY PERMISSION.

around 4236 B.C., to revamp their calendar, to one based on a 365-day cycle.

Elsewhere, the Mayan culture that flourished in Central America between 2600–1500 B.C. had 260 and 365-day calendars that were based on the Sun, Moon, and planet Venus. Portions of their calendars were sculpted into large calendar stones, which have survived to the present day.

The Julian calendar was introduced by Julius Caesar in 45 B.C. This version was a calendar based on the daily passage of the Sun across the sky. Each month was equal in length. Every fourth year a day was added to keep the calendar year in synch with the seasonal year. Today, principles of this calendar such as the fourth year added day and the January 1 beginning of a new year are still in use.

The year 45 B.C. is also known as the 'year of confusion' since Caesar inserted 90 days into the year to bring the calendar months back into synch with the seasons. It must have been a confusing year, indeed!

There are others examples of calendars. Indeed, even today there are approximately 40 different kinds of calendars in use.

Real-life Applications

As is evident that part of the math behind calendars is the segregation of time into units. For example, the Gregorian calendar that guides the days of many of us is another 365-day based design (except for every fourth year, the so-called leap year, when an extra day is added on to the month of February). Each 365-day period is divided into collections of days, usually 30 and 31, except for the 28 or 29 days of February, that are called months. In turn, the days in each month are organized into groups of seven, each of which represents a week. Fine-tuning things further, each day is divided into the 24 equal splits of time called hours and each hour into 60 minutes. Further divisions are possible.

The math at the heart of calendars organizes and at least gives the sense of controlling time. So, the math connects people to the world and even to the universe. It is not surprising that calendars assume such central and even sacred importance to societies throughout recorded history and even back into the mists of time.

The daily division of time in many calendars is based on astrological events. One is the daily cycle that results from the rotation of the Earth on its axis. Because the Earth is moving, relative to the sun, any particular portion of the globe will light and dark periods. The Gregorian and Julian calendars are solar calendars.

The monthly calendar cycle is based on the revolution of the Moon around the Earth. This is a lunar calendar. An example of a lunar-based calendar is the Islamic calendar. Because the phases of the moon do not match up with the months of the year, the Gregorian and Islamic calendars do not 'match up.'

Calendars also track the length of the four seasons. The basis of seasons is also astrological. The Earth's north-south axis is not oriented at 90 degrees to the Sun. Rather, the axis is tilted, with the North Pole being further away from the Sun than the South Pole. The result is that, as the Earth revolves around the Sun, the sunlight is more intense over certain regions of the planet at different times of year.

This tropical year is incorporated into a third type of calendar that is a blend of the solar and lunar calendars. The Hebrew and Chinese calendars are examples of this blend, which is called a lunisolar calendar. A lunisolar calendar has a sequence of months that are based on the cycle of the moon. But, every few years a month is added in, to bring the calendar back in synch with the tropical year.

LEAP YEAR

As mentioned earlier, the Julian and Gregorian calendars have some years that are one day longer. In the Gregorian calendar these leap years are 366-days long, rather than the usual 365-day year. The determination of when a leap year occurs is a straightforward mathematical process. Every year that can be divided evenly by 4 is a leap year, except for years that can also be divided evenly by 100. The latter, those years that mark the end of a century, can be leap years, but only when they can be divided evenly by 400.

Using these rules, the year 2000 is a leap year, since it can be divided exactly by 4 (to yield 500), by 100 (to yield 20) and by 400 (to yield 5). But the year 1900 is not a leap year. This is because it can be divided evenly by 4 (to yield 475) and by 100 (to yield 19), but cannot be divided exactly by 400 (1900/400 = 4.75). The year 2100 is also not a leap year.

The 400-year cycle of the Gregorian calendar comprises 146,097 days. Dividing the number of days by the number of years results in 365.24 (just about the number of days in each year). Multiplying this number by 2 or 3 does not produce a whole number. But, when 365.24 is multiplied by 4, the result is very close to a whole number. This is the basis of the 4-year cycle of leap years.

MATHEMATICAL ORIGIN OF THE GREGORIAN CALENDAR

The Gregorian calendar was devised to recalculate the dates of Easter. Centuries ago, the March 21 date of Easter coincided with the spring equinox; one of two days each year when the length of daytime and nighttime are the same at 12 hours. But, by the thirteenth century, people became aware that Easter was falling earlier in the month than the equinox. Popes Pius V and Gregory XIII worked on readjusting things. The solution implemented by Gregory was to delete October 5 through October 14, 1582, from the calendar. In that year, October 4 was followed by October 15, which put occurrence of the spring equinox back around March 21.

MATH AND THE ISLAMIC AND CHINESE CALENDARS

In the Islamic calendar, the months correspond to the lunar cycle. Twelve lunar cycles comprises a period in the Gregorian calendar equivalent to about 33 years.

The Chinese calendar is a lunisolar calendar whose months depend on the positions of the Sun and Moon. The pattern of 29- or 30-day months forms the basis for a 60-year cycle of names. A year name consists of a name from a group of celestial names and terrestrial names. The latter are names of animals and is the basis of Chinese years such as 'Year of the Rat' and 'Year of the Pig'.

For all the different calendar systems that have and still exist, several fundamentals are common. One is the intent of a calendar to organize time. The other is the vital relationship of math to the structure of the calendar. As in many other aspects of life, calendars are all about real-life math.

Where to Learn More

Books

Bourgoing, J.D. *Discoveries: The Calendar History, Lore, and Legend.* New York: Harry N. Abrams, 2001.

Judge, M. *The Dance of Time: The Origins of the Calendar – A Miscellany of History and Myth.* New York: Arcade Publishing, 2004.

Richards, E.G. *Mapping Time: The Calendar and Its History.* New York: Oxford University Press, 2000.

Web sites

Doggett, L.E. "Calendars." <http://astro.nmsu.edu/~lhuber/leaplist.html> (November 3, 2004).

Cartography

Overview

Cartography—the making of maps and charts—relies on mathematical principles to produce accurate representations of features or attributes distributed in space and time. The primary uses of mathematics in cartography are to accurately transform the spatial relationships among features on a curved surface onto a plane such as a piece of paper or a computer monitor and to determine the precise locations of features. Maps can depict physical features such as cities and roads, the type of bedrock exposed at the surface, or the elevation of the ground surface. They can also depict non-physical attributes such as the likelihood of damage during a strong earthquake or the average income of residents. The word chart is usually restricted to highly specialized maps showing coastlines, water depths, navigational aids such as buoys, and navigational hazards such as reefs in great detail.

Fundamental Mathematical Concepts and Terms

SCALE

The scale of a map is the mathematical relationship between real distances and those shown on the map. If two buildings located 1 mi (1.6 km) apart are shown 1 in (2.5 cm) apart on a map, the map scale is 1 in : 1 mi or, using consistent units, 1:63,360. If the same two buildings are located 0.1 in (0.25 cm) apart on the map, then the scale becomes 1:633,600. Scale can also be written in a fractional form, for example 1/100,000, or shown graphically by a scale bar printed on the map. A scale of 1:100,000 is said to be larger than a scale of 1:1,000,000 because the fraction 1/100,000 is larger than 1/1,000,000. The concept of large and small scale can sometimes be confusing because a small scale map will cover a larger area than a large scale map of the same size. Large scale maps, however, show more detail than small scale maps covering the same area. Although there is no universally accepted definition of the difference between large, intermediate, and small scale maps, those with scales larger than 1:25,000 are usually considered large scale and those with scales less than 1:250,000 are usually considered small scale. Maps with scales between 1:25,000 and 1:250,000 are generally considered to be intermediate scale. To eliminate the possibility of confusion, it is always best to specify the scale numerically rather than just qualitatively using words like large or small.

MAP PROJECTION

One of the most difficult problems facing cartographers is the development of methods to transfer Earth's

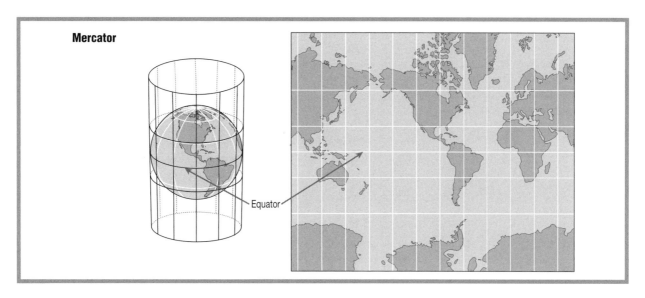

Mercator

Equator

A Mercator map. MAP BY XNR PRODUCTIONS, INC. THE GALE GROUP.

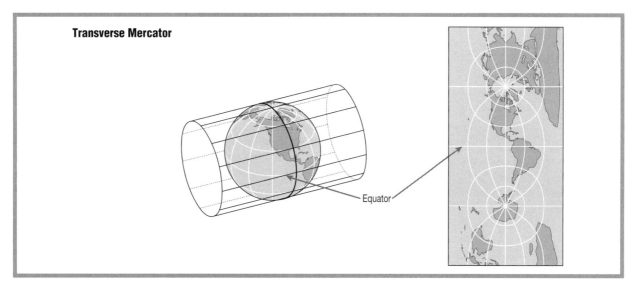

Transverse Mercator

Equator

A transverse mercator map. MAP BY XNR PRODUCTIONS, INC. THE GALE GROUP.

spherical shape onto flat pieces of paper. This is done using an application of geometry known as map projection. Cartographers have developed dozens of different projections over the years, each of which has its advantages and disadvantages. Regardless of the projection used, however, all maps produced on flat surfaces will include some distortion.

The Mercator projection, invented by the cartographer Gerardus Mercator in 1569, is useful for navigation because a straight line drawn on a Mercator map represents a straight line on Earth's surface. Because Earth is spherical, however, the straight line does not necessarily represent the shortest distance between two points.

Mercator projections are projections of the spherical Earth onto a vertical cylinder tangent to the Equator. Therefore, a Mercator projected map will occupy an unbroken rectangle representing an unrolled cylinder. Its primary disadvantage is that the Mercator projection distorts areas and shapes to a degree that increases as one moves away from the Equator. Therefore, land masses at mid- to high latitudes appear disproportionately large. A variation on the Mercator projection, the transverse Mercator projection, was created by Johann Heinrich Lambert in 1772. It is a projection of the spherical Earth onto a horizontal cylinder. The Mercator projection was brought into the space age during the 1970s, when cartographers

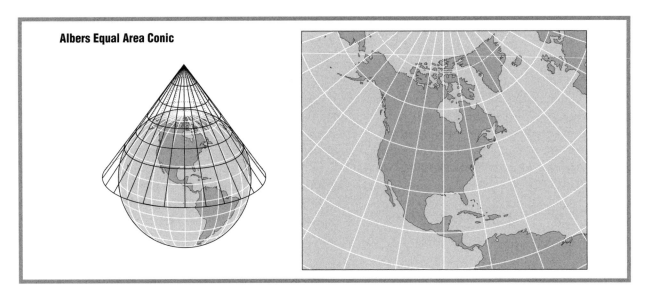

Albers Equal Area Conic

An Albers equal area conic map. MAP BY XNR PRODUCTIONS, INC. THE GALE GROUP.

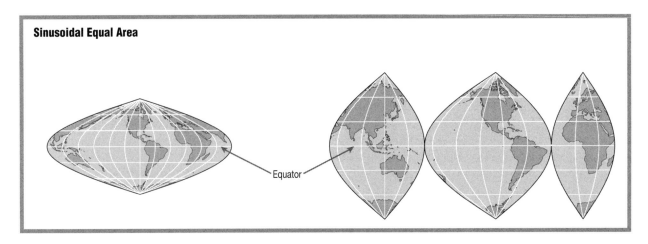

Sinusoidal Equal Area

Equator

A sinusoidal equal area map. MAP BY XNR PRODUCTIONS, INC. THE GALE GROUP.

A.P. Colvocoresses, J.P. Snyder, and J.L. Junkins invented the space oblique Mercator projection in order to project images obtained from Landsat satellites orbiting Earth.

As the name implies, conic projections use a cone that is tangent to Earth's surface rather than a cylinder. One kind of conic projection, the Albers equal area conic projection, is used by the U.S. Geological Survey for maps of the conterminous United States because it is well-suited for areas that have a large east-west extent. Another conic projection, the Lambert conformal conic projection, is also well-suited to areas with large east-west extents and is used for many maps of the United States.

Sinusoidal equal area projections, which have been in use since 1570, avoid the distortion of Mercator projections and are often used for maps in which it is important to compare the sizes or shapes of features in different parts of the world. An example of this would be a map showing the distribution of oil fields around the globe. A Mercator projected map would exaggerate the sizes of oilfields far from the Equator, but a sinusoidal equal area projection does not. A disadvantage of the sinusoidal equal area projection is that its projected shape is a series of lozenges or pods rather than a simple rectangle.

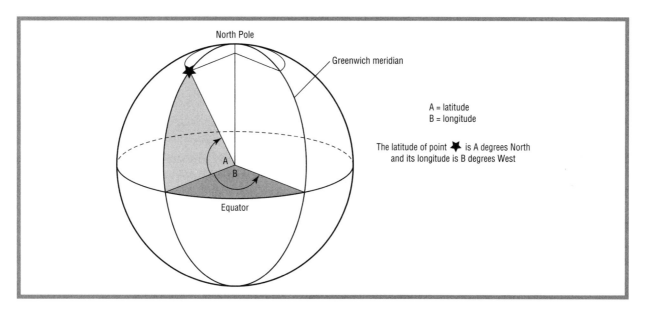

Figure 1.

COORDINATE SYSTEMS

Coordinate systems provide a way to record the position of a feature on the ground or on a map. The most widely known geographic coordinate system consists of lines of latitude and longitude on a sphere, which are measured as angles north or south of the Equator for latitude and east or west of the Greenwich (or Prime) Meridian for longitude (see Figure 1).

Angles of latitude range from 0 at the Equator to 90 degrees at the North and South Poles, and angles of longitude can range from 0 to 360 degrees. It is important to specify whether latitude is north or south of the Equator and whether longitude is east or west of the Greenwich Meridian. Lines of latitude are parallel to each other and each degree of latitude is equivalent to about 70 mi (112 km). Lines of longitude, which are known as meridians, converge at the North and South Poles and the distance between them decreases away from the Equator. The distance between meridians is about 70 mi (112 km) at the equator and decreases to zero at the two poles. Because there is such a large distance between lines of latitude and longitude, it is convenient to subdivide each degree into smaller parts. Degrees (°) have traditionally been divided into 60 minutes (′) of latitude or longitude, and minutes into 60 seconds (″) of latitude or longitude. For example, the location of the Seattle-Tacoma International Airport is 47° 26′ 56″ North latitude and 122° 18′ 34″ using degrees, minutes, and seconds. Because it can be difficult to perform arithmetic with latitude and longitude values given in degrees, minutes, and seconds, latitude and longitude can also be specified in decimal degrees. The location of the Seattle-Tacoma airport in decimal degrees is 47.45° North latitude and 122.31 ° West longitude.

Latitude and longitude are well suited for locating points on spheres and global navigation, but can be inconvenient to use for small areas. Therefore, other coordinate systems have been developed over the years. One of these is the Universal Transverse Mercator (UTM) grid system, which divides the globe into 60 zones each 6° of longitude wide and divided into North and South halves. Each zone has its own Transverse Mercator projection, which allows the curved surface of Earth to be accurately projected onto a flat map with grid lines that are parallel and perpendicular to each other. Because Earth is covered by 60 projections, distortion within any one of the zones is minimal. The UTM coordinates of a point are given by specifying the zone number, an east-west distance known as the easting, and the distance north or south of the Equator, which is known as the northing. UTM positions are always given in meters and never in feet or miles. Using UTM coordinates, the location of Seattle-Tacoma International Airport is Zone 10 N, 552,058 E, 5,255,280 N. One disadvantage of UTM coordinates is that is that they cannot be extended beyond zone boundaries. Therefore, latitude and longitude remain the standard for global mapping and navigation.

Each state within the United States also has its own coordinate system, known as a State Plane Coordinate System, that is used by surveyors and government agencies. State plane coordinate systems give distances east and north of a specified point in each state, can can be divided into parts for large states. Unlike UTM coordinates, state plane

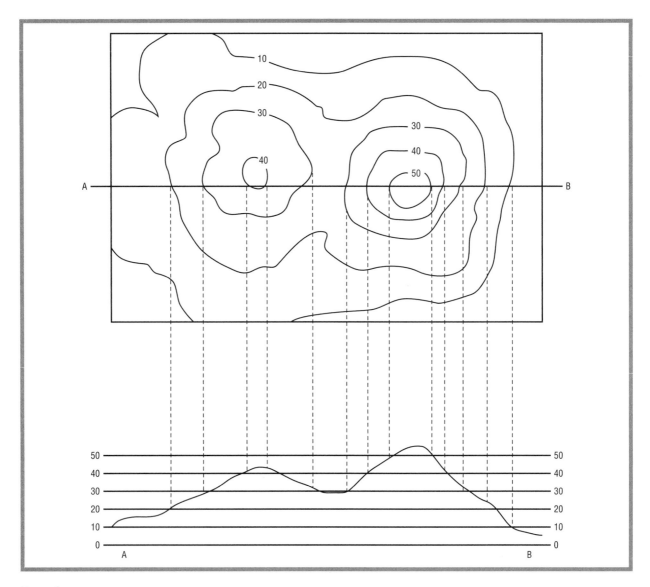

Figure 2.

coordinates can be given in either meters or feet (although it is important to specify which units are being used).

TOPOGRAPHIC MAPS

Topographic maps (an example is shown in Figure 2) are specialized maps that show the elevation of Earth's surface using contour lines, which connect points of equal elevation, or shading. They are especially useful for route finding, search and rescue operations, construction site selection, and scientific studies because of their accurate representation of landforms such as hills and valleys. Skilled map readers can easily interpret patterns of contour lines and visualize the landscape. Topography can also be represented using digital elevation models, which

are computer files containing the elevation of Earth's surface at thousands or even millions of known points. Digital elevation models can be used to create contour maps, shaded relief maps, or three-dimensional surfaces that are useful for many applications.

A Brief History of Discovery and Development

Maps have been important in warfare, agriculture, trade, and the growth of civilizations for thousands of years. The oldest map discovered to date is a small clay tablet unearthed in 1930 at an archeological excavation in

Iraq, at the ancient city of Ga-Sur about 200 mi (320 km) north of present-day Baghdad. The tablet is thought to date from the time of Sargon of Akkad (2334–2279 B.C.) and shows two hilly areas on either side of a stream. More than two millenia passed before the Greek mathematician Eratosthenes (276–194 B.C.) estimated the circumference of Earth by measuring the lengths of noon shadows cast at two distant points at midsummer. The Greek astronomer Ptolemy (ca. 100–170, exact dates unknown) produced the first maps showing Earth as a sphere and included lines of latitude and longitude, but his maps were not widely used until the fifteenth century. Scientific cartography languished through the Middle Ages but blossomed again in the Renaissance and the Age of Exploration, when accurate maps became so important that they were considered to be military, diplomatic, and commercial secrets. It was at this time that Mercator developed his projection and produced a widely known map of the world. Map production technology continued to advance through the centuries, spurred on by events such as the introduction of aerial photography in the early twentieth century, satellite-based remote sensing in the 1960s and 1970s, and the widespread use of geographic information systems (GIS) software in the 1980s and 1990s.

Real-life Applications

GPS NAVIGATION

Maps are an important part of global positioning system (GPS) navigation systems installed in personal automobiles, commercial vehicles, aircraft, and ships. GPS receivers obtain signals from a system of satellites and, in the best of circumstances, can calculate locations with an accuracy of a few feet. In order to be useful for aviation or marine navigation, the location provided by the GPS receiver must be combined with an accurate map showing navigational hazards such as mountain ranges or submerged reefs.

GIS-BASED SITE SELECTION

GIS software allows map users to combine different kinds of maps in order to select the best locations for everything from new stores to hazardous waste storage facilities. Maps showing topography, transportation routes, population, natural hazards such as flood plains, and many other factors can be combined and sites selected using sophisticated mathematical algorithms that weigh the importance of information contain on each map or determine the most economical route between two locations. Maps and GIS systems are also used to understand the spread of diseases, evaluate patterns of

> ## Key Terms
>
> **Cartographic projection:** A geometric transfer of patterns, shapes, and locations from a spherical globe to a flat surface.
>
> **Scale:** The ratio of the size of an object to the size of its representation.

criminal activity, and distribute aid in the wake of natural disasters such as floods, hurricanes, and earthquakes.

NATURAL RESOURCES EVALUATION AND PROTECTION

Scientists and engineers use maps on a daily basis to record the distribution of natural resources ranging from ore deposits to endangered species habitat, to depict the locations of hazards such as landslides or earthquake faults, and to regulate activities such as commercial logging and urban growth. Some of their cartographic products include geologic maps showing the distribution of rock types, soil survey maps useful for agriculture and land use planning, and land cover maps that can be used to help assess the likelihood for erosion during heavy rainstorms.

Potential Applications

Maps will remain vital tools as long as humans continue to travel and explore. In the future, maps may help to guide the exploration of nearby planets and moons. Maps and GIS software will continue to integrated into cellular phones and portable computers used in day-to-day activities such as shopping and vacation travel. Computer mapping software and databases will allow users to create maps custom tailored to their unique needs.

Where to Learn More

Books

Crane, Nicholas. *Mercator: The Man Who Mapped the Planet.* New York: Henry Hold and Company, 2003.

Thrower, N.J.W. *Maps and Civilization: Cartography in Culture and Society,* 2nd ed. Chicago: University of Chicago Press, 1999.

Wilfor, J.N. *The Mapmakers (Revised).* New York: Knopf, 2000.

Web sites

Aber, J.S. "Brief History of Maps and Cartography". 2004. <http://academic.emporia.edu/aberjame/map/h_map/h_map.htm> (February 12, 2005).

Cartography Associates. "David Rumsey Map Collection". 2003. <http://www.davidrumsey.com/> (February 12, 2005).

U.S. Department of the Interior. "National Atlas Home Page". January 26, 2005. <http://nationalatlas.gov/> (February 12, 2005).

U.S. Geological Survey. "Map Projection Publications", Fact Sheet 087-99, May 1999. February 19, 2004. <http://erg.usgs.gov/isb/pubs/factsheets/fs08799.html> (February 12, 2005).

U.S. Geological Survey. "A Tapestry of Time and Terrain: The Union of Two Maps—Geology and Topography." March 29, 2002. <http://tapestry.usgs.gov/two/two.html> (February 12, 2005).

Woodward, David. "The History of Cartography." January 11, 2005. February 19, 2004. <http://feature.geography.wisc.edu/histcart/> (February 12, 2005).

Overview

Charts are a graphical representation of quantitative data, usually laid out in two-dimensional form. A chart is a picture of related data, or, with line charts, a picture of an equation.

Charts and graphs are used extensively to help people both communicate numerical information and understand the relationship between different data sets. Charts are a way to view numbers and/or math, in picture form.

Charting grids are called a planes because of their two-dimensional qualities, usually represented in x-axis and y-axis forms, where the x axis is the horizontal plane and the y axis is the vertical plane. Each point, or coordinate, represents the x-value location relative to the y-value location. Charts are also commonly called graphs because the exercise of plotting these x and y coordinates is known as graphing. This mapping process follows a diagram where the variation of a dependent variable is in comparison with another, independent variable.

Fundamental Mathematical Concepts and Terms

There are three basic chart formations: line charts, column charts, and pie charts.

Line charts are pictures plotted, connected dots, often representing equations. Column charts are pictures of data clusters, while pie charts are pictures of percentages.

BASIC CHARTS

Although the line chart and the column chart are similar, each has unique characteristics that help the user better understand the data. The pie chart represents parts of a whole (percentages) and shows how parts of a data set are combined to create a complete picture.

There are many variations of these three basic charts, but the motivation between choosing which chart to use depends on how to communicate the information.

LINE CHARTS

Line charts are commonly used to graph time horizon data such as stock prices, economic data, or sales of specific companies. Many different types of data can be plotted in a line chart but the main benefit is to show how a data set trends, usually over a time horizon. Figure 1 shows a line chart in which the line goes up or down.

A line chart is a series of connected dots. Each dot represents a coordinate value of independent and dependent

Attorney William Lerach, representing Amalgamated Bank, uses a timeline chart of the Enron Corp. daily share price at a New York news conference in 2001. AP/WIDE WORLD PHOTOS. REPRODUCED BY PERMISSION.

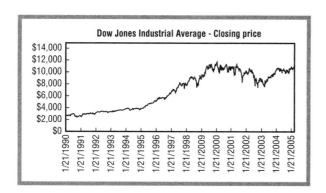

Figure 1.

variables. Usually the independent variable is plotted on the x axis (horizontal), and the dependent variable is plotted on the y axis. Each coordinate of the x value is plotted in relation to its y value. The dots are then connected to create a line.

When plotting changes over time, the x axis usually represents the time, the independent variable, and the y axis represents the dependent variable, the data values.

With all line charts, the order of the values on both the x axis and the y axis is relational.

In line charts, the slope of the line between two points is easily recognized simply by observation. No calculations are necessary to get a good idea of the relative change from one segment to the next. Another quality about line charts is that it is easy to see if the values are increasing or decreasing along different segments. This change is often referred to as volatility.

Multiple-line charts can be plotted on one graph to show the relationship of different data sets with similar parameters relative to each other. A good example of multiple-line charts is seen daily when looking at the price changes of two similar stocks trading on the New York Stock Exchange. Figure 2 shows how to plot both lines on the same graph and quickly see which stock yields the better return for its investors.

In a double-axis line chart, one or more lines is plotted on the left axis, and one or more lines is plotted on the right y axis. This is useful when looking at the relative changes of different data that have similar independent variables (e.g., time horizon), but the data does not have similar dependent variables. Using the stock price example, Figure 3 shows how to compare the price change of two different companies over the same time horizon, but they have very different stock prices. One set of data is plotted on the left y axis, and the other on the right y axis.

X-Y SCATTER GRAPHS

At the fundamental level, all line charts are x-y scatter graphs. The x-y scatter graph is a plotting of all the coordinated points in a set of data. It does not usually have connecting lines. Consequently, the line chart is the next step after plotting the relationships between the x values and the y values. By simply connecting the dots, the XY scatter graph turns into a line chart.

X-y scatter graphs are useful when there are numerous variables to plot and a connecting line is not important. With the x-y scatter graph, clusters of data are easy to follow. These types of charts are used extensively in statistics to see the correlation of dependant and independent variables. As shown in Figure 4, a trend line could be drawn through the midpoint of all the data to show the aforementioned trending of different variables.

COLUMN AND BAR CHARTS

Column charts are useful when sizing different categories of data and comparing them to each other. Column charts are usually used for fewer observations than seen in a line chart. For instance, to measure points scored per starting basketball player in a particular game, a column chart, such as Figure 5, would have a column (category) to represent each player. The players would be labeled along the x axis, and the points scored for each measured on the y axis.

To measure hours of sunlight per day over a one-year period, a column chart would need 365 x-axis categories. It is easier to show this with a line graph and have 12 x-axis categories, each measuring one month with the appropriate number of days between each category. Having 365 columns on one graph is usually unrealistic.

BAR CHARTS

Bar charts are constructed in the same way as column charts with the difference being that the categories are plotted on the y axis and the data values are plotted on the x axis. This results in the columns protruding out horizontally, rather than up vertically. Because the column is now horizontal, it is referred to as a bar, like a parallel bar, rather than a column. Figure 6 represents a bar chart.

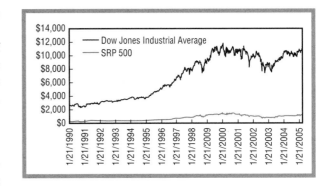

Figure 2.

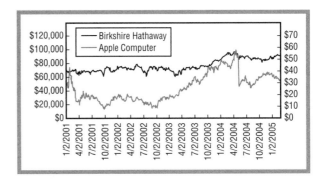

Figure 3.

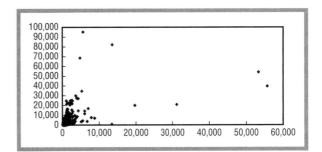

Figure 4.

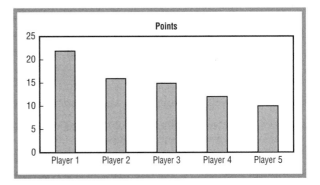

Figure 5.

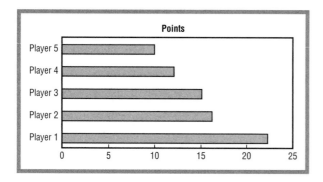

Figure 6.

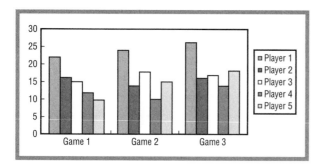

Figure 7.

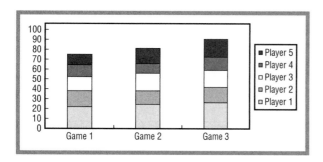

Figure 8.

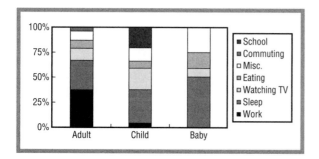

Figure 9.

CLUSTERED COLUMN CHARTS

Clustered columns are charts with the columns adjacent to each other, that is, they are touching, with multiple columns per category. For instance, when charting the points scored per player per game, each category would represent one game and there would be columns per category, all adjacent to each other. The next category would also have the same number of columns and would represent a different game. As Figure 7 shows, this type of chart needs a legend explaining what each column, or data series, represents.

Clustered column charts are very useful in comparing different categories as well as the series within the category. Often the individual columns are color coded as another way to separate and compare the data.

STACKED COLUMN CHARTS

The stacked column chart puts the different values per category on top of each other. This type of chart gives a relative sizing of each value per category. The stacked chart is good for comparing data both within categories and across categories. As in Figure 8, sometimes lines are drawn from the tops of each value between categories to help show the rate of change from one category to the next.

100% STACKED COLUMN CHARTS

Similar to the stacked chart, the 100% stacked chart gives a relative measurement categories, but sets the sum of all category values to 100%. The end result is a percentage measure of each value per category, rather than the actual value. The 100% stacked chart is useful when the overall category is a fixed amount. As an example, since there are 24 hours in the day, Figure 9 shows how those hours are spent on a percentage basis.

PIE CHARTS

Pie charts—much like the 100% stacked chart—convert all the data to percentages. That is, a pie chart sums all the different data categories and accepts this total to be 100% of the data, as in the whole pie. Then, each category is measured as a slice of the pie and represents therefore the percentage of the pie.

The formula for taking hypothetical data of 95 12th graders, 115 11th graders, 150 10th graders, and 180 9th graders is $95 + 115 + 150 + 180 = 540$. Therefore the pie calculations for 12th graders would be $95 / 540 = 18\%$. Figure 10 represents doing the same calculation for each grade, keeping the denominator at 540 and changing the numerator to represent the total students in each grade.

Pie charts are useful to help show how data sizes up to other data in one category. They also help to rank data quickly because they show which data category is largest, which is smallest, and all others in between.

A Brief History of Discovery and Development

Charts are sometimes referred to as Cartesian planes. This is the early name for a two-dimensional grid representation of numbers, first developed by French mathematician, philosopher, and scientist René Descartes (1596–1650).

Real-life Applications

People in finance and business often use charts, consequently they have develop elaborate charting techniques and different ways of displaying data in the three basic charting types.

In the field of finance there are people who make their living trading securities based solely on analyzing the charts of stocks, bonds, and commodities. This business is known as technical analysis, and the people who work in this field are known as chartists.

LINE CHARTS

Line charts are useful for charting large quantities of data, with numerous categories that would be impractical to view individually. For instance, to look at the daily closing price of a stock over a five-year period, it would require viewing approximately 1,260 individual data points (there are approximately 252 trading days in each calendar year). By using a line chart, each point can be graphed individually with tiny lines connecting the dots to give the visual quality of a continuous line. This will show trends over the five-year period and offers a sense of how the price has behaved during the entire time horizon or during specific time periods.

A more technical application for line charts is to plot a few data items and then calculate the slope of the line between each data point, or selective data points. This is useful in measuring volatility, and it also reveals trends and periods of drastic change more precisely.

COLUMN/BAR CHARTS

The term stack-up is pertinent in column charts because that is exactly what happens: data are stacked on

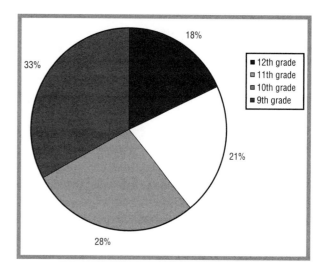

Figure 10.

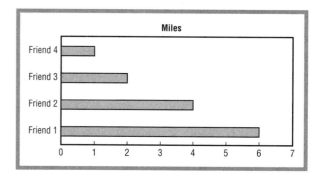

Figure 11.

top of each other or next to each other, for comparative purposes.

Bar charts are often used to measure distance or growth. Since distance of travel is usually viewed from left to right, the length of each bar is a perfect way to visualize how far the data series goes. To compare distance traveled from one person's house to four other people's houses, Figure 11 is a bar chart that shows how far each trip is. It is easy to see which distance is furthest, which is shortest, and others in between.

PIE CHARTS

Pie charts are used to see how a particular data set is partitioned. A pie chart assumes that the data set is the whole universe of data, and it will show the individual percentages that make up that whole. Pie charts are most useful when the total is a fixed quantity.

When looking at total points scored in a football game, each player's points scored is shown as slices of the

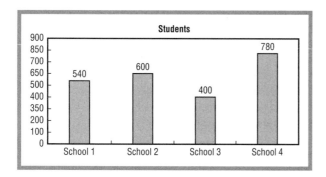

Figure 12.

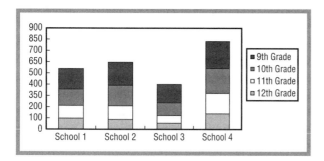

Figure 13.

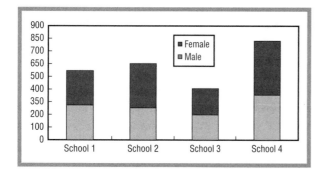

Figure 14.

pie, and the whole pie would represent the total points scored. Alternatively, to look at points scored by just three players, a pie chart is not useful, because other points could have been scored by different players, and the players do not represent the whole, they are only a fraction of the whole.

USING THE COMPUTER TO CREATE CHARTS

There are many computer programs that quickly do most chart plotting. The most common is Microsoft Excel,

which has many different predetermined chart templates, based on the three basic charts, and formats data into a chart.

Excel and other charting programs have created pre-formatted charts to represent data in as many ways as possible, but at the root of all these charts are the three basic chart formats. One area where they have made significant changes in appearance is in area charts, or other three-dimensional chart types. While the basic charting procedure is basically the same, these charting programs have tried to add a third dimension, depth, to the basic two-dimensional chart. While this is helpful with very specific types of data, the two-dimensional charts are still the most commonly used.

CHOOSING THE RIGHT TYPE OF CHART FOR THE DATA

Organization of data is an important part of telling a story, and conveying that story to others. Charts are a quick way of showing the relational aspects of different categorized data sets; charts take the quantitative aspects of information and create a picture to make it easier for the viewer to quickly see relationships. Therefore, choosing the correct chart to represent data sets is a key element of conveying the story, and communicating how the data looks.

For example, at the beginning of the semester the math teacher makes the following announcement: the school administrators want to analyze the demographics of this high school relative to three other high schools in neighboring states. Furthermore, the administration has made the analysis a contest, and everyone in any math class is welcome to participate. All entries will be voted on fairly and independently. The teacher also states: if the winner is in a particular class, that participating student will receive an A for the course.

After collecting the data, the student ends up with the following information for all four schools: total students, broken out by grade; number of male and female students; total square feet of each school; number of teachers; number of classes offered; and the number of students who took the SAT tests, per state, over a 25-year period.

Using line, column, and pie charts, the data is organized in the following way: First, a basic column chart is created showing the total students for each school, as in Figure 12. Secondly, in Figure 13, a stacked bar chart is created, each with four columns, so each segment is representing one grade and each column is representing each school. Figure 14 represents this same concept used to show the distribution of males and females for each school.

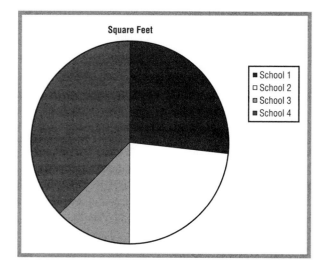

Figure 15.

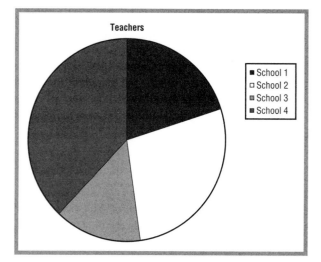

Figure 16.

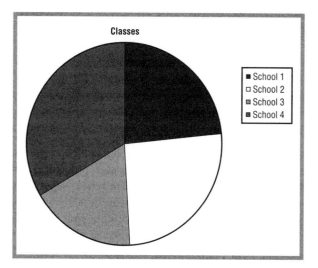

Figure 17.

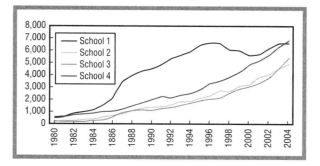

Figure 18.

Lastly, Figure 18 is a line chart used to plot the average SAT scores over the 25-year period. With 25 categories on the x axis, and the scores on the y axis, the data points are plotted, the dots connected, and a line chart is created that spans the 25-year period.

Using a pie chart to plot the square feet per school, the pie chart has four segments, one for each school, and each segment of pie represents the percentage of square feet as a portion of the whole, as shown in Figure 15. Figure 16 represents a pie chart to plot the number of teachers for each school, and Figure 17 is the third pie chart that has the number of classes per school.

Where to Learn More

Books
Excel Charts. Somerset, NJ: John Wiley & Sons, 2005.

Key Terms

Dependant variable: What is being modeled; the output.

Independent variable: Data used to develop a model, the input.

Computers and Mathematics

Overview

Mathematics is integral to computers. Most computer processes and functions rely on mathematical principles. The word "computers" is derived from computing, meaning the process of solving a problem mathematically. Large complex calculations (or computing) in engineering and scientific research often require basic calculators and computers.

Computers have evolved greatly over the years. These days, computers are used for practically anything under the Sun, education, communication, business, shopping, or entertainment. Mathematics forms the basis of all these applications.

Applications of mathematical concepts are seen in the way computers process data (or information) in the form of bits, bytes, and codes, store large quantities of data by compression, and send data from one computer to another by transmission. With the advent of the Internet, communication has become extremely easy. Every computer is assigned a unique identity, using mathematical principles, making communication possible. In addition, mathematics has also found other applications in computers, such as security and encryption.

Fundamental Mathematical Concepts and Terms

BINARY SYSTEM

All computers or computing devices think and process in binary code, a binary number system. In a binary number system, everything is described using two values—on or off, true or false, yes or no, one or zero, and so on. The simplest example of a binary system is a light switch, which is always either on or off. A computer contains millions of similar switches. The status of each switch in the computer represents a bit or binary digit. In other words, each switch is either on or off. The computer describes one as "on" and zero as "off."

Any number can be represented in the binary system as a combination of zeros and ones. In the binary number system, each number holds the value of increasing powers of two, e.g., 2^0, 2^1, and so on. This makes counting in binary easy. The binary representation for the numbers one to ten can be shown as follows:

- $0 = 0$
- $1 = 1$
- $2 = 10$
- $3 = 11$

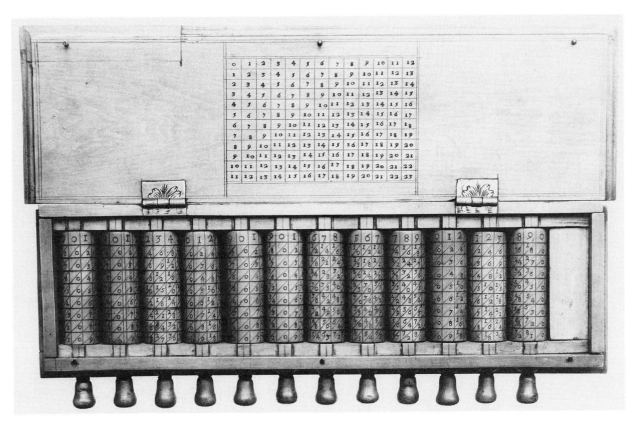

A calculating device created by Scottish mathematician John Napier in 1617 which consists of cylinders inscribed with multiplication tables. It's also known as "Napier's Bones." BETTMANN/CORBIS.

- 4 = 100
- 5 = 101
- 6 = 110
- 7 = 111
- 8 = 1000
- 9 = 1001
- 10 = 1010.

ALGORITHMS

The key principle in all computing devices is a systematic process for completing a task. In mathematics, this systematic process is called an algorithm. Algorithms are common in daily life as well. For example, when building a house, the first step involves building the floor base (or foundation), followed by the walls, and then the ceiling or roof. This systematic procedure to solve the problem of building a house is an example of an algorithm.

In a nutshell, algorithms are a list of step-by-step instructions. In mathematical terms, these are also sometimes known as theorems. A computer program, or application, is made up of a number of such algorithms. Besides, every process in a computer also depends on a specific algorithm. For example, when switching on the computer, the computer does what is known as "booting." Booting helps in properly loading the operating system (Windows, Mac, Dos, UNIX, and so on). During booting, the computer follows a set of instructions (defined by an algorithm). Similarly, while opening any program (say, MS Word), the computer is again instructed to follow a set of tasks so that the program opens properly.

Like complex mathematical problems, even the most complex software programs are based on numerous algorithms.

A Brief History of Discovery and Development

Although the modern computer was built only in the twentieth century, many primitive forms of the computer were used in ancient times. The early calculators can also be considered as extremely basic computers based on similar mathematical concepts. The word calculator, is derived from the Latin word *calculus* (or a small stone). Early human civilizations used small stones for counting. Counting boards made up of stones were used for basic arithmetic tasks such as addition, subtraction, and multiplication.

This led to development of devices that enabled calculation of more complex numbers, and in quick time. With the progress of civilization, man saw the development of the abacus, the adding machine, the Babbage, and the prototype mainframe computers.

Modern computers, however, were invented in the twentieth century. In 1948, the mathematician Claude Shannon (1916–2001), working at Bell Laboratories in the United States, developed computing concepts that would form the basis of modern information theory. Shannon is often known as the father of information science. Computers were earlier only used by government institutions. Home or personal computers (known as PCs) came much later in the late 1970s and 1980s.

Today, personal computers and servers with a microprocessor chip (a small piece of computer hardware) are embedded in almost all lifestyle electronic products, from the washing machine and television to calculators and automobiles. Many of these chips are capable of computing in the same capacity as some basic computers. The advancement of mathematical concepts and theories has made it possible to develop sophisticated computers in smaller and smaller sizes, such as those found in handheld computers like the PDA (personal data assistant) and PMP (personal media player).

Ciphers, codes, and secret writing based on mathematical concepts have been around since ancient times. In ancient Rome, they were used to communicate secrets over long distances. Such codes are now used extensively in the field of computer science.

Real-life Applications

BITS

The bit is the smallest unit of information in a computer. As discussed earlier, a bit is a basic unit in a binary number system. A bit or binary digit stands for true or false, one or zero, on or off. The computer is made up of numerous switches. Each switch has two states (on and off). The value of each state represents a bit.

Bits are the basic unit of storage in computers. In other words, all data is stored in the form of bits. The reason for using a binary number system rather than decimal system for storage (and other purposes) is that with prevailing technology, it is much easier to implement the binary system in computers. Implementing the binary system is significantly cheaper, as well.

The speed of the computer (processor speed) in terms of processing applications is related to many factors, including memory space (also known as random access memory, or RAM). Most home computers are either 32-bit or 64-bit; 32-bit and 64-bit are the sizes of the memory space.

BYTES

In computers, bits are bundled together into manageable collections called bytes. A byte consists of eight bits. Bits and bytes are always clubbed together like atoms and molecules. Computers are designed to store data and process instructions in bytes. To handle large quantities of information (or bits), other units such as kilobytes, megabytes, and gigabytes are used. One kilobyte (KB) = 1,024 bytes = 2^{10} bytes (and not 1,000 bytes as commonly thought). Similarly, 1 megabyte (MB) = 1,048,576 bytes = 2^{20} bytes, and 1 gigabyte (GB) = 1,073,741,824 bytes = 2^{30} bytes.

The first computers were 1-byte machines. In other words, they used octets or 8-bit bytes to store information, and they represented 256 values (2^8 values, integers zero to 255).

The latest computing machines are 64-bit (or eight bytes). This type of representation makes computing easier in terms of both storage and speed. Bits and bytes form the basis of many other computer processes and functions. These include CD storage, screen resolution, text coding, data comparison, data transmission, and much more.

TEXT CODE

All information in the computer is stored in the form of binary numbers. This includes text, as well. In other words, text is not stored as text, but as binary numbers. The rule that governs this representation is known as ASCII (American Standard Code for Information Interchange). The ASCII system assigns a code to every letter of the alphabet (and other characters). This code is stored as a seven digit binary number in computers. Moreover, the ASCII code for a capital letter is different than the code for the small letter. For example, the ASCII code for "A" is 10, whereas that for "a" is 97. Consequently, the value of "A" is stored as 0001010 (its binary representation), whereas "a" is 1100001.

Every character is stored as eight bits (a leading bit in addition to the seven bits for the ASCII code), or one byte. Thus, the word "happy" would require five bytes. An entire page with 20 lines and 60 characters per line would require 1,200 bytes.

The main benefit of storing text code as binary numbers is that it makes it easier for the computer to store and process the data. Besides, mathematical operations can be performed on binary representations of text.

PIXELS, SCREEN SIZE, AND RESOLUTION

A pixel is derived from the words picture and element. The smallest and the most basic unit of images in computers is the pixel. A pixel is a tiny square block. Images are made up of numerous pixels. The total number of pixels in a computer image is known as the resolution of the image. For example, a standard computer monitor displays images with the resolution 800×600. This simply means that the image (or the entire computer screen) is 800 pixels wide and 600 pixels high.

Each pixel is also stored as eight bits (or one byte). Again, its representation is in the form of binary numbers. Storing the value of the color of a pixel is far easier in binary format, as compared with other formats. The maximum number of combinations of zeros and ones in an 8-bit number is 256 (2^8). Each combination represents a color. Simply put, every pixel can have one of 256 different colors.

This kind of computer display is called an "8-bit" or "256-color" display, and was very common in computers built in the 1990s. In contrast, newer computer monitors built after the year 2000 have a significantly higher number of colors (in millions). These are the 16-bit and 24-bit monitors.

The color of every pixel in a computer image is a combination of three different colors—red, green, and blue (RGB). RGB is common terminology used in computer graphics and images, and simply means that every color is a combination of some portion of red, green, and blue colors. The value of each of these colors is stored in one byte. For example, the color of a pixel could be 100 of red, 155 of green, and 200 of blue. Each of these values is stored in binary format in a byte. Note that the color values can range from zero to 255. Thus, every color pixel has three bytes. Subsequently, a computer monitor with the resolution 800×600 would need $3 \times 800 \times 600$, or 1,440,000 bytes.

IP ADDRESS

Every computer on a network has a specific address. A number, known as the Internet protocol address, or IP address, indicates this. The reason for having an IP address is simple. To send a packet or a letter through regular mail, the address of the recipient is required. Similarly, for communicating with a computer (from another

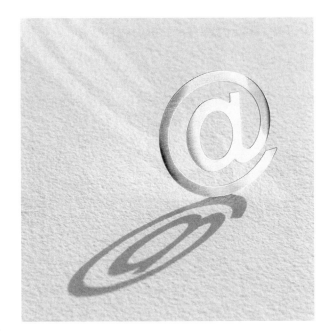

Internet mathematics translates binary code into web addresses and other information. ROYALTY-FREE/CORBIS.

computer), the address of that computer is required. Every computer has a unique IP address that clearly distinguishes it from other computers. The concept of the IP address is based on mathematical principles, and there are rules that govern the value of the IP address. For example, an IP address is always a set of four numbers separated by dots (e.g., 204.65.130.40).

Remember, the computer only understands binary numbers. Consequently, the IP address is also represented as a binary number. The binary representation is octet (equivalent to the representation of a byte). Technically, every IP address is a 32-bit number divided into four bytes, or octets (eight bites). Each octet represents a specific number. For example, in the above case, 204 would be stored in one octet, 65 in another octet, and so on. The binary representation (as stored in the computer) for the above-mentioned IP address would be: 11001100 .01000001.10000010.0101000.

Communication between computers becomes far easier with binary representation. The IP address consists of two components, the network address and the host address. The network address (the first two numbers) represents the address of the entire network. For example, if a computer is part of a network of computers connected into an entire company, the first two numbers would represent the IP address of the company. In other words, for all computers connected to the company network, the first two numbers would remain the same.

The host address (the last two numbers) represents the address of a computer specifically. For example, the third number might represent a particular department within a company, whereas the last number would represent a particular computer in that department. Consequently, two computers within the same department (and part of the same company) would have the same first three numbers. Only the last number would be different. Similarly, two computers that are part of different departments would have the same first two numbers.

As each number in the IP address is allowed a maximum of one octet (or eight bites), the maximum value the number can have is 255. In other words, the values of every number in the IP address ranges from zero to 255. An IP address that contains a number higher than this range would be incorrect. For example, 204.256.12.0 is incorrect, as 256 is not valid.

SUBNET MASK

With the advent of the Internet, the number of computers that are connected worldwide is quickly rising. The Internet is a huge network of computers. Subsequently, each computer has an IP address that helps it communicate with the rest. For example, to send an email, the email address must be entered. This email address is translated to a specific IP address, that of the recipient. As of 2005, there are millions of computers connected to the Internet. As mentioned earlier, IP addresses have a limitation. Each number can only have a value within a specific range (zero to 255).

The IP address given to any computer on the Internet is temporary. In other words, as soon as a computer connects to the Internet, it receives a unique IP address. As soon as the Internet is disconnected, this IP address is free and can be used by another computer. When the same computer connects again, it would get another IP address. With the high number of computers connected to the Internet simultaneously, it is difficult to accommodate every computer within this range. This is where the concept of Subnet mask comes in.

Subnets, as the name suggests, are sub-networks. The host address (from the IP address) is divided into further subnets to accommodate more computers. This is done in such a way that a part of the host address identifies the subnet. The subnet is also shown as a binary number. Communication becomes easier because of the binary representation.

Take, for example, the IP address 204.65.130.40. Its binary equivalent is 11001100.01000001.10000010 .00101000.

The subnets would have the same network address (first two numbers). The first four bits of the host address (third number) would be the same as well, to identify the host of the subnet. In this case, 1000 would be unchanged. The remaining four bits of the host address would be unique to each subnet. Every subnet, in turn, can have numerous computers. Every computer on the subnet would have a unique fourth number in the IP address. Consider the following scenario:

The main IP address is 11001100.01000001 .10000010.00101000. This could have many subnets such as 11001100.01000001.10000111.00111010, 11001100 .01000001.10000101.0100010, and so on. Note that the first four digits of the third number (host address) are same but the remaining are different, indicating different subnets on the same host. The fourth number indicates a specific computer on the subnet. For computers on the same subnet, the first three numbers would remain the same.

Simply put, the subnet mask ensures that more computers can be accommodated within a network. Every subnet mask number identifies the network address, the host, the subnet, as well as the computer.

COMPRESSION

Computers store (and process) data that include numbers, arithmetic calculations, and words. In addition, the data may also be in the form of pictures, graphics, and videos. In computers, data is stored in files. File sizes, depending on the type of data, can be huge. Many times the size of a file becomes unmanageable. In such cases, better ways of storing and process data, must be used. Given below are some comparisons to provide a better understanding of sizes of different files on a computer.

One alphabetic character is represented by one byte, one word is equivalent to eight to ten bytes or so, a page averages about two kilobytes, an entire book averages one megabyte or more, twenty seconds of good quality video occupy anywhere from two to ten megabytes, and so on. Similarly, a compact disc (CD) has 600–800 megabytes of data.

Storing such huge amounts of information in a computer can often be difficult. Besides, it is almost impossible to send large data from one computer to another through e-mail or other similar means. Moreover, downloading a significant amount of data from the Internet (such as movie files, databases, application programs) can be extremely time consuming, especially if using a slow dial up connection. This is where compression of the data into a manageable size becomes important.

Certain applications based on mathematical algorithms compress the data. This allows the basic data that a computer sees in binary format, to be stored in a compressed format requiring much lower storage space. Compressed data can be uncompressed using the same application and algorithm.

Compression is extremely beneficial, especially when a large file has to be sent from one computer to another. In case of e-mail, sending a one-megabyte (MB) file through a dial up connection, would take considerable time, anywhere from fifteen to thirty minutes. Bigger files would take even longer. Besides, e-mails might not have the capacity of sending (or receiving) bigger files. In such cases, sending zipped files that are much smaller is useful. Similarly, downloading compressed files from the Internet rather than the large original ones is a better option.

There are also other types and methods for compressing. Run length compression is another type that is used widely. In run length compression, large chunks, or runs, of consecutive identical data values are taken, and each of these is replaced by a common code. In addition to the code, the data value and the total length are also recorded. Run length compression can be quite effective. However, it is not used for certain types of data such as text, and executable programs. For these types of files, run length compression does not work. Without going into the technical specifics of run length compression, this method works quite well on certain types of data (especially images and graphics), and is subsequently applied to many data compression algorithms. Most compressed files can be uncompressed to obtain the original. However, in almost all cases, some data is lost in the process. For visual and audio data, some loss of quality is allowed without losing the main data. By taking advantage of limitations of the human sensory system, a great deal of space is saved while creating a copy that is very similar to the original. In other words, although compression results in some data loss, this loss can be insignificant and the naked eye usually cannot usually discern the difference between the original and the un-compressed file. The defining characteristics of these compression methods are their compression speed, the compressed size, and the loss of data during compression.

Apart from computers, compression of images and video is also used in digital cameras and camcorders. The main purpose is to reduce the size of the image (or video) without compromising on the quality. Similarly, DVDs also use compression techniques based on mathematical algorithms to store video.

In audio compression, compression methods remove non-audible (or less audible) components of the signal while compressing. Compression of human speech is sometimes done using algorithms and tools that are far more complex. Audio compression has applications in Internet telephony (voice chat through the internet), audio CDs, MP3 CDs, and more.

DATA TRANSMISSION

In computing, data transmission means sending a stream of data (in bits or bytes) from one location to another, using different technologies. Two of these technologies are coding theory and hamming codes. These are both based on algorithms and other mathematical concepts.

Coding theory ensures data integrity during transmission. In other words, it ascertains that the original data is safely received, without any loss. Messages are usually not transmitted in their original form. They are transmitted in coded or encrypted form (described later). Coding theory is about making transmitted messages easy to read. Coding theory is based on algorithms. In 1948, the mathematician Claude Shannon presented coding theory by showing that it was possible to encode in an effective manner. In its simplest form, a coded message is in the form of binary digits or bits, strings of zero or one. The bits are transmitted along a channel (such as a telephone line). While transmitting, a few errors may occur. To compensate for the errors, more bits of information than required are generally transmitted.

The simplest method (part of the coding theory developed by Shannon) for detecting errors in binary data is the parity code. Concisely, this method transmits an extra bit, known as the parity bit, after every seven bits from the source message. However, the parity code method can merely detect errors, not correct them. The only method for correcting them is to ask for the data to be transmitted again.

Shannon developed another algorithm, known as the repetition algorithm, to ensure detection as well as correction of errors. This is accomplished by repeating each bit a specific number of times. The recipient sees which value (zero or one) occurred more often and assumed that was the actual value. This process can detect and correct any number of errors, depending on how many repeats of each bit are sent. The disadvantage of the repetition algorithm is that it transmits a high number of bits, resulting in a considerable amount of repetitive bits. Besides, the assumption that a bit that is received more often, is the actual bit, may not hold true in all cases.

Another mathematician, Richard Hamming (1915–1998), built more complex algorithms for error correction. Known as Hamming codes, these were more efficient, even with very low repetition. Initially, Hamming produced a code (based on an algorithm) in which four data bits were followed by three check bits that allowed the detection and the correction of a single error. Although, the number of additional bits is still high, it is without a doubt lower than the total number of bits transmitted by the repetition algorithm. Subsequently, these additional bits (check bits) were reduced even further by improving the underlying algorithms. Hamming codes are commonly used for transmitting not just basic data (in the form of simple email messages), but also more complex information.

One such example is astronomy. The National Aeronautics and Space Administration (NASA) uses these techniques while transmitting data from their spacecrafts back to Earth (and vice versa). Take, for example, the NASA *Mariner* spacecraft sent to Mars in the 1960s. In this case, coding and error correction in data transmission was vital, as the data was sent from a weak transmitter over very long distances. Here the data was read perfectly using the Hamming code algorithm. In the late 1960s and early 1970s, the NASA *Mariner* sent data using more advanced versions of the Hamming and coding theories, capable of correcting seven errors out of thirty-two bits transmitted. Using this algorithm, over 16,000 bits per second of data was successfully relayed back to Earth.

Similar data transmission algorithms are used extensively for communication through the Internet since the late 1990s. The Hamming codes are also used in preparing compact discs (CDs). To guard against scratches, cracks, and similar damage, two overlapped Hamming codes are used. These have a high rate of error correction.

ENCRYPTION

Considerable confidential data is stored and transmitted from computers. Security of such data is essential. This can be achieved through specialized techniques known as encryption. Encryption converts the original message into coded form that cannot be interpreted unless it is de-coded back to the original (decryption). Encryption, a concept of cryptography, is the most effective way to achieve data security. It is based on complex mathematical algorithms.

Consider the message abcdef1234ghij56789. There are several ways of coding (or encrypting) this information. One of the simplest ways is to replace each alphabet by a corresponding number, and vice versa. For example, "a" would become "1", "b" would be "2", and so on. The above original message can, thus be encrypted as 123456abcd78910 efghi. The message is decrypted using the same process and converted back in the original form.

Complex mathematical algorithms are designed to create far more complex encryption methods. The information regarding the encryption method is known as the key.

Cryptography provides three types of security for data:

- Confidentiality through encryption—This is the process mentioned above. All confidential data is encrypted using certain mathematical algorithms. A key is required to decrypt the data back into its original form. Only the right people have access to the key.
- Authentication—A user trying to access coded or protected data must authenticate himself/herself. This is done through his/her personal information. Password protection is a type of authentication that is widely used in computers and on the Internet.
- Integrity—This type of security does not limit access to confidential information, as in the above cases. However, it detects when such confidential is modified. Cryptographic techniques, in this case, do not show how the information has been modified, just that it has been modified.

There are two main types of encryption used in computers (and the Internet)—asymmetric encryption (or public-key encryption), and symmetric encryption (or secret key encryption). Each of these is based on different mathematical algorithms that vary in function and complexity.

In brief, public key encryption uses a pair of keys, the public key, and the private key. These keys are complimentary, in the sense that a message encrypted using a particular public key can only be decrypted using a corresponding private key. The public key is available to all (it is public). However, the private key is accessible only by the receiver of a data transmission. The sender encrypts the message using the public key (corresponding to the private key of the receiver). Once the receiver gets the data, it is decrypted using the private key. The private key is not shared with anyone other than the receiver, or the security of the data is compromised.

Alternatively, symmetric secret key encryption relies on the same key for both encryption and decryption. The main concern in this case is the security of the key. Subsequently, the key has to be such that even if someone gets hold of it, the decryption method does not become too obvious. For this purpose, encryption and decryption algorithms for secret key encryption are quite complex.

The key, as expected, is shared only by the receiver and the sender (unlike public key encryption, where everyone knows the public key). The key can be anything ranging from a number, a word, or a string of jumbled up letters and other characters. In simple terms, the original

Key Terms

Bit: The smallest unit of storage in computers. A bit stores a binary value.

Byte: A byte is a group of eight bits.

Encryption: Using a mathematical algorithm to code a message or make it unintelligible.

Pixel: Short for "picture," a pixel is the smallest unit of a computer graphic or image. It is also represented as a binary number.

data is encoded using a simple or complex technique defined by a mathematical algorithm. The key also holds the information on how the algorithm works. The same algorithm can then be used to decode the message back into its original form.

Encryption is used frequently in computers. Most data is protected using one of the above mentioned encryption techniques. The Internet also widely applies encryption. Most websites protect their content using these methods. In addition, payment processing on websites also follows complex encryption algorithms (or standards) to protect transactions.

Where to Learn More

Books

Cook, Nigel P. *Introductory Computer Mathematics.* Upper Saddle River, NJ: Prentice Hall, 2002.

Graham, Ronald H., et al. *Concrete Mathematics: A Foundation for Computer Science.* Boston, MA: Addison-Wesley, 1994.

Conversions

Conversion is the process of changing units of measurement from one system to another. The ability to convert units such as distance, weight, and currency is an increasingly important skill in an emerging global economy. In area of research and technological applications such as science and engineering, the ability to convert data is crucial.

No better example of how critical a role conversion math can play can be found in the destruction of NASA's *Mars Climate Orbiter* in 1999. The *Mars Climate Orbiter* was one of a series of NASA missions in a long-term program of Mars exploration known as the Mars Surveyor Program. The orbiter mission was designed to have the orbiter fire its main engine to enter into orbit around Mars at an altitude of about 90 miles (about 140 km). However, a series of errors caused the probe to come too close to Mars and, as a result, the probe was only about 35 miles (57 km) from the Martian surface when it attempted to enter orbit—an altitude far below the minimum safe altitude for orbit. As a result the *Mars Climate Orbiter* is presumed to have been destroyed as it reentered the Martian atmosphere.

Engineering teams contracted by NASA used different measurement systems (English and metric) and never converted the two measurements. As a result, the probe's attitude adjustment thrusters failed to fire properly and the probe drifted off course toward its fatal demise.

Fundamental Mathematical Concepts and Terms

In addition to traditional English measurements, International System of Units (SI) and MKS (meter-kilogram-second) units are part of the metric system, a system based on powers of ten. The metric system is used throughout the world—and in most cases provides the standard for measurements used by scientists. On an everyday basis, nearly everyone is required to convert values from one unit to another (e.g., the conversion from kilometers per hour to miles per hour).

This need for conversation applies widely across society, from fundamental measurement of the gap in spark plugs to debate and analysis over sports records.

When values are multiplied or divided, they can each have different units. When adding or subtracting values, however, the values must added or subtracted must have the same units. A notation such as "ms^{-1}" is simply a different way of indicating m/s (meters per second).

Units must properly cancel to yield a proper conversion. If an Olympic sprinter runs 200-meter race in 19.32 seconds, he runs at an average speed of average speed of 10.35 meters per second [200 m / 19.32 s = 10.35 m/s]. If a student wishes to convert this to miles per hour the conversion should be carried out as follows: (10.35 m/s) (1 mile / 1,609 m) (3,600 s / 1 hr) = 23.2 miles/hr. The units cancel as follows: (10.35 ^{m}/$_{s}$) (1 mile / 1,609 ^{m}) (3,600 ^{s} / 1 hr) = 23.2 miles/hr.

Students should remember to be cautious when dealing with units that are squared, cubed, or that carry another exponent. For example, a cube that is 10 cm on each side has a volume that is expressed as a cube value (e.g., m^3 that is determined from multiplying the cube's length times the width times the height: V = (10 cm) (10 cm)(10 cm) = 1,000 cm^3.

Many conversions are autoprogrammed into calculators—or are easily made with the use of tables and charts.

THE METRIC UNITS

The SI starts by defining seven basic units: one each for length, mass, time, electric current, temperature, amount of substance, and luminous intensity. ("Amount of substance" refers to the number of elementary particles in a sample of matter. Luminous intensity has to do with the brightness of a light source.) However, only four of these seven basic quantities are in everyday use by non-scientists: length, mass, time, and temperature.

The defined SI units for these everyday units are the meter for length, the kilogram for mass, the second for time, and the degree Celsius for temperature. (The other three basic units are the ampere for electric current, the mole for amount of substance, and the candela for luminous intensity.) Almost all other units can be derived from the basic seven. For example, area is a product of two lengths: meters squared, or square meters. Velocity or speed is a combination of a length and a time: kilometers per hour.

Because the meter (1.0936 yd) is much too big for measuring an atom and much too small for measuring the distance between two cities, we need a variety of smaller and larger units of length. But instead of inventing different-sized units with completely different names, as the English-American system does, metric adaptations are accomplished by attaching a prefix to the name of the unit. For example, since kilo- is a Greek form meaning a thousand, a kilometer is a thousand meters. Similarly, a kilogram is a thousand grams; a gigagram is a billion grams or 10^9 grams; and a nanosecond is one billionth of a second or 10^{-9} second.

Odometers sit in a shop that legally converts odometers from kilometers to miles in used cars imported from Canada. AP/WIDE WORLD PHOTOS. REPRODUCED BY PERMISSION.

THE ENGLISH SYSTEM

In contrast to the metric system's simplicity stands the English system of measurement (a name retained to honor the origin of the system) that is based on a variety of standards (most completely arbitrary).

There many English units, including buckets, butts, chains, cords, drams, ells, fathoms, firkins, gills, grains, hands, knots, leagues, three different kinds of miles, four kinds of ounces, and five kinds of tons. There are literally hundreds more. For measuring volume or bulk alone, the English system uses ounces, pints, quarts, gallons, barrels and bushels, among many others.

THE INTERNATIONAL SYSTEM OF UNITS (SI)

The metric system is actually part of a more comprehensive International System of Units, a comprehensive set of measuring units. In 1938, the 9th General [International] Conference on Weights and Measures, adopted the International System of Units. In 1960, the 11th General Conference on Weights and Measures modified the system and adopted the French name *Système International d'Unités*, abbreviated as SI.

Nine fundamental units make up the SI system. These are the meter (abbreviated m) for length, the kilogram (kg) for mass, the second (s) for time, the ampere (A) for electric current, the Kelvin (K) for temperature, the candela (cd) for light intensity, the mole (mol) for quantity of a substance, the radian (rad) for plane angles, and the steradian (sr) for solid angles.

DERIVED UNITS

Many physical phenomena are measured in units that are derived from SI units. As an example, frequency is measured in a unit known as the hertz (Hz). The hertz is the number of vibrations made by a wave in a second. It can be expressed in terms of the basic SI unit as s^{-1}. Hertz units are used to describe, measure, and calibrate radio wavelengths and computer processing speeds.

Pressure is another derived unit. Pressure is defined as the force per unit area. In the metric system, the unit of pressure is the Pascal (Pa) and can be expressed as kilograms per meter per second squared, or $kg/m\,s^2$. Measurements of pressure are important in determining whether gaskets and seals are properly placed on automobile motors or properly functioning in air-conditioning units.

Even units that appear to have little or no relationship to the nine fundamental units can, nonetheless, be expressed in terms of those units. The absorbed dose, for example, indicates that amount of radiation received by a person or object. In the metric system, the unit for this measurement is the "gray." One gray can be defined in terms of the fundamental units as meters squared per second squared, or m^2/s^2.

Many other commonly used units can also be expressed in terms of the nine fundamental units. Some of the most familiar are the units for area (square meter: m^2), volume (cubic meter: m^3), velocity (meters per second: m/s), concentration (moles per cubic meter: mol/m^3), and density (kilograms per cubic meter: kg/m^3).

As previously mentioned, a set of prefixes is available that makes it possible to use the fundamental SI units to express larger or smaller amounts of the same quantity. Among the most commonly used prefixes are milli- (m) for one-thousandth, centi- (c) for one-hundredth, micro- (μ) for one-millionth, kilo- (k) for one thousand times, and mega- (M) for one million times. Thus, any volume can be expressed by using some combination of the fundamental unit (liter) and the appropriate prefix. One million liters, using this system, would be a megaliter (ML) and one millionth of a liter, a microliter (μL).

UNITS BASED ON PHYSICAL OR "NATURAL" PHENOMENA

In the field of electricity the charge carried by a single electron is known as the elementary charge (e) and has the value of $1.6021892 \times 10^{-19}$ coulomb. This is termed a "natural" unit.

Other real-world or "natural" units of measurement include the speed of light (c: 2.99792458×10^8 m/s), the Planck constant (6.626176×10^{-34} joule per hertz), the mass of an electron (m_e: $0.9109534 \times 10^{-30}$ kg), and the mass of a proton (m_p: $1.6726485 \times 10^{-27}$ kg).

Each of the above units can be expressed in terms of SI units, but they are often also used as basic units in specialized fields of science.

A Brief History of Discovery and Development

Because the United States is the world's leading producer in many items, regardless of the near universal acceptance of the SI, the most frequent conversions between units are between the English system of weights and measures to those of the metric system. The metric system of measurement, first advanced and adopted by the France in the late eighteenth and early nineteenth century, has grown to become the internationally agreed-upon set of units for commerce, science, and engineering.

The United States is the only major economic power to yet fully embrace the metric system. The history of the metric system in the United States is bumpy, with progress toward inevitable metrification coming slowly over two centuries.

As early as 1800, U.S. government agencies adopted metric meter and kilogram measurements and standards. In 1866, the U.S. Congress first authorized the use of the metric system. Although internal progress is halting at best, the United States is one of the 17 original signers of the treaty establishing the International Bureau of Weights and Measures that was intended to provide worldwide metric standards. Most Americans do not know, for example, that since 1893, the units of distance (foot, yard), weight (pound), and volume (quart), have been officially defined in terms of their relation to the metric meter and kilogram.

After the modernization and international expansion of the metric system in the 1960s and 1970s following adoption of the SI, the United States soon stood alone among modern industrialized nations in failing to make full conversion. The English system was abandoned by the English as early as 1965 as part of Great Britain's integration into the European Common Market (a forerunner of the modern European Union) and countries such as Canada completed massive metrification efforts throughout the 1970s.

Following Congressional resolutions and studies that recommended U.S. conversion to the metric system by 1980, an effort toward voluntary conversion began with the 1975 Metric Conversion Act that established a subsequently short-lived U.S. Metric Board. The

American public simply refused to embrace and use metric standards.

It was not until 1988 the Congress once again tried to spur metric conversion with the Omnibus Trade and Competitiveness Act of 1988. The Act specified that metric measurements are to be considered the "preferred system of weights and measures for U.S. trade and commerce." The Act also specified that federal agencies use the metric measurements in the course of their business.

Regardless of the efforts of leaders in science and industry, early into the twenty-first century, U.S. progress remains spotty and slow. However, the demands of global commerce and the economic disadvantages of the use of non-metric measurements provide an increasingly powerful incentive for U.S. metrification.

Although the SI is the internationally accepted system, elements of the English system of measurement continue in use for specialized purposes throughout the world. All flight navigation, for example, is expressed in terms of feet, not meters. As a consequence, it is still necessary for a mathematically literate person to be able to perform conversion from one system of measurement to the other.

Real-life Applications

There are more than 50 officially recognized SI units for various scientific quantities. Given all possible combinations there are millions of possible conversions possible. All of these require various conversion factors. However, in addition to metric conversions, a wide range of conversions are used in everyday situations—from conversion of kitchen measurements in recipes to the ability to convert mathematical data into representative data found in charts, graphs, and various descriptive systems.

Historical Conversions

Historians and archaeologists are often called upon to interpret text and artifacts depicting ancient systems of measurement. To make a realistic assessment of evidence from the past they must be able to convert the ancient measurements into modern equivalents.

For example, the Renaissance Italian artist, Leonardo da Vinci used a unit of measure he termed a *braccio* (English: arm) in composing many of his works. In Florence (Italian: Firenze) *braccio* equaled two *palmi* (English: palms). However, historians have noted that the use of such terms and units was distinctly regional and that various conversion factors must be used to compare drawings and manuscripts. In Florence, a *braccio* equaled about 23 in. (58 cm), but in other regions (or among different professional classes) the *braccio* was several inches shorter. In Rome, the *piede* (English: foot) measured near it modern equivalent of 12 in. (30 cm) but measured up to 17 in. (34 cm) in Northern Italy.

Conversion of Temperature Units

Temperature can be expressed as units of Celsius, Fahrenheit, Kelvin, Rankin, and Réaumur.

The metric unit of temperature is the degree Celsius (°C), which replaces the English system's degree Fahrenheit (°F). In the scientists' SI, the fundamental unit of temperature is actually the kelvin (K). But the kelvin and the degree Celsius are exactly the same size: 1.8 times as large as the degree Fahrenheit. One cannot convert between Celsius and Fahrenheit simply by multiplying or dividing by 1.8, however, because the scales start at different places. That is, their zero-degree marks have been set at different temperatures.

The measurement of thermal energy involves indirect measurement of the molecular kinetic energies of a substance. Rather than providing an absolute measure of molecular kinetic energy, thermal measurements are designed to determine differences that result from work done on, or by, a substance (e.g., heat added to, or removed from, a substance). Temperature differences correspond to changes in thermal energy states, and there are several analytic methods used to measure differences in thermal energy via measurement of temperature. When dealing with the terminology associated with the measurement of thermal energy, one must be mindful that there is no actual substance termed "energy" and no actual substance termed "heat." Accordingly, when speaking of energy "transfer" or heat "flow" one is actually referring to changes in functions of state that can only be raised or lowered within a body or system. Neither energy or heat can really be "transferred" or "flow."

In thermodynamics, temperature is directly related to the average kinetic energy of a system due to the agitation of its constituent particles. In practical terms, temperature measures heat and heat measures the thermal energy of a system.

In meteorological systems, for example, temperature (as an indirect measure of heat energy) reflects the level of sensible thermal energy of the atmosphere. Such measurements use thermometers and are expressed on a given temperature scale, usually Fahrenheit or Celsius.

The common glass thermometer containing either mercury or alcohol uses the property of thermal expansion of the respective fluid as an indirect measure of the increase or decrease in the thermal energy of a body or system. Other types of thermometers utilize properties such as electrical resistance, magnetic susceptibility, or light emission to measure temperature.

Electrical thermometers (e.g., thermoprobes, thermistor, thermocouples, etc.) relate changes in electrical properties (e.g., resistivity) to changes in temperature are extensively used in scientific research and industrial engineering.

Because energy is commonly defined as the ability to do work, the thermal energy of a system is directly related to a system's ability to translate heat energy into work. Correspondingly, the measurement of the thermal energy of a system must be interpreted as the measurement of the changes in the ability of a system or body to do work. Absolute zero Kelvin—notice that Kelvin is not expressed as "degrees Kelvin"—($-459.69°F$, $-273.16°C$, $0°R$ on the Rakine scale)—is the lowest temperature theoretically possible. At absolute zero there is a minimum of vibratory motion (not an absence of motion) and, by definition, no work can be done by a system on its surrounding environment. In this regard, such a system (although not motionless) would be said to have zero thermal energy.

In 1714, the German physicist Daniel Gabriel Fahrenheit (1686–1736) created a thermometer using liquid mercury. Mercury has a uniform volume change with temperature, a lower freezing point and higher boiling point than water, and does not wet glass. Mercury thermometers made possible the development of reproducible temperature scales and quantitative temperature measurement. Fahrenheit first chose the name "degree" (German: grad) for his unit of temperature. Then, to fix the size of a degree (°), he decided that it should be of such size that there are exactly 180° between the temperature at which water freezes and the temperature at which water boils. (180 is a "good" number because it is divisible by one and by 16 other whole numbers. That is why 360, or 2×180, which is even better, was originally chosen as the number of "degrees" into which to divide a circle.) Fahrenheit now had a size for his degree of temperature, but no standard reference values. Where should the freezing and boiling points of water fall on the scale? He eventually decided to fix zero at the coldest temperature that he could make in his laboratory by mixing ice with various salts that make it colder. (Salts, when mixed with cold ice, lower the melting point of ice, so that when it is melting it is at a lower temperature than usual.) When he set his zero at that point, the normal freezing

point of water turned out to be 32° higher. Adding 180 to 32 gave 212°, which he used for the normal boiling point of water. Thus, freezing water falls at 32° and boiling water falls at 212° on the Fahrenheit scale. The normal temperature of a human being is about 99°.

In 1742, the noted Swedish astronomer Anders Celsius (1701–1744), professor of astronomy at the University of Uppsala (Sweden), proposed the temperature scale which now bears his name, although for many years it was called the centigrade scale. As with the Fahrenheit scale, the reference points were the normal freezing and normal boiling points of water, but he set them to be 100° apart instead of 180. Because the boiling point and, to a lesser extent, freezing point of a liquid depend on the atmospheric pressure, the pressure must be specified: "normal" means the freezing and boiling points when the atmospheric pressure is exactly one atmosphere. These points are convenient because they are easily attained and highly reproducible. Interestingly, Celsius at first set boiling as zero and freezing as 100, but this was reversed in 1750 by the physicist Martin Strömer, Celsius's successor at Uppsala.

Defined in this way, a Celsius degree (°C) is 1/100 of the temperature difference between the normal boiling and freezing points of water. Because the difference between these two points on the Fahrenheit scale is 180°F, a Celsius degree is 1.8 times (or 9/5) larger than a Fahrenheit degree. You cannot convert between Fahrenheit and Celsius temperatures simply by multiplying by 1.8, however, because their zeroes are at different places. That would be like trying to measure a table in both yards and meters, when the left-hand ends (the zero marks) of the yardstick and meter stick are not starting at the same place.

One method to convert temperature from Fahrenheit to Celsius or vice versa, is to first account for the differences in their zero points. This can be done very simply by (step 1) adding 40 to the temperature you want to convert. That is because -40° (40 below zero) happens to come out at the same temperature on both scales, so adding 40 gets them both up to a comparable point: zero. Then (step 2) you can multiply by 1.8 (9/5) convert Celsius to Fahrenheit or divide by 1.8 (9/5) to convert Fahrenheit to Celsius to account for the difference in degree size, and finally (step 3) subtract the 40° originally added.

WEATHER FORECASTING

An understanding of the daily weather forecast, especially in areas outside the United States requires the ability to convert temperatures between Celsius and Fahrenheit temperature scales. The standard conversion from Fahrenheit to Celsius is expressed as $°C = (°F - 32) / 1.8$.

Accordingly a 72°F expected high temperature equates to approximately 22.2°C.

COOKING OR BAKING TEMPERATURES

To convert a temperature used for cooking (the expected oven temperature) for an French recipe for baking bread one might be called on to convert °C to °F and that conversion is obtained via °F = (°C × 1.8) + 32. So if an oven should be set at 275 °C in France to produce a crispy baguette (the traditional French long an thin loaf of bread) then an oven calibrated in °F should be set to approximately 525°F (275°C × 1.8) + 32 = 527°F.

Canceling Units

Notice that we are performing simple conversions, without the formality of labeling the units that must cancel to make the transformation. In the above example regarding oven temperature, the conversion factor 1.8 really represents 1.8°F / 1°C, read as 1.8 degrees Celsius to 1 degree Fahrenheit. This allows the units to cancel (275°C × 1.8 °F / 1 °C) + 32°F = 527°F.

In the prior example related to weather, the factor reciprocal of the factor 1.8 is used in the conversion formula °C = (°F − 32) / 1.8 equals 1°C per 1.8 °F or 1°C / 1.8°F and so the °F cancels as 22.2°C = (72 − 32) °F / 1.8 °C / °F.

ABSOLUTE SYSTEMS

About 1787 the French physicist Jacques Charles (1746–1823) noted that a sample of gas at constant pressure regularly contracted by about 1/273 of its volume at 0°C for each Celsius degree drop in temperature. This suggests an interesting question: If a gas were cooled to 273° below zero, would its volume drop to zero? Would it just disappear? The answer is no, because most gases will condense to liquids long before such a low temperature is reached, and liquids behave quite differently from gases.

In 1848 William Thomson (1824–1907), later Lord Kelvin, suggested that it was not the volume, but the molecular translational energy, that would become zero at about –273°C, and that this temperature was therefore the lowest possible temperature. Thomson suggested a new and more sensible temperature scale that would have the lowest possible temperature—absolute zero—set as zero on this scale. He set the temperature units as identical in size to the Celsius degrees. Temperature units on Kelvin's scale are now known as Kelvins (abbreviation, K); the term, degree, and its symbol, °, are not used. Lord

Conversion of measurements in recipes if often necessary.
ALEN MACWEENEY/CORBIS.

Kelvin's scale is called either the Kelvin scale or the absolute temperature scale. The normal freezing and boiling points of water on the Kelvin scale, then, are 273K and 373K, respectively, or, more accurately, 273.16K and 373.16K. To convert a Celsius temperature to Kelvin, just add 273.16.

The Kelvin scale is not the only absolute temperature scale. The Rankine scale, named for the Scottish engineer William Rankine (1820–1872), also has the lowest possible temperature set at zero. The size of the Rankine degree, however, is the same as that of the Fahrenheit degree. The Rankin temperature scale is rarely used today.

Absolute temperature scales have the advantage that the temperature on such a scale is directly proportional to the actual average molecular translational energy, the property that is measured by temperature. For example, if one object has twice the Kelvin temperature of another object, the molecules, or atoms, of the first object actually have twice the average molecular translational energy of the second. This is not true for the Celsius or Fahrenheit scales, because their zeroes do not represent zero energy. For this reason, the Kelvin scale is the only one that is used in scientific calculations.

A traffic sign near the U.S. border in Quebec. OWEN FRANKEN/CORBIS.

ARBITRARY SYSTEMS

On the Réaumur scale, almost forgotten except in parts of France, freezing is at 0 degrees, and the boiling point is at 80 as opposed to 100° Celsius, or 212° Fahrenheit. The gradation of temperature scales is, however, arbitrary.

Conversion of Distance Units

Distance conversions are common to hundreds of everyday tasks, from driving to measuring. Conversion factors for distance are uncomplicated and easily obtained from calculators and conversion tables (e.g., 1 inch = 2.54 centimeters, 1 yard = 0.9144 meter, and 1 mile = 1.6093 km).

The meter was originally defined in terms of Earth's size; it was supposed to be one ten-millionth of the distance from the equator to the North Pole, going straight through Paris. However, because Earth is subject to geological movements, this distance cannot be depended upon to remain the same forever. The modern meter, therefore, is defined in terms of how far light will travel in a given amount of time when traveling at—naturally—the speed of light. The speed of light in a vacuum is considered to be a fundamental constant of nature that will never change, no matter how the continents drift. The standard meter turns out to be 39.3701 inches.

10K and 5K walks and races (measuring 10 and 5 kilometers, properly abbreviated km, or 10,000 and 5,000 meters) are popular events, often used for local charitable fund raising and well as sports competition. A 10K race is about 6.21 miles and a 5K race is, of course, half that distance (about 3.11 miles, with rounding). One kilometer = .6214 mile and so 10,000 km × .6214 miles/km = 6.21 km.

Other units of measurement related to distance encountered include: Admiralty miles, angstroms, astronomical units, chains, fathoms, furlongs (still used in horse racing), hands, leagues, light years, links, mils (often used to measure paper thickness), nautical miles (with different U.K. and U.S. standards), parsecs, rods, Roman miles (*milia passuum*), Thous, and Unciae (Roman inches).

Conversion of Mass Units

The kilogram is the metric unit of mass, not weight. Mass is the fundamental measure of the amount of matter in an object. For example, the mass of an object will not change if you take it to the Moon, but it will weigh less—have less weight—when it lands on the Moon because the Moon's smaller gravitational force is pulling it down less strongly.

Regardless, in everyday terms on Earth, we often speak loosely about mass and weight as if they were the same thing. So you can feel free to "weigh" yourself (not "mass" yourself) in kilograms. Unfortunately, no absolutely unchangeable standard of mass has yet been found to standardize the kilogram on Earth. The kilogram is therefore defined as the mass of a certain bar of platinum-iridium alloy that has been maintained since 1889 at the International Bureau of Weights and Measures in Sèvres, France. The kilogram turns out to be approximately 2.2046 pounds.

To convert from the pound to the kilogram, for example, it is necessary to multiply the given quantity (in pounds) by the factor 0.45359237. A conversion in the reverse direction, from kilograms to pounds, involves multiplying the given quantity (in kilograms) by the factor 2.2046226.

For large masses, the metric ton is often used instead of the kilogram. A metric ton (often spelled tonne in other countries) is 1,000 kilograms. Because a kilogram is about 2.2 pounds, a metric ton is about 2,200 pounds—ten percent heavier than an American ton of 2,000 pounds.

Some remnants of English weights and measures still exist in popular culture. It is not uncommon to have weights of athletes in football (American soccer) and rugby matches quoted by commentators in terms of "stones." A stone is the equivalent of 14 pounds, so a 15-stone goalkeeper or rugby forward would weigh a formidable 210 pounds.

Other units of mass encountered include carats (used for measuring precious stones such as diamonds), drams, grains, hundredweights, livre, ounces (Troy), pennyweights, pfund, quarters, scruples, slus, and Zentners.

Conversion of Volume Units

For volume, the most common metric unit is not the cubic meter, which is generally too big to be useful in commerce, but the liter, which is one thousandth of a cubic meter. For even smaller volumes, the milliliter, one thousandth of a liter, is commonly used.

Other units of volume include acre-feet, acre-inches, barrels (used in the petroleum industry and equivalent to 42 U.S. gallons), bushels (both United States and United Kingdom), centiliters, cups (both U.S. and metric), dessertspoons (U.S., U.K., and metric, and in the U.S. about double the teaspoon in volume) fluid drams, pecks, pints, quarts, tablespoons, and teaspoons.

Units such as tablespoons and teaspoons are among the most common of hundreds of units related to cooking where units can be descriptive (e.g., a "pinch" of salt). Most cookbooks carry conversions factors for units described in the book.

In the United States, gasoline is sold and priced by the English gallon, but in Europe gasoline is sold and priced by the liter. The unsuspecting tourist may not take immediate notice at the great difference in price because roadside signs advertising the two can sometime be very similar. Aside from differences in currency value explained below, a price of $2.10 per gallon is far less than 1.30 € (Euros) per liter. There are more than 3.78 liters per gallon and so the price of 1.30 €/liter must be multiplied by 3.78 to arrive at a gallon equivalent cost of approximately 4.91 Euros per gallon.

Currency Conversion

The price difference in the above fuel purchase example is exacerbated (increased not for the better) by the need to convert the value of the two currencies involved. As of mid-2005, 1 Euro equaled $1.25 (in other words, it took $1.25 to purchase 1 Euro). And so the actual price of the fuel in the above example was 1.30 Euro/liter × 1.25 $/Euro = 1.625 $/liter and thus a gallon equivalent price of $6.14 per gallon (1.625 $/liter × 3.78 liter/gallon).

Although currency values (and thus conversion factors) can change rapidly—over the years between 2001 and 2005 the Euro went from being worth only about 75 U.S. cents to more than $1.30—such price differences for fuel are normal, because fuel in Europe is much more expensive than in the United States.

Non-standard Units of Conversion

Another often-used, non-standard metric unit is the hectare for land area. A hectare is 10,000 square meters and is equivalent to 0.4047 acre.

Other measurements of area include Ares, Dunams, Perches, Tatami, and Tsubo.

Key Terms

English system: A collection of measuring units that has developed haphazardly over many centuries and is now used almost exclusively in the United States and for certain specialized types of measurements.

Derived units: Units of measurements that can be obtained by multiplying or dividing various combinations of the nine basic SI units.

Kelvin: The International System (SI) unit of temperature. It is the same size as the degree Celsius.

Mass: A measure of the amount of matter in a sample of any substance. Mass does not depend on the strength of a planet's gravitational force, as does weight.

Matter: Any substance. Matter has mass and occupies space.

Metric system: A system of measurement developed in France in the 1790s.

Natural units: Units of measurement that are based on some obvious natural standard, such as the mass of an electron.

SI system: An abbreviation for Le Système International d'Unités, a system of weights and measures adopted in 1960 by the General Conference on Weights and Measures.

Temperature: A measure of the average kinetic energy of all the elementary particles in a sample of matter.

Conversion of Units of Time, an Exception to the Rule

The metric unit of time, the second, no longer depends on the wobbly rotation of Earth (1/86,400th of a day), because Earth is slowing down; with days keep getting a little longer as time passes. Thus, the second is now defined in terms of the vibrations of the cesium-133 atom. One second is defined as the amount of time it takes for a cesium-133 atom to vibrate 9,192,631,770 times. This may sound like a strange definition, but it is a superbly accurate way of fixing the standard size of the second, because the vibrations of atoms depend only on the nature of the atoms themselves, and cesium atoms will presumably continue to behave exactly like cesium atoms forever. The exact number of cesium vibrations was chosen to come out as close as possible to what was previously the most accurate value of the second.

Minutes are permitted to remain in the metric system for convenience or for historical reasons, even though they do not conform strictly to the rules. The minute, hour, and day, for example, are so customary that they are still defined in the metric system as 60 seconds, 60 minutes, and 24 hours—not as multiples of ten.

Where to Learn More

Books

Alder, Ken. *The Measure of All Things: The Seven Year Odyssey and Hidden Error that Transformed the World.* New York: Free Press, 2002.

Hebra, Alexius J. *Measure for Measure: The Story of Imperial, Metric, and Other Units.* Baltimore: Johns Hopkins University Press, 2003.

Periodicals

"The International System of Units (SI)." *United States Department of Commerce, National Institute of Standards and Technology, Special Publication* 330 (1991).

Web sites

Bartlett, David. *A Concise Reference Guide to the Metric System.* <http://www.bms.abdn.ac.uk/undergraduate/guidetounits.html> (2002).

Coordinate systems are grids used to label unique points using a set of two or more numbers with respect to a system of axes. An axis is a one-dimensional figure, such as a line, with points that correspond to numbers and form the basis for measuring a space. This allows an exact position to be identified, and the numbers that are used to identify the position are called coordinates. One example of the use of coordinates is labeling locations on a map. Street maps of a town, or maps in train and bus stations allow an overview of areas that may be too difficult to navigate if all features of the area were to be shown. Without a coordinate system, these maps would represent no sense of scale or distance.

The most common use of coordinate systems is in navigation. This allows people who cannot see each other to track their positions via the exchange of coordinates. In a complex transport system, this allows all the components to work together by exchanging coordinates that reference a common coordinate system. An example is an aviation network, where air traffic control must constantly monitor and communicate the positions of aircraft with radar and over radio links. Without a coordinate system, it would be impossible to monitor distances between aircraft, predict flight times, and communicate direction or change of direction to aircraft pilots over the radio.

DIMENSIONS OF A COORDINATE SYSTEM

Coordinate systems preserve information about distances between locations. This allows a path in space to be analyzed or areas and volumes to be calculated. For example, if a position coordinate at one point in time is known and the speed and direction are constant, it is possible to calculate what the position coordinate will be at some future time.

The number of unique axes needed for a coordinate system to work is equal to the number of unique dimensions of the space, and is written as a set of numbers (x,y,z). In ordinary day-to-day life, there are three unique directions, side-to-side, up and down, and backwards and forwards. It was the German-born American physicist Albert Einstein (1879–1955) who suggested that there is a fourth dimension of time. This suggestion led to

Einstein's famous theory of relativity. However, these effects are normally not visible unless the velocities are very close to the speed of light or there is a strong gravitational field. Therefore, the dimension of time is not usually used in geometric coordinate systems.

Sometimes it is sensible to reduce the number of dimensions used when constructing a coordinate system. An example is seen on a street map, which only uses two axes, (x,y). This is because changes in height are not important, and locations can be fixed in two of the three dimensions in which humans can move. In this case, a coordinate system based on a two-dimensional flat surface (a map) is the best system to use.

CHANGING BETWEEN COORDINATE SYSTEMS

Coordinate systems denote the exact location of positions in space. If two or more sets of coordinates are given, it is possible to calculate the distances and directions between them. To see this, consider two points on a street map that uses a two-dimensional Cartesian coordinate system. A line can be drawn between the two points that extend from a reference point, say a building where a friend is staying, located at (a,b) on the map, to the point where you are standing (x,y). This line has a length, called a magnitude, and a direction, which in this case is the angle made between the line and the x axis. In Cartesian coordinates, the magnitude is given by Pythagoras' theorem:

$$\text{Magnitude} = \sqrt{(x-a)^2 + (y-b)^2}$$

The angle that this line makes with the x axis moving anticlockwise is given by:

$$\text{Angle} = tan^{-1}\left(\frac{y-b}{x-a}\right)$$

If you were to walk toward your friend along the line, the magnitude would change, but the angle would not. If you were to walk in a circle around your friend, the angle would change, but the magnitude would not.

You may have noticed that the magnitude (radius of the circle around your friend) and the angle taken together form a coordinate in the polar coordinate system, (*radius, angle*). These equations are an example of how it is possible to convert between coordinate systems. The Cartesian coordinates of your position can be redefined as a polar coordinates. The reverse is also possible.

VECTORS

This example also leads to the concept of vectors. Vectors are used to record quantities that have a magnitude and a direction, such as wind speed and direction or the flow of liquids. Vectors record these quantities in a manner that simplifies analysis of the data, and vectors are visually useful as well. For example, consider wind speed and direction measured at many different coordinates. A map can be made with an arrow at each coordinate, where each arrow has a length and direction proportional to the measured speed and direction of the wind at that coordinate. With enough points, it should be possible just by looking at this map to see patterns these arrows create and hence, patterns in the wind data.

CHOOSING THE BEST COORDINATE SYSTEM

Coordinate systems can often be simplified further if the surface being mapped has some sort of symmetry, such as the rotational symmetry of a radar beam sweeping out a circular region around a ship. In this case, the coordinate system with axes that reflect this circular symmetry will often be simpler to use. Coordinates can be converted from one system to another, and this allows changing to the simplest coordinate system that best suits each particular situation.

CARTESIAN COORDINATE PLANE

A common use of the Cartesian coordinate system can be seen on street maps. These will quite often have a square grid shape over them. Along the sides of the square grid, numbers or letters run along the horizontal, bottom edge of the map and the other along the vertical, left hand side of the map. In this example, assume that both sides are labeled with numbers. These two sides are called the axes and for Cartesian coordinate systems, they are always at 90 degrees to each other.

By reading the values from these two axes, the location of any point on the map can be recorded. The values are taken from the horizontal x axis, and the vertical y axis. The value of the x axis increases with motion to the right along the horizontal axis, and the value of the y axis increases with motion up along the vertical axis.

By selecting a point somewhere on the map, two lines are drawn from the point that crosses both the x axis and y axis at 90 degrees. The values along the two axes can then be read to give coordinates. The exact opposite technique will define a point on the map from a pair of coordinates. Two lines drawn at 90 degrees to the x axis and y axis will locate a point on the map where the two lines cross.

The coordinates for a point on the map are often written as (x,y). The order of expressing the coordinates is important; if they are mixed up the wrong point will be defined on the map.

Figure 1 shows an example of a two-dimensional Cartesian coordinate system. In three dimensions, a Cartesian system is defined by three axes that are each at 90-degree angles to each other. There is some freedom in the way three axes in space can be represented, and an error could invalidate the coordinate system. The usual rule to avoid this is to use the right-handed coordinate system. If you hold out your right hand and stick your thumb in the air, this is the direction along the z axis. Next, point your index finger straight out, so that it is in line with your palm; this is the direction along the x axis. Finally, point your middle finger inwards, at 90 degrees to your index finger; this is the y axis. The fingers now point along the directions of increasing values of these axes. A point is now located in a similar way to two-dimensional coordinates. From a set of coordinates, written as (x,y,z), a point is located where three planes, drawn at 90-degree angles to these axes, all cross.

POLAR COORDINATES

The polar coordinate system (see Figure 2) is another type of two-dimensional coordinate system that is based on rotational symmetry. The reason this system is useful is that many systems in nature exhibit rotational symmetry, and when expressed in these coordinates, they will often be simpler and more enlightening than using two-dimensional Cartesian coordinates.

The two coordinates used to define a point in this system are the radius and the polar angle. To understand this, imagine standing at the center of a round room that has the hours of a clock painted around the walls. Elsewhere in the room is a dot painted on the floor. The distance between you and the dot is the radius. The angle is a bit more involved. Standing facing 3 o'clock, the polar angle is given by the number of degrees you turn your head counter-clockwise to face the dot. For example, if the dot is at the 12 o'clock mark, it has a polar angle of 90 degrees with respect to you; if it is at 9 o'clock, it has an angle of 180 degrees; and if it is at 6 o'clock, it has an angle of 270 degrees. The line at 0 degrees, the 3 o'clock mark, is defined to coincide with the horizontal, or the x axis in the Cartesian system.

A Brief History of Discovery and Development

Humans have been mapping their location and travels since the dawn of human history. Examples are seen throughout history, such as the mapping of land in the

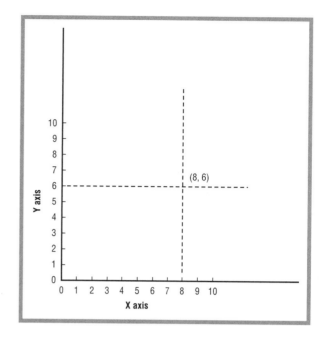

Figure 1: Rectangular coordinates.

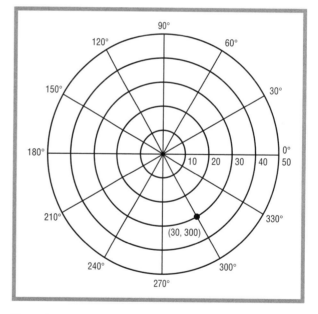

Figure 2: Polar coordinates.

valley of the Nile in ancient Egypt, and recording journeys of global exploration such as those of the Spanish explorer Christopher Columbus (1451–1506) and others.

Today, the management of the world's natural and economic resources requires the availability of accurate and consistent geographic information. The methods for storing this data may have changed, with computer-based storage replacing paper maps, yet the underlying principals for ensuring compatibility has remained the same.

With coordinate systems, locations can be placed on maps and navigation can be achieved. Such systems allow a location to be unambiguously identified through a set of coordinates. In navigation, the usual coordinates in use are latitude and longitude, first introduced by the ancient Greek astronomer Hipparchus around 150 B.C.

Like so many mathematical ideas in history, coordinates may have existed in many forms before they were studied in their own right. French philosopher and mathematician René Descartes (1596–1650) introduced the use of coordinates for describing plane curves in a treatise published in 1637. Only the positive values of the x and y coordinates were considered, and the axes were not drawn. Instead, he was using what is now called the Cartesian coordinate system, named after him. The polar coordinate system was introduced later by the English mathematician and physicist Isaac Newton (1642–1727) around 1670. Nowadays, the use of coordinate systems is integral to the development and construction of modern technology and is the foundation for expressing modern mathematical ideas about the nature of the universe.

Real-life Applications

COORDINATE SYSTEMS USED FOR COMPUTER ANIMATION

Films makers and photographers use computers to manipulate images in a computer. Some common applications include photo manipulation, where images can be altered in an artistic manner, video morphing, where a computers morph an image into another image, and other special effects. Blue screen imaging is an effect where an actor acts standing in front of a screen, which is later replaced with an image. This would allow an actor dressed as Superman in front of a blue screen to later be seen flying over a town in the film, for example.

Leaps in computing power and storage have allowed animators to use computers to design and render breathtaking artistic works. Rendering is a process used to make computer animation look more lifelike. Some of these animations are works in their own right, and others can be combined with real life film to create lifelike computer generated effects.

All of these techniques require coordinate systems, as a computer's memory can only store an image as a sequence of numbers. Each set of coordinates will be associated with the position, velocity, color, texture, and other information of a particular point in the image. As an example, consider animating the figure of a dog in a cartoon. If the dog was featured in many scenes, it would be inefficient to redraw

each movement of the dog. To simplify the animation, each part of the picture is split up into objects that can be animated individually. In this case, a coordinate system can be set up for each moving part of the dog.

For the finished animated picture, all the objects will be drawn together on some background image all at once, maybe with some objects rotated, shifted, or enlarged to refine the final effect. Vectors can be used to make this process more efficient and flexible. In two-dimensional animation and computer graphics design, this is often called vector graphics. In three-dimensional graphics, it is usually referred to as wire frame modeling.

COORDINATE SYSTEMS USED IN BOARD GAMES

Some games use boards that are divided up into squares. An example of this is chess, an ancient and sophisticated game that is played and studied widely. By defining a coordinate system on the board, the positions of the individual pieces can be located. Examples of this are found in books on the game and even in some newspapers, where rows of letters and numbers define the position and movements of the pieces. In this way, many famous games of chess have been recorded and a student of the game can replay them to learn tactics and strategies from masters of the game.

In computer chess simulators, the locations of the pieces have to be stored as coordinates as numbers in the computers memory. Once in the computer's memory, various algorithms calculate the movements of the pieces, which are then displayed on the computer screen.

Even without computers, if two chess players are separated by vast distances, the coordinate system allows the game to be played by the transmission of the coordinates of each move. There are many games of chess that have been played over amateur radio or by mail in this manner. In this case, the players can be separated by many thousands of miles and still play a game of chess.

PAPER MAPS OF THE WORLD

Assuming that the terrain one wishes to cross is flat, a coordinate system based on two dimensions and a Cartesian grid can be used for a paper map. This is suitable in shipping for maps of coastlines and maps of areas up to the size of large islands. However, the world is not flat, but curved, and for maps with areas larger than about 4 mi^2 (10 km^2), a Cartesian map of the surface will not be accurate.

One way to make an accurate map that covers most of the world on paper is to use a Mercator projection (a two-dimensional map of the Earth's surface named for Gerhardus Mercator, the Flemish cartographer who first created it in 1569). This projection misses the North and South Poles, as well as the international date line. At the equator, the map is a good approximation of the Cartesian system, but because of Earth's curved shape, no two axes can perfectly represent its surface. Toward the poles, the image of the Earth's surface becomes more and more distorted. It is impossible to accurately project a spherical surface onto a flat sheet, as there is no way to cut the sphere up so that its sections can be rolled out flat. No matter what projection is used, flat paper maps of Earth's surface will always have some distortion due to the curved nature of Earth.

COMMERCIAL AVIATION

Coordinate systems allow a location to be transmitted over a radio link if two people have a map with a common coordinate system. Shipping is one example of this, but another important commercial use of coordinate systems is in aviation. In the skies, positions can be communicated as a series of coordinates verbally or electronically over radio links that allow many planes to be flown into or out of airports. In commercial aviation, there will often be many planes in the sky at one time coming in from all different directions toward an airport. At busy airports, sometimes there will not be enough runways to deal with all the traffic, and airplanes will often be put into a holding pattern while awaiting clearance to land. Positions of the aircraft are continually monitored by air traffic controllers with coordinates given both verbally by pilots and mechanically by radar.

As air traffic increases each year, it becomes more critical that coordinates and other information are relayed quickly and clearly. Air traffic controllers must make sure that coordinates are correct and understood clearly. Apart from all of the sophisticated technological safeguards, a simple misunderstanding of a spoken coordinate could be enough to cause a disaster. To avoid this, all commercial pilots must communicant in English, and flight terminology is common and standard across countries.

LONGITUDE AND JOHN HARRISON

In navigation, some point of reference is needed before a coordinate can be found. On a street map, a person could look for a street name or some other landmark to pinpoint their position. However, on the open seas and without fixed landmarks, it was not always simple for a ship to find a point of reference. To fix a position on Earth's surface requires two readings, called latitude and longitude. If the Earth is pictured as a circle, with the North Pole at the top and the South Pole at the bottom, and the ship is on the edge of the circle, the latitude is the angle between the ship, the center of the Earth, and the equator. Longitude can then be pictured as the circle when looking down from on top of the Earth, with the North Pole at the center of the circle. The angle between the ship and Greenwich, England is the longitude. Finding latitude is quite simple at sea using the angle between the horizon and the North Star or noon Sun. A device called a sextant was commonly used for this, but finding an accurate reading for longitude was more problematic.

Calculating longitude was a great problem in the naval age of the seventeenth and eighteenth century, and occupied some of the best scientific minds of the time. The British announced a prize of £20,000 for anyone who could solve the problem. It was finally solved by the invention of a non-pendulum clock that could kept accurate time at sea. It was invented by the visionary English clock maker John Harrison (1693–1776), who spent a great part of his life trying to construct a clock that was thought by many to be impossible with the technology of the time. It contained several technological developments that allowed it to work and keep time in the rough conditions at sea. During this time, John Harrison was constantly battling with the Royal Society, England's preeminent scientific organization. Ironically, while the members of the Royal Society were still debating if his clock really did work, it was already being used at sea for navigation by the navy. Eventually, after a long battle, John Harrison received the money and recognition he deserved. With the invention of this clock, calculating longitude at sea became simple. The clock is set to a standard time, taken as the time of Greenwich and called Greenwich Mean Time (GMT). If a person looks at the clock at noon, when the sun is directly overhead, and it reads 2 P.M., then two hours ago it was noon in Greenwich, as the sun rotates 360 degrees around the Earth every 24 hours. The equation is:

$$\frac{360°}{24 \text{ hours}} \quad \frac{2 \text{ hours}}{\text{difference}} = \frac{30° \text{ Longitude}}{\text{from Greenwich}}$$

MODERN NAVIGATION AND GPS

In the twenty-first century, most navigation is based on the global positioning system (GPS). This is a network of 24 American satellites that orbit the Earth, allowing a position coordinate to be read off the screen of a special radio receiver anywhere on Earth, and is accurate to within 16.4 yd (15 m). Interestingly, this system requires use of a special coordinate system based on Einstein's theory of

Key Terms

Axis: Lines labeled with numbers that are used to locate a coordinate.

Coordinate: A set of two or more number or letters used to locate a point in space. For example, in 2 dimensions a coordinate is written as *(x,y)*.

Cartesian coordinate: A coordinate system were the axes are at 90 degrees to each other, with the x axis along the horizontal.

Dimension: The number of unique directions it is possible for a point to move in space. The world is normally thought of as having three. Flat surfaces have two dimensional and more advanced physical concepts require the use of more than three dimensions such as spa.

Polar angle: The angle between the line drawn from a point to the center of a circle and the x axis. The angle is taken by rotating counterclockwise from the x axis.

Polar coordinate: A two-dimensional coordinate system that is based on circular symmetry. It has two coordinates, the radius and the polar angle.

Radius: The distance from the center of a circle to its perimeter.

Vector: A quantity consisting of magnitude and direction, usually represented by an arrow whose length represents the magnitude and whose orientation in space represents the direction.

relativity called spacetime. In spacetime, time itself becomes a coordinate axis added to the normal three-dimensional world. The four-dimensional spacetime may seem strange, and the effects of it are far too small to be seen unless scientists or mathematicians are dealing with very high velocities or gravitational fields. However, the GPS satellites must give a very accurate time signal for the calculation of a coordinate. To do this, the satellites have small on-board atomic clocks. Relativistic effects from the high velocity of the satellites orbit relative to the Earth's surface distort this time signal and this distortion must be accounted for. If these effects were not taken into account, the resulting coordinates would be off by more than 6.2 miles (10 km) per day. This is all accomplished with an internal computer that returns the corrected map reading to the user.

3-D SYSTEMS ON ORDINANCE SURVEY MAPS

Some examples of three-dimensional coordinate systems can be found on ordinance survey maps. In this case, a two-dimensional Cartesian system is modified by the addition of lines to map height above sea level. These maps are used by surveyors and in sports, such as climbing and hiking, to map terrain with valleys and mountains. To define the height of the ground above sea level, two coordinates would not be enough. The basic map is a Cartesian system with a grid that gives two coordinates, but the third dimension for height is represented by curved lines drawn on the map. Each one of these lines

represents a height in meters above sea level, giving the third dimension.

RADAR SYSTEMS AND POLAR COORDINATES

Modern radar systems are based on a device called a magnetron that produces a highly focused beam of microwaves. The beam can be rotated so that a radar operator can see all of a ship. A radar system that uses this method is seen on ships as a rotating parabolic aerial attached somewhere on top of the ship. This radar system is used to detect ships and other large solid objects in the sea, as the beam sweeps around the ship in a circular path. The radar screen will look like the familiar radar screen seen in movies, shaped as a round monitor with a line from the center sweeping around it in a circular path. Objects on the screen will show up as points as the beam sweeps over them.

The beam rotates in a two-dimensional fixed plane, so in order to locate objects, changes in height can be ignored, and a two-dimensional coordinate system can be used. The two-dimensional Cartesian coordinate system is not the best coordinate system to use in this case. Consider the operator's screen, for example. Although one might cover the round screen in a square mesh and put the round screen into a square box to draw the x and y axis, this would be impractical. The length from the center of the screen to a point to the edge of the round screen is constant, and is related to the maximum range the radar system can physically detect. As the edge of the round screen is at maximum range, there would be areas dead areas between this and the square box used to define the Cartesian coordinate system. Another problem comes with the calculation of the distance and angles of objects in relation to the ship.

A better coordinate system to use in this example is the polar coordinate system, which reflects the circular nature of the sweeping beam. The radius axis is the distance along a line, drawn from the detected object to the center of the screen. The polar angle is measured between the horizontal line that crosses the center of the screen and the beam line. To draw a reference grid for the radius of this coordinate system, the screen is divided up into a number of concentric circles, or circles that get bigger with equal spacing, and are all centered at the screen center. Each of these circles is at a different fixed radius so the distance of the detected object can be read on the screen. A number of lines drawn at equal angles emanating from the center of the screen, like the spokes of a bicycle wheel, allow the polar angle to be read off, giving the angle between the ship and the detected object.

The center of the screen is always the location of the ship. If the radar operator sees a flash on the screen, the polar coordinate of the object is identified by the finding the circle and line that meet at the detected object. If each circle is labeled as 1km and each line labeled in 1-degree increments of angle, with the right hand side of the horizontal line representing the front of the ship, a polar coordinate made from the twentieth circle and the ninetieth line counter-clockwise from the horizontal instantly tells the radar operator that the object is 20 km away and 90 degrees to the right of the ship. More importantly, this information is read from the screen without using any mathematical conversion to find these figures, which would have been needed had a Cartesian system been used.

Where to Learn More

Books

Sobel, Dava, and William J. H. Andrewes. *The Illustrated Longitude.* New York: Walker & Company, 2003.

Web sites

Dana, Peter H. "Coordinate Systems Overview" *The Geographer's Craft*. University of Colorado. <http://www.colorado.edu/geography/gcraft/notes/coordsys/coordsys_f.html> (accessed March 18, 2005).

Stern, David P. "Navigation." <http://www-istp.gsfc.nasa.gov/stargaze/Snavigat.htm> (accessed March 18, 2005).

Decimals

Decimals can precisely indicate amounts, time speed to the hundredths or even thousandths of a second, precisely indicate the passage of time, accurately represent measurements of parameters that include weight, height, temperature and distance, and even help nab drivers who are speeding down the highway.

This article will consider decimals: what they are, how numbers are represented, and how decimals form a vital part of real-life math.

Fundamental Mathematical Concepts and Terms

The simplest way to answer this is visually: suppose that there are ten boxes on a table, as depicted in Figure 1.

Three of the boxes in Figure 1 are black in color and the remaining seven boxes are white. An ideal way to describe this relationship nonverbally is to use the language of math. A central part of a mathematical description can revolve around decimals. In order to write the preceding sentence using math language instead of words. The black colored boxes can be denoted as $1/10 + 1/10 + 1/10 = 3/10$. Another way to mathematically write the same information is in decimal form, expressed as 0.3.

This particular decimal consists of three components. The zero is in the ones column. Although other numbers are not present to the left of the zero, if they were, they would be in the familiar tens, hundreds, thousands, etc. columns. In other words, these numbers would be increasing from zero in $10\times$ increments. The number three is located immediately to the right of the period (the decimal point), in the column that depicts tenths (1/10).

If there was a number to the right of the three, that number would be in the hundredths (1/100) column. In the present example, 0.3, there are zero ones and three tenths. The number is pronounced as 'zero point three'.

Thus decimals can be seen as a short way of expressing certain types of fractions, namely those whose denominator are sums of powers of ten (tenths, hundredths, thousandths, etc.).

As an example, consider the number 8.53479. The number can be written in fractional form in terms of the place values of its various digits: $8.53479 = 8/1 + 5/10 + 3/100 + 4/1,000 + 7/10,000 + 9/1,000,000$. However, it is certainly a lot easier and more understandable to write this number in the decimal form (also called decimal

Figure 1.

notation) of 8.53479 than in the long and cumbersome fractional form.

A Brief History of Discovery and Development

Interestingly, although decimals are relatively new to numbering systems, base numbering systems like base 10 and base 60 have been around for thousands of years. In 1579, a book written by an Italian/French mathematician named François Viete contains a quote that argues for the use of the base 10 decimals (the tenths, hundredths and thousandths pattern seen above) instead of a more complex base 60 (sexagesimal) system that was then in vogue.

Viete argued, 'Sexagesimals and sixties are to be tested sparingly or never in mathematics, and thousandths and thousands, hundredths and hundreds, tenths and tens, and similar progressions, ascending and descending, are to be used frequently or exclusively.'

Just a few years later, in 1585, a book entitled *De Thiende* (The Tenth) popularized the concept and structure of decimals. However, the structure was a bit different than the decimals known today. The present day format of decimals came about in seventeenth century Scotland, courtesy of mathematician John Napier. It was Napier who introduced the decimal point as the boundary between the place values on ones and tenths. In some areas of the world a decimal comma is still used instead of a point.

Real-life Applications

As noted in the previous section, decimals numbers are easier to write and comprehend than numbers as represented in a fractional format, especially larger numbers. This ease of use and understanding has made decimals a centerpiece of disciplines including medicine, finance, and construction that call for the precise representation of distance, mass, and currency.

SCIENCE

In science, virtually all measurements are recorded and expressed as decimals. This accuracy is important to

the scientific method, since it makes it possible for someone to repeat the reported experiments. Repetition of experiments and the resulting confirmation or refuting of the reported results is the cornerstone of science.

MEASUREMENT SYSTEMS

In countries that use the metric system, such as Canada and most of Europe, decimals predominate. Glancing at the digital thermostat might reveal a temperature of 68°F (20.17°C). A glance at the cereal box might reveal that a 1 cup (0.25 liter) serving of cereal contained 8.5 grams of protein and 2.7 grams of fat. A coffee bought at the local drive-through java emporium costs $3.00 plus a 15% tax (another $0.45).

Sports

There are many others examples of decimals in our everyday lives. Watch just about any sporting event in which timing of the game or the race is involved and a digital clock will inevitably be in use. Indeed, in track and field events like the 100-, 200- and 400-meter runs, the finish line clock is capable of measuring to the hundredths of a second. That is why a winning 100-meter time will be reported as 9.89 seconds, for example.

In the sport of baseball, a common practice for a team is to position one of their personnel in the stands to monitor the speed of the pitches thrown by the team's starting pitcher. Compiling this information can help the coach know at about what point in the game the pitcher starts to get tired and the velocity of his or her pitches begins to decrease. The timing device is used to record the speed of the pitches. This device is essentially the same as the one that police officers use to record the speed of vehicles zooming along a highway. These 'speed guns' display the speed digitally. So, when a coach sees the pitches drop to 75.5 miles per hour, or the police officer times a car moving at 80.3 miles per hour, action is likely to be taken.

GRADE POINT AVERAGE CALCULATIONS

Another example of one of the thousands of uses of decimals strikes motivating fear into the hearts of students, calculating their grade point average or GPA. The GPA is a cumulative score of the individual grades attained for the various courses taken. As high school seniors are well aware, universities, colleges and other institutions can place great emphasis on GPA when deciding on admittance of students.

Letter grade	Points
A	4.00
A−	3.67
B+	3.33
B	3.00
B−	2.67
C+	2.33
C	2.00
C−	1.67
D+	1.33
D	1.00
D−	0.67
F	0.0

Figure 2.

Bob		John	
A	4.00	A	4.00
B+	3.33	A	4.00
B	3.00	B+	3.33
C	2.00	B	3.00
C−	1.67	D	1.00

Figure 3.

GPA is based on the points that are assigned to a course. The points are usually based on a four-point grading scale similar to those in Figure 2.

In this example, Bob and John have received the following grades for the five courses taken: Bob received an A, B+, B, C, and a C−. John received two As, a B+, B, and a D. Using the grade point scale, the points for each of the courses is expressed in Figure 3.

In order to calculate the GPAs for the Bob and John, each student's individual scores are totaled and that number is divided by the number of courses. In other words, the average score is determined. Bob's GPA is (4.00 + 3.33 + 3.00 + 2.00 + 1.67) / 5, or 2.80. John's GPA is (4.00 + 4.00 + 3.33 + 3.00 + 1.00) / 5, or 3.066.

Where to Learn More

Books

De Francisco, C., and M. Burns. *Teaching Arithmetic: Lessons for Decimals and percents, Grades 5-6.* Sausalito: Math Solutions Publications, 2002.

Mitchell, C. *Funtastic Math! Decimals and Fractions.* New York: Scholastic, 1999.

Schwartz, D.M. *On Beyond a Million: An Amazing Math Journey.* New York: Dragonfly Books, 2001.

Web sites

BMCC Math Tutorials "Introduction to Decimals." <http://www.bmcc.org/nish/MathTutorials/Decimals/> (October 30, 2004).

Overview

Demographics is the mathematical study of populations, and groups within populations.

Demographics uses characteristics of a population to develop policies to serve the people, to guide the development and marketing of products that will be popular, to conduct surveys that reveal opinions and how these opinions vary among various sectors of those surveyed, and of continuing news interest, to analyze polls and results related to elections.

Math lies at the heart of demographics, in the methods used to assemble information that is accurate and representative of the population. Without the accuracy and precision that mathematics brings to the enterprise, the demographic analysis will not provide meaningful information.

But demographics is not entirely concerned with math. Because demographics is also concerned with factors like cultural characteristics and social views, factors such as how people think about the issue at hand are also measured. Or, even less precisely, demographics can be concerned with how people 'feel' about something. These sorts of factors are more difficult to put into numbers and they are described as being qualitative (measuring quality) as opposed to quantitative (measuring an amount). Qualitative and quantitative aspects are often combined to form a 'demographic profile.'

Some of the mathematical operations that can be useful in the analysis of demographic information include the mean (the average of a set of numbers that is determined by adding some aspect of those numbers and dividing by some aspect of the numbers), the median (the value that is in the middle of a range of values) and the distribution (the real or theoretical chances of occurrence of a set of values, usually patterned with the most frequently-occurring values in the middle with less frequently-occurring values tailing off in either direction.)

Demographic information can be very powerful. It can reveal previously unrecognized aspects of a population and can be used to predict future trends. Part of the reliability of the demographic information comes from the mathematical operations used to derive the data.

Real-life Applications

ELECTION ANALYSIS

The analysis of the 2004 general election (also called the Presidential election) in the United States offers an example of the use of demographics to analyze the voting patterns. By asking people questions about their beliefs and opinions on a variety of issues, and by utilizing databases

Artists (such as hip hop artists jace, Buckshot, and Freddie Foxxx, shown here) and other activists use demographics to identify specific areas and populations where advertising and money will be most effective. AP/WIDE WORLD PHOTOS. REPRODUCED BY PERMISSION.

that yield information on aspects such as age, gender, and income (more on this sort of information is presented below), a more complete picture can be built of the characteristics of those who voted for a certain candidate.

For example, exit polls (asking people questions after they have voted) were used to determine voter preferences and what issues were important in deciding how to cast votes in various races.

These characteristics can be considered along with information on employment, geographic residence, home-owner status, and other factors, to build up a profile of a 'typical' person who will vote for a particular politician.

These demographic patterns were known beforehand to campaign organizers, who conducted their own surveys of the public. So, aware of the characteristics of a certain segment of the population and the percentage of total voters who fit this demographic, candidates target specific groups with specific messages and promises.

CENSUS

Many countries periodically undergo a process known as a census. Essentially, a census is an organized gathering of information about the adult population of the country. Citizens and other eligible residents of the country complete a form or participate in an interview. Many questions are asked in a census. Example categories include age, gender, employment status, income range, educational background, marital status, number of dependents, ethnic background, place of residence (both geographically and in terms of whether a residence is owned or rented), history of residence change, and record of military service.

These categories of information can be analyzed to provide details of the characteristics of the population, and the proportions of the populations that make up each of the characteristic groups.

The demographic information in a census is used by governments to develop policies that will hopefully best

Cohort	Dates of birth	Events	Example characteristics
Great Depression	1912–1921	Depression, high unemployment, hard times	Need for financial security and comfort, Conservative
World War II	1922–1927	War, women working, a common enemy	The common good, patriotism, teamwork
Generation X	1965–1976	Space disasters, AIDS, safe sex, Berlin wall	Need for emotional security and independence, importance of money
Generation N	1977–present	September 11, Iraq wars, Internet	Need for physical safety, patriotism, increased fear, comfortable with change

Table 1.

serve their constituents. As well, the information represents a wonderful database for marketers to sell their wares. For example, it would not make sense for car company to target a region of high unemployment as a market for its top-of-the-line luxury car.

Demographics and the Marketplace

Demographics such as contained in a census have long been a tool of those who make and sell products. Knowing the characteristics, likes and dislikes of the buying public is obviously important when trying to sell a product.

The baby boom that occurred during the 1950s and 1960s provides a prime example of an identified demographic group. The increased birth rate in North America during those decades will have a number of effects that have and will continue to ripple through the ensuing decades. In the first few years, there was an increased demand for products to do with infants (baby food, diapers). Savvy entrepreneurs took advantage of the knowledge that an increasing number of new parents identified strongly with environmental protection to market organic baby foods and re-popularize nondisposable diapers. In the following few years as infants became youngsters, adolescents and young adults there was a succession of increased demands for children's toys and clothes, better educational facilities, housing and furniture. In the last decade, as the baby boomers have reached middle age, there has been an increased demand for certain types of vehicles such as SUVs, for health clubs and weight loss centers to help trim sagging waistlines, and for expertise in investment help as retirement draws closer. In the coming decades, as the baby boomers become infirmed, there will be a demand for more health-care services and funeral services.

Baby boomers came into the world at about the same time and, as they age, experience similar things and have similar demands. This generation is a perfect example of what was termed, way back in the 1920s, a 'generational cohort.' The designation has roots in mathematics. In statistical analysis, it can be advantageous and more meaningful to group items in cohorts that are similar in whatever aspect(s) is being studied. Historic examples of other demographic cohorts, and their associated characteristics, are given in Table 1.

GEOGRAPHIC INFORMATION SYSTEM TECHNOLOGY

Geographic information system (GIS) technology is the use of computers and computer databases to assemble information that have a geographical component. The information can come from reports, topographical maps that display elevation, land use maps, photographs, and satellite images of an area.

Knowledge of the geography can be combined with other data including information on age, gender, employment, health, and other aspects that are collected in a census, and data collected from other surveys. The aim is to provide a more complete picture of a region, in which demographic characteristics can be related to geographical features.

As an example, combining GIS data with population information could reveal that there is a higher incidence of fatal diseases in rural and mountainous areas. This could help health care providers in designing better ambulance service or telephone-based health advice.

The analysis and interpretation of geographic information can be a mathematical process. Equations can be applied to images to help sort out background detail from the more relevant information. Data can be statistically analyzed to reveal important associations between various data groups.

Where to Learn More

Books

Foote, D.K., and D. Stoffman. *Boom Bust & Echo: Profiting from the Demographic Shift in the 21st Century.* Toronto: Stoddart, 2000.

Rowland, D.T. *Demographic Methods and Concepts.* New York: Oxford University Press, 2003.

Wallace, P. *Agequake: Riding the Demographic Rollercoaster Shaking Business, Finance, and Our World.* London: Nicholas Brealey Publishing, 2001.

Discrete Mathematics

Discrete mathematics includes all types of math that deal with discrete objects, that is, things that are distinct, unconnected, or step-by-step in nature. For example, the natural numbers 0, 1, 2, 3, 4, . . . are discrete, and counting is a discrete process.

The increasing use of computers in science, engineering, mathematics, and daily life has led to fast growth in discrete mathematics over the last 30 years or so. Digital computers store numbers, words, and images in discrete form, that is, as ones and zeroes, so designing and programming computers involves discrete mathematics. Today, discrete mathematics is basic to most areas of computer science, including operating systems, algorithms, security, cryptography, networking, and database searching. Discrete math also used in engineering, chemistry, biology, operations research (the scientific management of large systems of people, machines, money, and material), and in many other fields.

The counterpart of discrete is continuous. Something is continuous if it changes smoothly from one place (or time) to another. A flight of stairs is discrete; a ramp is continuous. Calculus, which studies the behavior of curves and irregularly shaped areas, is an example of a branch of mathematics that is concerned mostly with continuous rather than discrete objects. However, discrete and continuous mathematics often overlap or influence each other.

Fundamental Mathematical Concepts and Terms

LOGIC, SETS, AND FUNCTIONS

The foundation of discrete mathematics is the study of logic, statements, sets, and functions.

Logic is the study of the rules of thinking. It helps mathematicians distinguish trains of thought that are valid (correct throughout) from ones that contain hidden errors. Logic can be considered a form of discrete mathematics because the steps in a logical train of thought occur one at a time, that is, in a discrete way.

Statements are claims about the way things are. True statements obey the rules of logic, and false statements break them. The statement $1 + 1 = 2$ is true, and the statement $1 + 1 = 5$ is false.

Set theory studies the ways in which statements, facts, or objects can be arranged into groups. These groups are called sets. For example, the two numbers 0 and 1 can be grouped into a set. Logic governs the making of statements about sets.

Finally, a function is a rule that connects the objects in one set with those in another. A function can be defined by a sentence like "Adding 1 to every number in set A gives a number in set B," by a formula like $f(x) = x + 1$, or by a computer program. Functions are basic to both discrete and non-discrete mathematics; they are the language that mathematics uses to describe changes and relationships.

BOOLEAN ALGEBRA

In 1854 the English mathematician George Boole (1815–1864) published a book called *The Laws of Thought*. In it, he described the rules for doing mathematics with the numbers 0 and 1. These are the rules of the type of discrete mathematics called Boolean algebra. In Boolean algebra, every operation (such as addition or multiplication) must give a result that is still a 0 or a 1. In ordinary arithmetic, $1 + 1 = 2$; but in Boolean algebra, $1 + 1 = 1$. Ordinary arithmetic can be approximately translated into Boolean algebra, which is how computers and calculators are programmed to do math.

NUMBER THEORY

Number theory is the study of integers, which are the counting numbers 0, 1, 2, 3, . . . and their negatives, –1, –2, –3, and so on. Numbers that cannot be written as whole numbers, like 1/2, 3/4, or 5.6, are not integers.

But how can there be a whole field of study devoted to something so simple? When you've counted 1, 2, 3, what else is there to say? Plenty, as it turns out. Prime numbers, for example, are one of the main interests of number theory. An integer is a prime number if it cannot be evenly divided by any number smaller than itself except 1. (Any number divided by 1 is just itself.) 4 is not a prime number because it can be evenly divided by 2, but 5 is prime because it cannot be evenly divided by 2, 3, or 4. One sure way to tell whether a number is prime is to try to divide it by every positive integer smaller than itself; if none of them divide the number evenly, it is a prime. Primes have surprisingly complicated and useful properties.

COMBINATORICS

Combinatorics is the mathematics of counting. Counting may also seem, at first glance, too simple to be a whole field of mathematics. When you count, you just point to each object to be counted and say "one, two, three . . ." until you run out of objects—right?

But this will not work if there is nothing to point at, or if there are too many things for one-by-one counting

to be practical. This is often the case when we are trying to count not objects, but *arrangements* of objects, also called "permutations." For example, we might be designing a computer password system to serve as many as a billion users. We don't want to require extremely long passwords, because this might annoy users and drive them away. However, if we use passwords that are too short, there will not be enough passwords to go around. For example, if the passwords were only one letter long, there would only be 26 passwords (A, B, C, . . . Z). Would five letters be enough? We could answer such a question by writing down all possible five-letter combinations and then counting them, but this would take too long— remember, we want at least a billion passwords. Combinatorics answers questions like this efficiently. In the case of the five-letter password, one of combinatorics's simplest rules says that there are 26^5 or 11,881,400 possible passwords. This is not enough for a billion users. To allow for more than a billion passwords, we must use at least seven letters. Combinatorics enables us to count possibilities in more complicated situations, too, with many applications in computer science and other real-world fields.

PROBABILITY THEORY

Probability theory is the study of how likely things are to happen. For instance, when a fair coin is flipped, the probability of getting heads is 1/2 and the probability of getting tails is 1/2. Probability theory often uses continuous variables, but it is rooted in discrete mathematics because the "events" it deals with are separate or discrete.

Probability theory is used throughout science and business. Whenever we have to make a guess about the future—or about past events of which we cannot have certain knowledge—we must think in terms of probability. Corporations deciding how many items to manufacture and where to send them must decide what the market will probably want; gamblers and betters try to make the most probable bets; testing memory chips and other manufactured items for quality control is done using probability-based methods; and many computer tasks, such as searching a database for a particular name or other item, are treated by designers as "probabilistic" (random) processes. Combinatorics is used heavily in probability theory because to know how probable a particular event or group of events is, we need to know how many possible events there are. We calculate this using combinatorics.

ALGORITHMS

An algorithm is a set of instructions for solving a problem or performing a task. The problem may be mathematical—like deciding whether a particular number is a prime number, or finding the square root of a number—or it may be non-mathematical, like baking a cake or finding a word in a computer database. Algorithms are used to tell computers how to do virtually everything that they do. When a digital camera focuses automatically, for example, its internal computer obeys an algorithm that tells it what part of the picture to focus on and how to know when that part is in focus. Large prime numbers, which are important for sending secret messages (cryptography), are found using algorithms. Algorithms are a part of discrete mathematics because the steps of an algorithm can be taken separately, one at a time.

CRYPTOGRAPHY

Cryptography is the science of making and reading secret (or "encrypted") messages—messages that look like completely random strings of symbols (letters, numbers, or bits). Cryptography is a sub-field of discrete mathematics because it deals with discrete (separate) symbols and words.

GRAPHS

In mathematics, graphs are drawings consisting of points (or circles) joined by lines. The circles are called nodes and the lines that join them are called edges. If you were to draw a five-pointed star, going directly from point to point with your pencil, and then put a circle at each point of the star, you would have drawn a graph with five nodes and five edges. Many real-world problems can be drawn as graphs. Nodes can stand for actual places (cities, say) connected by edges representing roads, railways, or telephone lines. Nodes can also be used to stand for states or conditions of a machine, with edges standing for possible changes from one state to another. For instance, two nodes connected by a single edge might stand for the ON and OFF states of a television, with the line between them standing for the fact that we can make the machine go from one state to another (turn it on and off).

MATRIX ALGEBRA

Matrix algebra gives the rules for handling matrices (plural of "matrix"). A matrix is a group of numbers or other symbols that have been arranged in a rectangular array, as if glued to the squares of a chessboard. Whenever we have a list of related mathematical equations (a "system of equations") and want to find a solution that satisfies all of them at once, we can write that list of equations as a matrix. Matrix algebra is used in computer programs designed to predict or mimic real-world events. Each number in computer memory can be treated as a number in a matrix, making it possible to solve large, difficult systems of equations efficiently.

Real-life Applications

SEARCHING THE WEB

You want to find something on the Web, so you call up the window of a favorite search engine such as AltaVista or Google and type in a word or phrase. In a few seconds or less, results appear—the first 10 or 20 out of what may be hundreds or even tens of thousands of matches, also called "hits." Somehow, in a fraction of a second, a computer (not yours, one belonging to the company that runs the search engine) has managed to comb through the contents of several billion Web pages to see which ones contain the word or words that you've entered. Each page may contain hundreds or thousands of words.

The search engine manages this trick by searching not the Web itself, but an index. A separate program is constantly "crawling" the Web, that is, automatically calling up hundreds of millions of Web pages. It then looks at all the words on each page and adds the words it finds to a large index or database along with information about how important each particular Web page might be. Such an index is huge—gigabytes or even terabytes (trillions of bytes)—but it is still far smaller than the Web itself. When you enter a word or phrase in the search engine, the engine searches the index. Structuring and searching large indexes and databases relies on the mathematics of graphs, especially that kind of graph called a "tree." Structuring an index as a tree makes searching it highly efficient. The result is that a search engine can dish up thousands of hits almost in almost as little time as it takes to get your request and send the results back.

COMPUTER DESIGN

When George Boole published his book *The Laws of Thought* in 1854, digital computers had not yet been thought of (though a few mechanical adding machines had been built). Boole's book, in which he laid out the rules of arithmetic using the simplest possible number system (0 and 1), was thought to be "pure" math, that is, math having no application to "reallife." But in 1938 the American mathematician Claude Shannon (1916–2001) showed that Boolean algebra could be used to design electrical circuits. It is easier and cheaper to build a circuit that represents 1 and 0 by switching itself on and off than to design a circuit that represents many numbers by switching between many in-between states. Today, all computer circuits are designed using Boolean algebra.

SHOPPING ONLINE AND PRIME NUMBERS

Cryptography is probably the most important application of number theory. It is, for example, basic to the functioning of the Internet. Without cryptography, millions of credit-card numbers could not be sent safely and automatically over the Internet every day. Whenever a web browser such as Explorer or Communicator announces that it has given you a "secure" connection after you have clicked on a link to make a credit-card purchase, a link using a type of encryption known as a "public-key cipher" is established. It is practically impossible for anyone (except, maybe, somebody at the National Security Agency, the United States government organization devoted to eavesdropping on communications and breaking codes) to read a message sent over such an Internet connection, even if they somehow manage to intercept the message somehow. Public-key ciphers depend on the fact that when very large two prime numbers are multiplied to give a third, larger number it is difficult—almost impossible, in practical terms—to discover the two primes from knowledge of their product (the large number made by multiplying them).

COMBINATORIAL CHEMISTRY

Many of the tastes and smells that we experience every day—whether in foods like french fries and gum, or wafting from toilet paper, shampoo, or makeup—are created in laboratories. Chemists are always looking for new taste and smell chemicals, "tastants and odorants" as they are called in the industry. One of the fastest-growing methods for finding new tastants and odorants, as well as drugs, pesticides, dyes, catalysts, and other chemicals, is combinatorial chemistry. In theory, scientists should be able to predict the properties of a complicated chemical just by looking at the shape and composition of its molecules. In practice, though, the only way to know how a complicated chemical will behave is to put it together and perform tests on it. Combinatorial chemistry, guided by the mathematics of combinatorics, creates small amounts of hundreds or thousands of similar chemicals all at once. These are then tested by computers at high speeds. Combinatorial chemistry has greatly speeded up the discovery of new drugs and other useful chemicals.

LOOKING INSIDE THE BODY WITH MATRICES

For about a hundred years, x-ray images of patients were taken by shining x rays (a type of high-energy light) through the body and capturing the shadows cast by body parts, especially bones, on a piece of photographic film. But in the 1980s, a new kind of x-ray came into being,

called CAT scanning (for computerized axial tomography). In CAT scanning, a narrow x-ray beam, like the beam of a flashlight, is moved all around the patient in a circle. It is turned to point inward as it moves so that it shines through the patient crosswise from every possible angle. On the far side of the patient, an instrument records the power of the x rays shining through the patient. Where the x-ray beam meets more bone or other tissue that absorbs it, the beam is weaker on the far side of the patient. This process produces a long series of numbers (beam brightness measurements) that do not look anything like a picture of the inside of the patient's body—but using matrix algebra, a computer makes them into a clear, sharp "cross-section" image resembling what you would see if you could slice the patient in half. CAT scans show pictures of fine details inside the body that doctors could never see before. Other modern imaging methods, such as nuclear magnetic resonance imaging, also use matrix algebra.

FINDING NEW DRUGS WITH GRAPH THEORY

The chemical industry has for decades been building up databases that record the three-dimensional (3-D) structures of millions of molecules. The 3-D structure or shape of a molecule helps determine its medical properties. Researchers designing new drugs often know what shape they want a drug molecule to have in order to produce a certain effect in the body, but searching through millions of 3-D molecule records by calling up each one on a screen and looking at it is too slow. Instead, since the early 1990s drug designers have been using graph theory and algorithms to search for molecules with useful shapes. In this method, each molecule is represented as a graph, with atoms for nodes and chemical bonds for edges. Fast algorithms have been designed that look for matches between a "query graph" (the molecule the drug designer is looking for) and the graphs of the molecules in the database.

COUNTING JAGUARS USING PROBABILITY THEORY

Jaguars live in the jungles of Central and South America. In the wild, jaguars—like other predators—roam over vast areas, making it hard to know how many jaguars there are in a given area. Yet it is important to know how many jaguars there are in an area such as Kaa-Iya National Park in Bolivia, in order to know how best to protect them from extinction.

The probability theory of discrete events provides an answers. (Counting jaguars is discrete because jaguars are discrete.) Researchers set up "camera traps" in the forest

Key Terms

Boolean algebra: The algebra of logic. Named after English mathematician George Boole, who was the first to apply algebraic techniques to logical methodology. Boole showed that logical propositions and their connectives could be expressed in the language of set theory.

Combinatorics: The study of combining objects by various rules to create new arrangements of objects. The objects can be anything from points and numbers to apples and oranges. Combinatorics, like algebra, numerical analysis and topology, is a important branch of mathematics. Examples of combinatorial questions are whether we can make a certain arrangement, how many arrangements can be made, and what is the best arrangement for a set of objects. Combinatorics can be grouped into two categories. Enumeration, which is the study of counting and arranging objects, and graph theory, or the study of graphs. Combinatorics makes important contributions to fields such as computer science, operations research, probability theory, and cryptology.

Function: A mathematical relationship between two sets of real numbers. These sets of numbers are related to each other by a rule which assigns each value from one set to exactly one value in the other set. The standard notation for a function $y = f(x)$, developed in the 18th century, is read "y equals f of x." Other representations of functions include graphs and tables. Functions are classified by the types of rules which govern their relationships.

Logic: The study of the rules which underlie plausible reasoning in mathematics, science, law, and other discliplines.

Matrix: A rectangular array of variables or numbers, often shown with square brackets enclosing the array. Here "rectangular" means composed of columns of equal length, not two-dimensional. A matrix equation can represent a system of linear equations.

Prime number: Any number greater than 1 that can only be divided by 1 and itself.

Set: A collection of elements.

that automatically photograph jaguars as they pass by. Since the pattern of spots on each jaguar is unique, like a fingerprint, these photographs tell the researchers how many individual jaguars they have seen at each camera trap. The whole population of jaguars cannot be expected to walk past the cameras, however—it would cost too much to build that many camera traps—so a mathematical model is used instead, along with a method called "maximum likelihood estimation," that guesses what the most probable or likely total number of jaguars is based on the number that have been photographed. In 2004, biologists announced that using cameras and probability theory they estimated that there were about 1,000 jaguars in Kaa-Iya Park—more than they had thought, which is good news for this endangered species.

Where to Learn More

Books

Bogart, Kenneth P. *Discrete Mathematics.* Lexington, MA: D.C. Heath and Co., 1988.

Matousek, Jiri, and Jaroslav Nesetril. *Invitation to Discrete Mathematics.* Midsomer Norton, Avon, UK: Oxford University Press, 1999.

Rosen, Kenneth H. *Discrete Mathematics and Its Applications.* New York: WCB/McGraw-Hill, 1999.

Periodicals

Friedman, Richard R. "The suggestibility of children: scientific research and legal implications." *Cornell Law Review,* Nov. 1, 2000.

Web sites

National Institute of Standards and Technology. "Advanced Encryption Standard: Questions and Answers." Computer Resource Security Center. March 5, 2001. <http://csrc.nist .gov/encryption/aes/round2/aesfact.html> (June 16, 2004).

Thomas, Rachel. "Cat Count." Plus Magazine. May 31, 2004. <http://plus.maths.org/latestnews/may-aug04/jaguar/ index.html> (Sep. 5, 2004).

Fundamental Mathematical Concepts and Terms

Division is the inverse operation of multiplication, and is used to separate a set quantity into several smaller equal quantities. Simple division involves three quantities. The beginning value in a division problem is called the dividend, and the amount by which it is divided is labeled the divisor. The solution to a division equation is called a quotient, so a simple division equation takes the form of dividend / divisor = quotient.

Two symbols are used to signify a division operation. The commonly used division symbol ($\div$) is called an obelus, though this name is rarely used. The fraction line (/) is also used to signify division, and this symbol is called either a diagonal or a solidus. A fraction, written as one value separated from another by a solidus, is in reality a division equation that has not yet been evaluated. The upper (left-hand) value, or dividend, in a fraction is called the numerator, and the lower value or divisor is called the denominator. In cases where the dividend does not divide evenly by the divisor, the quantity left over after the division is termed a remainder.

A Brief History of Discovery and Development

As the inverse of multiplication, division probably developed around the same time it did. However, the need for complex division calculations emerged only fairly recently, and early applications of division were probably simple equations used to evenly divide and distribute quantities of tangible objects.

As the need for multi-digit calculations became more common, the process of long division was gradually refined, and relatively complex procedures were developed to deal with these increasingly challenging problems. The ancient Egyptians developed a repetitive, but effective, method for calculating long division solutions using only simple multiplication. In this process, the divisor is repeatedly doubled until the product is more than the original dividend; at the point this occurs, specific intermediate values from this process are added to find the solution. While this process works well for even division, additional complications arise when the initial division leaves a remainder, an outcome for which additional procedures were developed to approximate the resulting fractional result.

A second ancient method of division is attributed to the Hindus and goes by several names, including galley division, *batello* division, and scratch division. This

Division

method, one of the most commonly used techniques prior to 1600, is well-suited to the abacus and counting table, but can quickly become confusing when done with pencil and paper due to the necessity of repeatedly crossing out or erasing numbers and replacing them with others. The basic methodology involves guessing a potential solution and replacing the existing value with this guess, then evaluating the outcome and making another guess. While tedious and needlessly complex in the twenty-first century, this method allowed lengthy equations to be evaluated many centuries before mathematics became a widely taught subject.

Ironically, one modern division technique is quite similar to the scratch method of repeatedly guessing and evaluating trial values. Trial division is a commonly used method of finding the prime factors of any number simply by evaluating different divisors or sets of divisors until a solution is found. While simple in application, this approach is computationally intensive, requiring hundreds or thousands of trials to produce a solution; however modern computers are ideally suited to such brute force methods, making this method useful in many situations.

Although powerful computers can solve virtually any division problem, equations with a divisor of 0 cannot be evaluated, and division by 0 produces an error message on pocket calculators, in spreadsheet formulas, and in computer programming applications. The reason for this error can be demonstrated by conducting a series of calculations which approach this value. For instance $1 / 1 = 1$, and each reduction in the divisor produces a corresponding increase in the result. Accordingly, if a sequence of calculations is evaluated, we find that $1 / .001 = 1,000$ and $1 / .0000001 = 10,000,000$. As this progression continues and the divisor gets smaller and smaller (approaches 0), the result of the equation will grow larger and larger. If taken to its ultimate extreme, this process would produce the equation $1 / 0 = $ infinity (or, $1 / 0 = $ undefined). Because infinity is a symbolic value with little practical meaning, this progression explains a computer's inability to perform calculations in which infinity is the result. Given this impossibility, most computers respond with an error code, informing the user that dividing by 0 is illegal or impossible.

Real-life Applications

DIVISION AND DISTRIBUTION

Three boys have six cookies to eat; how do they determine the most equitable way to distribute the treats? For many people, a simple situation like this one will be

their first exposure to one of the most common uses of division: distribution. In the case of boys and cookies, six cookies will divide evenly among three stomachs, and giving two cookies to each child will probably satisfy them all fairly well.

Of course if the boys are not all the same age and size, some may argue for a different distribution, with the largest boy arguing that he is entitled to three due to his larger appetite, or the smallest claiming he is due a larger share since his mother baked the cookies. While these childish disputes are relatively insignificant, they are different only in scale from the division disputes which take place daily in the world of commerce. For example, the 2004 major professional hockey season was cancelled, costing both players and owners millions of dollars in earnings, largely because the owners and the players could not agree how to equitably divide the profits of their joint venture. In this case, with the season cancelled, the income available to divide became 0, making the question of how to divide it much simpler.

Similar dilemmas arise when business ventures fail and companies declare bankruptcy. In such cases, a court will hear arguments from the company's management or their attorneys, and from claimants who have a legal right to receive some of the assets of the failed firm. In determining how to divide the assets, the judge is guided in part by a specific set of laws which govern bankruptcy and his or her own understanding of how strong each claimant's position is. In such cases, it is not unusual for stockholders to receive nothing, while bondholders and others receive twenty-five to seventy-five cents for each dollar they are owed.

Large corporations do not have an individual owner, per se, but are owned by their stockholders, a diverse group which can include individuals, pension funds and retirement plans, and other corporations. These owners are compensated for their investment by the payment of dividends, normally distributed on a quarterly schedule. A typical method of accounting for corporate earnings allows the firm to subtract all its costs of doing business, as well as the taxes it pays, to produce a final net earnings value. From these net earnings, the firm will take some funds to invest in future growth and other strategic goals; the remainder will typically be distributed to shareholders as a dividend, normally by paying an equal share to the owner of each share of common stock.

In the case of a large corporation such as Apple Computer, General Motors, or Wal-Mart, earnings for a given quarter could be hundreds of millions of dollars. However, the number of shares of stock is so large that a firm's dividend, or the amount that it passes along to shareholders, is usually fairly small, often in the range of $1.00 to $2.00 per share. In order to determine the value of this payment to an investor, analysts frequently calculate a value known as dividend yield, which equates to the percent of the share price which the annual dividend equals. Dividend yield is simple to calculate: annual dividend is divided by stock share price to give yield. This value provides an estimate of the first year return an investor might make if he or she purchased shares of the firm. Dividends are typically not paid by firms which are losing money, or by rapidly growing start-up companies which prefer to reinvest earnings in their current business. Microsoft, a leading software company, grew rapidly through the 1980s and 1990s, however it did not pay its first regular dividend until 2004.

Budgeting, whether dealing with money or any other limited commodity, provides many opportunities to apply division. An annual income of $48,000 can be budgeted across 12 months using division, giving a monthly spendable income of $4,000; in practical terms, many expenses do not appear evenly across the months of the year, so a prudent budget will include some unspent money for unforeseen or irregularly timed expenses.

At the corporate level, high-flying companies often start with a large pot of investor cash and race to become profitable before the pot runs dry; these firms, or the analysts who watch them, frequently assess their progress using the graphically labeled "burn rate." A company's burn rate is simply the amount of cash which it is burning, or spending in the course of a month, and is found by subtracting the current month's cash reserves from last month's figure to find out how much has been spent. Once the burn rate is calculated, the company's total cash nest egg can be divided by this figure top project an expected lifetime for the company before it runs out of cash and goes out of business. In the case of a firm with a $45 million cash start and a $9 million burn rate, the equation is 45 / 9 = 5, giving the firm's managers five months to either turn a profit, find additional investors, or look for other employment. Many of the internet startup firms of the early twenty-first century had burn rates of several million dollars per month, and in most cases, they never reached profitability.

Rationing is another form of budgeting, in which scarce resources such as water or food are distributed at a slower rate in order to prolong the supply. Survivors of a shipwreck might determine that they have eighteen gallons of water to divide among three survivors, and simple division will tell them that six gallons per person is all they have. Similarly, these same survivors might also wonder how long their limited water supply will last; dividing each individual's supply of six gallons by each person's minimum consumption of one gallon per day provides the bad news that the water will last less than one week.

DIVISION AND COMPARISON

Division is frequently used to compare two items to each other using a relationship known as a ratio. For example, a researcher might wish to know how accessible medical care is in different parts of the country. Common sense would tell her that a large city like Chicago should have more doctors than a small town like Whiteland, Indiana, but how can she compare these two, since they are so different? One approach to making this comparison involves the use of ratios. Dividing the number of doctors by the number of residents gives a ratio of doctors to residents, such as one doctor for each 1,400 residents. For example, if Whiteland has 2,700 residents and 3 doctors, dividing these values would produce a ratio of 1 doctor for every 900 residents. Other locations would have different ratios; some small towns might have ratios of 1 doctor for every 5,000 or more residents, while an urban area with a teaching hospital or an affluent suburb might have one doctor for every 100 residents. Using this ratio calculation, the researcher will be able to compare any locale with any other. In most cases, the conclusion will be that residents in areas with a higher ratio of doctors will have better access to medical care.

In numerous settings, division can provide a measure of the relationship between two quantities. Environmental scientists wishing to examine the rate of deforestation could choose to assess the density of trees in an area by dividing the number of trees in a forest by the number of acres covered, producing a ratio of trees per acre, which can be tracked over time. Similar methods can be used to assess the number of people living in an area by counting the population in a city, then dividing by the number of square miles in the urban area to produce a measure of residents per square mile.

Speed is commonly expressed in terms of miles or kilometers per hour, a simple ratio of distance covered in a set quantity of time which can be calculated for a given trip by dividing the number of miles covered by the number of hours required for the trip. Chemical concentrations are frequently expressed in ratio form using the measure parts per million (PPM), or the number of particles of a given substance which would be found in a million particles of the combined substance. Home swimming pools typically use low levels of chlorine to keep algae and bacteria from growing in the water, and the chlorine level in a swimming pool is typically measured using this ratio. While chlorine levels will vary depending on the weather and the number of swimmers, appropriate levels of 1 to 3 PPM will produce comfortable swimming and low maintenance requirements. At higher levels, chlorine stings the eyes of swimmers and tints hair, while lower levels are inadequate to control algae, causing the pool water to gradually shift from clear to green.

AVERAGES

Division is commonly used to produce averages. An average allows a large amount of data to be clearly and succinctly summarized in a single value. A simple average is found by adding the quantities involved and dividing by the number of items. If the daily temperatures for a week are 70, 72, 71, 75, 77, 71, and 45, the average is found by adding all the values, then dividing by seven, which is the number of data points included. In this case, the average would be 481 / 7 = 68.71, meaning the average temperature throughout the week was 68.71 degrees.

Because averages sum up a large amount of data, they frequently pay for this efficiency by obscuring individual values or trends in the data. In the temperature example above, a seasoned weather watcher would probably conclude from the raw data that a cold front blew into the region on the last day of the week, causing the temperature to drop rapidly, and significantly impacting the average. Recalculating the average for the first six days of the week produces an average temperature of 72.67, a value much closer to the temperatures actually experienced on the first six days of the week.

In this example, the accuracy of the average was improved by eliminating the final temperature reading from the calculation. Values which lie far from the rest, referred to as outliers, can reduce the accuracy of a simple average by skewing the results away from what it would otherwise be. For example, when evaluating average income in a city, one might find that 450 residents each earned an average of $40,000 per year while a single wealthy resident earned $9 million. While an estimate of $40,000 per resident would be quite accurate for a typical wage earner in this town, calculating the average income for the area gives a value of $59,866. This value provides a misleading picture of the area's income by suggesting that the average wage-earner in the area actually makes approximately 50% more than he actually does. In addition, no person in the city earns a salary anywhere close to this value, making its usefulness as a summary value suspect.

A second weakness of simple averages is that they sometimes produce values which are mathematically correct, but which actually provide a misleading picture of the underlying values. For example, a store owner might wish to assess the age of his customers in order to better serve their needs. To accomplish this, he assigns an employee to ask customers their ages and then calculate an average age. As soon as the manager looks at the average age of 45, he contemplates firing the employee, since he knows from observation that most of the customers in his store walk over from a nearby high school and a senior citizens center, and he has seen nobody near the age of 45. But further investigation reveals that the average is actually correct, since the ages of the customers were 14, 15, 16, 74, 75, and 76. In this case, where the total population is made up of two distinct groups, rather than a continuous distribution, the averaging process produces a solution which is mathematically correct, but practically misleading. In such cases, other measures of central tendency may be used, either alone or in concert with the average. One such measure, the mode, locates the most common or frequently occurring value in a distribution; this measure is useful when values are presented in whole numbers, such as ages. A second measure, the median value, locates the center-most value of a sequence. This assessment is useful for dividing a distribution in half, since it provides a line which lies below half the values in the distribution and above the other half.

Averages are often used to compare athletic performers and coaches. A baseball player's hitting is calculated by dividing his number of hits by his number of times at bat, though walks, fielder's choices, and other outcomes are counted differently than actual safe hits. This value, the player's batting average, tells how often a given hitter can expect to bat successfully in a given number of attempts. In the case of a batter who has hit fifteen times out of 100 attempts, the batting average would be 15 / 100 = .150. Batting averages are generally recorded to three decimal places and stated without the decimal point, meaning that these averages would typically be stated as "one-fifty."

Coaches are often assessed on their ratio of wins to attempts, or the number of wins, compared to the total number of games played. A successful coach might have compiled a record of 400 wins and 125 losses. This coach would have won 400 of his total 525 games, and dividing these values (multiplied by 100 to create a percentage) would produce a career winning percentage of 76%. Top-ranked coaches in professional basketball and football are frequently able to win far more games than they lose, producing high winning percentages.

In many cases, a simple average provides a useful summary of the values involved. But in some situations, one or more of the items being averaged is considered more important than the others; in these cases, a weighted average may be used to more accurately assess the data. A weighted average assigns a factor, or weight, to each value in the data set, giving some values a larger influence on the final result.

Consider the case of a new car buyer who wishes to compare three vehicles. The buyer has decided that his major criteria are price, efficiency, size, and sportiness, and he proceeds to rate each of the three vehicles on each of these four criteria using a scale from 1 to 10. However, the buyer is on a limited budget, so financial considerations are more important to him than other factors. For this reason, he assigns a weight of 1.0 to the factors of size and sportiness, but a weight of 2.0 to price and a weight of 1.5 to efficiency, which impacts the long-term cost of ownership.

Once these weights are multiplied by the raw scores to produce weighted scores, the resulting average will give the factor of price twice the influence of size and sportiness, while efficiency will be one and one-half times as influential. Weighted averages can be extremely helpful as a decision-making tool by allowing specific factors to influence the final score more highly than others.

PRACTICAL USES OF DIVISION FOR STUDENTS

Students are frequently encouraged to learn good time management skills, that in many cases requires making projections to determine how long a project might take. For example, a student who must read a 30-page chapter and has allotted one hour to read it can divide sixty minutes by thirty pages to find that he can devote two minutes to each page if he wants to finish in the allotted time. He can also determine that if he chooses to spend his first thirty minutes watching television, he will then have only thirty minutes left for his assignment, giving him one minute per page.

Similar choices exist when meal time arrives. The student eating on a budget of $15.00 per day can easily figure out using division that her choices will determine both how well and how often she eats. For instance, if she keeps her meal costs to $5.00 a piece, division tells her that she can plan to eat three meals per day. However if her meal cost climbs, the number of meals will fall; at $6.00 per meal, the number of meals will be 15 / 6, = 2.5, meaning she can afford 2 full meals and a snack, or perhaps a soft drink at mid-afternoon.

When a semester ends, one particular average takes center stage: the grade point average, or GPA. GPA is an average that summarizes a student's grades for a single semester or an entire academic career. GPA is found by assigning a scale to the letter grades, most commonly with a grade of A earning 4 points, B earning 3, and so on. These grades are then averaged by adding the values and dividing by the number of courses taken (or more frequently by the number of credit hours earned), and this ratio is the GPA. A perfect GPA of 4.0 would mean a student had earned only grades of A during the entire academic career, while a GPA of 1.0 would mean the student's average grade was a D.

Calculations of GPA are important to colleges evaluating applicants and to insurance companies. Students with high GPAs have also been found to experience fewer automobile accidents, a relationship that some insurers exploit by offering good student discounts to these young drivers.

OTHER USES OF DIVISION

Measuring a nation's affluence can be difficult, particularly when trying to compare it to another country with different characteristics. Consumer affluence is often measured by examining the availability of consumer goods, such as automobiles, televisions, and housing. While these goods are available in virtually every economy on Earth, the relationship between an item's price and a consumer's hourly wages may make it unaffordable.

Consider an automobile which sells for $20,000 in the United States, and a comparable car which sells for $8,000 in a less-developed nation. Before concluding that the car is more affordable in the other country, it is necessary to compare these prices to the wages of the citizens who might purchase them. Assuming that the average hourly wage in the U.S. is $10.00, we can divide the purchase price of the car by this value to conclude that a typical U.S. worker will spend 2,000 hours at work in order to pay for the car (and in reality much more because that income does not include deductions for taxes, etc.).

In the case of the second country, low prices are accompanied by much lower wages, in this case an average hourly rate of $1.00. Dividing the $8,000 price of the car by the hourly wage tells us that a worker in this country will have to work 8,000 hours to buy the car, or four times as long as the American worker. If similar relationships hold for other consumer goods, one can reasonably conclude that the American worker will be able to purchase more consumer goods for a given number of hours worked, and will enjoy a higher standard of living.

Retailing is a competitive industry, and in addition to the challenge of facing their competitors, retailers often face problems caused when their own employees either steal or give away merchandise, a crime known as pilferage. In many cases, employees assume that because the firm has such a large supply of hamburgers, pencils, or other items, a few will not be missed and will cause little harm. Unfortunately, this assumption is often incorrect.

Consider a young man working at a hamburger restaurant. After several months of work, he decides to do something nice for his best friend, so when the friend orders a meal, the young man simply does not charge him for his hamburger, which would normally sell for $1.65. What is the impact on the company of this theft? One way to determine the damage is to calculate how many hamburgers must now be sold in order to offset the loss caused by the employee's dishonesty. The company must first pay its owns costs before it earns any profits, and its expenses for rent, beef, wages, lettuce, and all its other costs come to $1.50 per hamburger, meaning that its net profit per sandwich is only 15 cents.

Because this 15 cents in profit is all that can be used to pay for the materials in the stolen burger, the company must divide the lost cost of the stolen burger by the profit earned on each additional burger in order to determine how many must be sold to recoup the loss. In this case, the $1.50 loss divided by 15 cents profit per sandwich means the restaurant must sell ten more burgers to make up the loss incurred on the stolen sandwich. These ratios are typical for most retailers, making employee theft a major threat to company profitability.

As digital cameras rapidly replace film cameras, users must determine what size memory card to purchase. One simple way to answer this question uses division to calculate how many photos will fit on each card. Imagine a photographer who wants to know how many shots he can take before he fills up his memory card and is forced to either download or delete shots. The memory card in this case is a 512 megabyte model, and the user's camera produces shots requiring 4 megabytes of storage apiece. Dividing the card capacity by the size per shot, the user finds that he will be able to store 128 digital pictures at a time.

Percentages are one of the most common methods of using division to represent and compare different values. For example, imagine that two extreme sports fans are arguing about which of their two favorite competitors is a better street luge racer. One racer has competed in 200 races and has won 150, while the other, who is in his rookie season, has only competed 38 times, of which he has won 30. Which racer is having a more successful career?

While many other aspects of the question could be argued, one simple comparison would be to calculate which racer has won more of his competitions. Obviously the first racer has won more races than the second has even entered, so how can these two performances be compared? Calculating a percentage allows both scores to be standardized for easy comparison.

A percentage expresses a ratio or division equation as if its divisor were 100. For example, in examining the human population of earth, we could easily determine that every living human being on the planet is breathing, meaning that the ratio of breathing humans to humans is around 7,000,000,000 / 7,000,000,000, or 1.0. Expressing this value as a percentage simply requires moving the decimal point two places to the right; thus 1.0 equals 100%, which is the percentage of living humans who are breathing.

In a similar manner, we could count and determine that of the 7 billion humans on Earth, about 3.4 billion are male, meaning that the ratio of males to humans is 3.4 / 7, and if we solved this equation we would find a decimal value of .486, or 48.6%. A similar process would tell us that the U.S. population of 300 million accounts for only 6% of the total world population.

Applying this percentage technique to our original question, we can calculate each racer's winning percentage in order to compare them. For the more experienced racer, 150 wins divided by 200 races produces a win ratio of 75%, meaning that in 100 races he can expect to win about 75. How does this compare to the newcomer's performance? Dividing his 30 wins by his 38 races gives a win ratio of 78.9%, meaning that in the same 100 races he would probably triumph in 79 of them, or about 4 more races than his competitor. While this comparison may not settle the debate over which extreme racer is better, it does provide a simple technique for comparing one racer's performance to another.

Many drivers get in a hurry, sometimes choosing to disobey posted speed limits in order to arrive at their destinations more quickly. What is the cost of this choice, assuming the driver is pulled over by a highway patrol officer? One way to assess the cost of this decision involves determining how much the driver paid for each mile over the speed limit he drove, and this can be found by dividing the fine by the number of miles over the limit. In this case, a fine of $75.00 for going 15 miles per hour over the limit would mean the driver paid $5.00 for each mile per hour that his speed exceeded the limit.

Car maintenance can extend an automobile's life; in particular, changing a car's oil regularly will improve its chance of a long life. Today's car owner faces many choices in the oil market, and division can help her determine the relative cost of each choice. Assuming this driver changes her own oil, she might wonder whether she should use standard motor oil, which costs $1.00 per quart, or synthetic oil, which costs $5.00 per quart.

Key Terms

Dividend: A mathematical term for the beginning value in a division equation, literally the quantity to be divided. Also a financial term referring to company earnings which are to be distributed to, or divided among, the firm's owners.

Percentage: From Latin for per centum meaning per hundred, a special type of ratio in which the second value is 100; used to represent the amount present with respect to the whole. Expressed as a percentage, the ratio times 100 (e.g., 78/100 = .78 and so .78 × 100 = 78%).

Stockholder: The partial owner of a public corporation, whose ownership is contained in one or more shares of stock. Also called a shareholder.

At first glance, synthetic oil appears far more expensive, since a typical oil change requires five quarts. However, the newest synthetic oils claim they will last a full year, while standard oil must be changed 3–4 times per year. How can a driver compare these options? A simple division equation allows a direct comparison of these two possibilities.

An oil change using five quarts of synthetic oil will cost $25.00, while each of the year's four changes using standard oil will cost $5.00 to drive the car for three months. Dividing the $25.00 cost of the synthetic change by four, we find an equivalent value for the synthetic change of $6.25, compared to $5.00 for standard oil. While this calculation demonstrates that synthetic oil does cost an average of about $1.25 more per oil change period (or about $5.00 more per year), this calculation does not take into account the other costs associated with the two options, such as the fact that one produces four times the quantity of used oil which must be disposed of, or that the time required to change the car's oil four times rather than one may be valuable to some owners.

Potential Applications

A traditional telephone system sends a person's voice as a continuous electrical signal over a pair of wires. But data sent over most digital data networks is actually divided into numerous small pieces, or packets, which are sent across the network independently, then reassembled at the destination. This method of dividing a message into numerous small pieces offers numerous advantages, including greater efficiency and much lower costs than the use of a dedicated line.

One of the most intriguing uses of this digital division process is the rapid emergence of VOIP (voice over internet protocol) telephone systems. The systems take the sound of a speaker's voice, break it into small packets of data, and send these packets across the Internet with all the other data packets traveling there; at their destination, the packets are reassembled to produce an audible signal with very little delay or system degradation. Because of the enormous cost savings and flexibility offered by VOIP, many technology analysts predict that by the year 2010, virtually all phone calls will be carried using VOIP technology. This same technology is also expected to be used for transmitting movies and other forms of entertainment.

Where to Learn More

Books
Seiter, Charles. *Everyday Math for Dummies.* Indianapolis: Wiley Publishing, 1995.

Web sites
A Brief History of Mechanical Calculators. "Leibniz Stepped Drum." <http://www.xnumber.com/xnumber/mechanical1.htm> (April 5, 2005).

History of Electronics. "Calculators." <http://www.etedeschi.ndirect.co.uk/museum/concise.history.htm> (April 6, 2005).

Homerun Web. "How to Calculate Earned Run Average (ERA)." <http://www.homerunweb.com/era.html> (April 12, 2005).

Mathworld. "Division." <http://mathworld.wolfram.com/division.html> (April 16, 2005).

Mathworld. "Trial Division." <http://mathworld.wolfram.com/trialdivision.html> (April 16, 2005).

The Math Forum at Drexel. "Dividing by 0." <http://mathforum.org/dr.math/faq/faq.divideby0.html> (April 16, 2005).

The Math Forum at Drexel. "Galley or Scratch Method of Division" <http://mathforum.org/library/drmath/view/61872.html> (April 16, 2005).

The Math Forum at Drexel. "Egyptian Division" <http://mathforum.org/library/drmath/view/57574.html> (April 16, 2005).

Domain and Range

Overview

A function is generally defined as a rule that partners to each number x in a set a unique number y in another set. The set of x values to which the rule applies is the function's domain, and the set of y values to which it applies is its range.

Calculus allows us to mathematically study rates of change and motion, something not precisely possible prior to discovery of the fundamental theorem of calculus. The domain and range of a function are the essence or foundation of algebraic equations and calculus formulas. Everyday uses include graphs, charts and maps.

Fundamental Mathematical Concepts and Terms

A function is a set of ordered pairs (x, y) such that for each first element x, there always corresponds one and only one element y. The domain is the set of the first elements and the range is the term given to name the set of the second elements. Often the domain is referred to as the independent variable and the range as the dependent variable.

The domain is the first group or set of values being fed or input into a function and these values will serve as the x-axis of a graph or chart.

The range is the second group or set of values being fed or input into a function with these values serving as the y-axis of a graph or chart.

A Brief History of Discovery and Development

The word function was first used mathematically by German philosopher and mathematician Gottfried Leibnitz (1646–1716) during his development of curve relationships. He used the term to describe a quantity relative to a curve such as a particular point of a curve or said curve's slope. Today the specific type of functions Leibniz referred to are called differentiable functions.

During the eighteenth century, Swiss mathematician and physicist Leonhard Euler (1707–1783) began using the word function to describe a formula involving various parameters. Over the next century, calculus and functions were being expanded upon and developed by German mathematician Karl Weierstrauss (1815–1897) who promoted developing calculus based upon arithmetic or number theory rather than geometry. By the end of the nineteenth century, mathematicians began defining mathematics using set theory thus seeking to describe

Key Terms

Dependent variable: What is being modeled; the output.

Function: A mathematical relationship between two sets of real numbers. These sets of numbers are related to each other by a rule that assigns each value from one set to exactly one value in the other set. The standard notation for a function y = f(x), developed in the eighteenth century, is read "y equals f of x." Other representations of functions include graphs and tables. Functions are classified by the types of rules which govern their relationships.

Independent variable: Data used to develop a model, the input.

every mathematical object as a set. It was German mathematician Johann Peter Gustav Lejeune Dirichet (1805–1859) and Russian born mathematician Nikolai Ivanovich Lobachevsky (1792–1856) who almost simultaneously gave a formal definition of function. They defined a function as a special case of a relation, with a relation being described by such concepts as "is greater than" or "is equal to" in arithmetic.

Real-life Applications

COMPUTER CONTROL AND COORDINATION

All modern applications and technology use functions to determine the domain and range of a given problem. Every time you observe a graph, use a calculator, turn on a computer, drive an automobile, or even watch television, you are interacting with calculus and the concepts of domain and range. An example of this would be the computer components found in modern aircraft. Equations, formulas, and functions are all utilized and working in onboard computer systems to increase the safety of modern aviation. These computer systems help compensate for the instability of the aircraft, weight vs. wingspan length disparity, and a host of other variables related to flight.

CALCULATING ODDS AND OUTCOMES

Another example of common uses would be calculating risk or determining odds. Share values can be depicted as functions with domain and range and analysis of such data differentiate a wise from a foolish investment. Insurance companies use calculus formulas and functions to determine the risk associated with insuring. Analysts and researchers use set theory and ratios as a method of evaluation.

Having a "working understanding" of functions—and of the components of domain and range (especially

which are dependent and independent variables) allows deeper understanding of graphs, charts, maps, finance, and business strategies.

PHYSICS

Physics relies heavily on calculus. English physicist and mathematician Sir Isaac Newton (1642–1727), who independently developed calculus about the same time as did Liebniz, used the concepts of domain and range of a function in advancing a Law of Gravity. In 2005, physicists use the domain and range of functions in designing and solving equations relative to nuclear energy development, nuclear fission processes, and other scientific experiments. Quantum mechanics, a rudimentary physical theory which refers to discrete units that the theory assigns to certain physical quantities, is the underlying structure in many fields of chemistry and physics. The fundamental beginnings of quantum mechanics stem from functional analysis.

ASTRONOMERS

Astronomers use the domain and range of functions to plot trajectories, calculate distances, measure space and speed of objects, and more. One example would be calculating the future path of asteroids, comets, and falling space debris. A complex set of functions are used to determine where to search for stars, planets, black holes, comets, asteroids, and other objects in our own galaxy as well as in other galaxies. Data sorted into tables by domain and range helped prove the existence of the planets Neptune and Pluto.

ENGINEERING

Determining the design and construction of highways, roads, wiring layouts, inventing new molecules, or developing molecular structures all require the use of domain and range in order to achieve the necessary

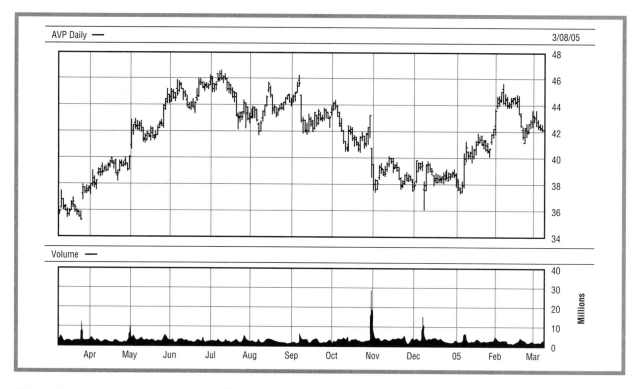

Figure 1. A one year chart of stock prices. (Share price is represented in U.S. dollars on the y axis.)

results. Structural engineers use functions to determine tensile strength of metals, to study aerodynamics of vehicles, and in the design of the subtle aerodynamic curves of an automobile.

COMPUTER SCIENCE

Computer hardware and software all use domain and range in the design and implementation of their programs. Computer programming languages such as Functional programming, is structured upon the use of functions. This language is a programming paradigm that relates computation as the evaluation of mathematical functions. Inductive logic programming also uses functions in a declarative programming paradigm that is concerned with finding general rules based on a sample of facts.

GRAPHS, CHARTS, MAPS

The ability to interpret graphs, charts and maps requires the understanding of the domain and range of functions. The x axis of a graph is the domain or all input elements of a function, and the y axis is all the actual elements derived from the function. The same is true for charts; the horizontal line represents the larger set, and the vertical line represents the specific information

derived from the larger set. (See Figure 1, a typical chart of stock prices.)

Satellite technology, space exploration, computer programming and languages, automobile design, nuclear physics and countless other scientific fields all rely on the interrelation of domain and range.

In this digital information age, privacy concerns are of increasing importance. Cryptography is a field which uses the domain and range of functions in order to develop keys and ciphers to hide important or sensitive information transmitted over the Internet or other media. For example, based upon a random number, the independent and dependent variables can construct an elaborate system of codes and keys.

Where to Learn More

Books

Krantz, Stephen G. *Calculus Demystified*. New York, NY: McGraw-Hill, 2003.

Websites

Husch, Lawrence S. Visual Calculus. 2001 <http://archives.math .utk.edu/visualcalculus/> (March 1, 2005).

Overview

Elliptic equations are a type of mathematical expression that is related to the geometric shape called an ellipse. An ellipse is a shape like a flattened circle. To draw an ellipse, one picks two points (called "foci") and encloses them with a curve drawn so that the sum of the distances from the foci to every point on the curve is always the same. An ellipse is flatter or more stretched if its foci are farther apart.

Fundamental Mathematical Concepts and Terms

An ellipse can also be defined using an equation. However, the equation that defines an ellipse is not what mathematicians mean when they speak of an "elliptic equation." An elliptic equation is a particular type of equation that arises from calculating the length of part of an ellipse. The lengths of curves are calculated using the mathematical technique called "integration." Integration of part of an ellipse produces a type of function (a function is a rule for going from one set of numbers to another) called an "elliptic integral." An elliptic equation is the opposite or inverse of an elliptic integral, the "inverse" of a function being a second function that undoes the first one. Elliptic equations are thus related to ellipses, but in a rather roundabout way.

Technically speaking, an elliptic function is a function in the complex plane that is periodic in two directions. This statement needs some explanation. The "complex plane" is a flat space like the *x-y* plane that is used in ordinary geometry and in graphing, except that one of the directions or axes is reserved for so-called "imaginary" numbers, which have special properties. A function is "periodic" if it repeats itself after some distance. A zigzag line is an example of a function that is periodic in one direction; the rings of a bull's-eye target are an example of a function that is periodic in all directions.

Elliptic functions are an important tool for mathematicians and physicists, because they crop up during the solution of many larger, more complex problems. For example, they appear in the exact mathematical description of pendulums and tumbling objects.

Real-life Applications

CONFORMAL MAPS

Flat maps have useful properties: they can be shown on computer screens and take up little room when printed on paper. However, most of the things that we are interested in making maps of—like the surface of the Earth—are curved. And whenever a curved surface is mapped to a flat surface, there is distortion or error. The Mercator projection of the Earth's surface (any flat map of a curved surface is called a "projection") is the most widely used flat map of the world, and it distorts the Earth by grossly expanding shapes near the North and South Poles. However, the Mercator projection does have one useful property: it is "conformal." Elliptic functions are involved in the mathematics of conformal projections. A conformal projection does not squeeze or stretch small shapes. Thus, a small circle or triangle on the globe is still a circle or triangle on the Mercator map. Conformal projections are used to map the surfaces of the Earth and of other planets. They are also used by researchers who want to make flat maps of the complexly folded surface of the brain, and are being studied by plastic surgeons as a way to describe and predict the effects of surgery on the nose.

E-MONEY

Many companies in Europe, Japan, and the United States are developing forms of e-money, also called digital cash. E-cash is electronic money on a card. A certain amount of money value is programmed into a memory circuit in the card, the card is swiped through a machine at the store when you buy something, and the cost of the purchase is subtracted from the value on the card. Unlike a debit card, you don't have to have a bank account to use an e-money card. Value is loaded directly into the card.

To keep people from simply programming more money into their e-money cards and becoming instant millionaires, a secret code or "cryptosystem" is used to check that e-money is real when it is spent. The kind of cryptosystem proposed for e-money is "public-key cryptography," and the particular kind of public-key cryptography that is proposed is elliptic-curve cryptography. Elliptic-curve cryptography doesn't require long, complicated calculations, which makes it ideal for use with the relatively simple circuits that can be embedded in e-money cards. Elliptic functions are involved in the mathematics of elliptic-curve cryptography.

THE AGE OF THE UNIVERSE

In 1915, the German physicist Albert Einstein (1879–1955) published his theory of general relativity. According to this theory (which has been checked against experiment many times, and is used every day in the global positioning system [GPS]), space does not go on forever. In a way that cannot be pictured in the mind but can be described mathematically, it bends around on itself. Space is finite or limited in size. Furthermore, it is expanding—and if it is finite and expanding, there must have been a time when it began expanding from zero size. In other words, the Universe must have had a beginning. Calculating the age of the Universe from the theory of general relativity involves the use of elliptic functions. Combining such calculations with astronomical observations has shown that the Universe is approximately 14.5 billion years old.

Where to Learn More

Books

Straffin, Philip, ed. *Applications of Calculus*, Vol. 3. Washington, DC: Mathematical Association of America, 1997.

Web sites

Wolfam. "Elliptic function." <http://mathworld.wolfram.com/EllipticFunction.html> (Feb 13, 2005).

Overview

Estimation is the act of approximating an attribute such as value, size, or amount. Estimation has applications in all walks of life—from how much salt to put on popcorn, to using complex mathematical methods to predict the economy. Many common ideas, such as estimating the distance from Earth to the Sun, would be inconceivable without mathematical estimation.

Fundamental Mathematical Concepts and Terms

Estimation is an essential tool in many mathematical situations. For example, many people have to set an alarm clock to make sure that they wake up in time to get ready and get to school. The first time they set their alarm, they probably had to think about the different tasks they needed to accomplish in the morning in order to get to school on time. Those tasks may be broken down in the following manner:

- School starts at 8:00 a.m.
- It takes about 20 minutes to walk to school.
- It takes about 30 minutes to make and eat breakfast.
- It takes 20 minutes to shower, brush their teeth, and get dressed.
- They may plan to press the snooze bar twice, so they set the alarm for about 20 minutes before they actually need to get up.

Adding all these amounts of time together gives an estimation that it takes about an hour-and-a-half to get ready and get to the destination. According to this estimation, these people need to set the alarm for about 6:30 a.m to be able to arrive on time. The ability to make a good estimate considerably depends on known information and past experience. Final estimations also often depend on previously established estimations. For instance, the amounts of time that they spend performing the various tasks are estimates in themselves. In this case, the final estimation is based on:

- Information provided—classes begin at 8:00 in the morning.
- Past experience—people have showered and brushed their teeth before; they have made and eaten breakfast before; they know if they are snoozers and that they usually push the button twice before they get out of bed.
- Rough estimation—they know how far school is from their house and about how long it takes to walk that distance; because they prefer to get to school early rather than late, so they overestimate the time just a bit.

Estimates are usually refined as they are tested. Each time a person wakes up, gets ready, and walks to school, they may make adjustments to their estimate. For instance, sometimes they may want to get to school earlier, so they set their alarm even earlier for that day. On the other hand, they may have the first class of the day canceled, and they decide to set the alarm for later so that they can sleep a little longer.

Defining acceptable levels of error is another key concept in making meaningful estimates. There are many different theories and methods for analysis of error in estimation.

Throughout history, multiple methods of estimation have been proposed and scrutinized. Controversy exists among the many estimators and analysts about which methods yield the most accurate results.

A Brief History of Discovery and Development

The word estimate is derived from a late sixteenth century Latin word meaning to determine, appraise, or value. However, various methods of estimation have been used throughout history.

An early account of mathematical estimation involves a question posed by Greek mathematician Archimedes (born c. 287 B.C.), in which he contemplated how many grains of sand would be needed to match the volume of Earth. In ancient times, the issue of understanding and labeling very large (and small) numbers posed a serious problem that hindered the capabilities of mathematicians. This issue is at least partly attributed to the limitations of the existing numbering system, which was much like the Roman numeral system. By utilizing his own numbering system (similar to the exponential numbering system adopted later), Archimedes was able to grasp numbers large enough to approximate key values for determining a reasonable estimation of the amount of sand required to fill the volume of the planet. His new notation also allowed him to convey his ideas to his peers and to effectively convince them of the relative accuracy of his estimate.

Archimedes employed existing geometric theory (the equation for finding the volume of a sphere), a commonly accepted estimation (an approximate value of the radius of Earth, the distance from the center to the crust), and an observed measurement (he estimated that an average grain of sand was basically spherical and had a diameter of about 1/100th of an inch). Using these tools, Archimedes was able to estimate that it would take over 10^{32} grains of sand to match Earth's volume. He was aware of many imperfections in his calculations, including the fact that every grain of sand is not perfectly spherical. Earth is not a perfect sphere either—it has mountains and canyons (and is squashed at the poles). The values for the radius of Earth and the average diameter for a grain of sand are also only approximate values. Nevertheless, he was able to derive an estimate that was substantial enough to give a manageable account of the magnitude of the answer to his question. This was an important step toward mankind becoming comfortable with previously unfathomable numbers.

In 1773, Benjamin Franklin found that a drop of oil placed on the surface of water will spread out across the water until it forms a layer that has the thickness of a single molecule of the oil (known as a monomolecular film). In addition to Franklin's immediate observations from this experiment (his notes describe the oil spreading quickly and causing wavy water to become calm almost immediately), his discoveries would have a profound influence in the first estimation of the thickness of a molecule more than a century later. Scientists first estimated the thickness of a molecule by recording the volume of a drop of oil and then placing the drop onto the surface of some water. Once the oil had spread to a thickness of a single molecule, the surface area of the floating film was estimated using an approximate value for the radius of the somewhat circular layer of film. Then the formula for volume was used to determine the thickness of the film. The volume of the oil is equal to the area of the film multiplied by the thickness of the film. So to determine the thickness of the film, which is the thickness of a single molecule, the volume of the drop of oil was divided by the approximate area of the film.

In 1801, Carl Frederick Gauss, also commonly viewed as one of the most important figures in the history of mathematics, made the first applicable estimation of the orbit of planets. His first subject was a newly discovered planet named Ceres. Using his method of least square, which remains an important contribution to the development of estimation methods, Gauss was able to enhance prior theories about the orbits of planets, incorporating calculations that represent imperfections in orbital paths due to factors such as interference caused by other celestial bodies.

From tracking and maintaining populations of endangered species, to predicting genetic abnormalities in unborn babies, estimation remains an essential concept in many mathematical and scientific procedures and discoveries.

Real-life Applications

BUYING A USED CAR

For most people, the purchase of their first car can be an overwhelming event. There is much more to consider than whether they like the color or the rims on the wheels. With so many factors that can affect the value of an automobile, people have to be careful that they don't get cheated.

Often, the seller will begin by asking for much more money than the car is actually worth, or keep any problems a secret until they officially sell the car. So, how can one know if they are getting a good deal or being cheated? First, one should be aware that issues might exist, such as engine trouble, damaged upholstery, or body damage that lower the true value of the car. Similarly, any enhancements that can raise the value must be considered, including custom parts, limited edition features, audio or video equipment, global positioning system (GPS) tracking devices, safety features, and so on.

The most significant obstacle in coming to an agreement between the seller and the buyer is that there is no true value of a used car. Too many factors influence each car to be defined generally. Luckily, one thing that the buyer and the seller will probably agree on is that they both want to finalize the transaction as quickly and smoothly as possible. Therefore, both people have to use estimation involving available information to find an agreeable price range. Most buyers and sellers alike refer to one of many periodically released publications, such as the *Kelley Blue Book*, as a basis for what the car should cost depending on the year, make, model, mileage, and general condition. Using this base price, both parties attempt to factor in as many positive and negative characteristics as they can find to determine what they think the car is worth and to arrive at lower and upper bounds for an acceptable price. From there, everything depends on keen negotiation skills.

GUMBALL CONTEST

Jen has been entered into the critical-thinking contest at the annual mathematics fair, in which the top math students from around the region compete to solve difficult problems. The first problem posed to the contestants involves the estimation of the number of gumballs contained in a glass case that is 4 feet long, 4 feet wide, and 8 feet tall. Each contestant is expected to use mathematical reasoning to decide whether they think that the number of gumballs inside the glass case is less than or greater than 25,000.

Jen examines the glass case and thinks about how she can make a good approximation of the number of gumballs inside. The first thing she does is collect as much information as she can about the problem at hand. Jen takes note of the following information:

• The glass case is transparent, so Jen can approximate the size of the gumballs. As far as she can see, the gumballs are all about the same size—somewhere between 1 inch and 2 inches in diameter.

• The volume of the glass case is equal to the product of its dimensions. Since she will be estimating the volume of the gumballs in cubic inches, Jen multiplies each dimension of the glass case by 12 to convert to inches. The glass case is 48 inches by 48 inches by 96 inches. Multiplying these values together, she finds that the volume of the glass case is 221,184 cubic inches.

Her estimate of the diameter of each gumball consists of an upper bound and a lower bound. In this way, she hopes to simplify the problem without concerning herself too much with the true size of one gumball—let alone all of them! If the total estimated volume that she finds using the lower bound for the size of a gumball is more than the volume of the case, then she will know that 25,000 gumballs will not fit into the case. If the estimate she finds using the upper bound for the size of a gumball is less than the volume of the case, then she can safely conclude that 25,000 gumballs will fit into the glass case. However, if the volume of the glass case is between her lower and upper estimates, then she cannot make a confident conclusion using this information, and she will have to attempt to refine her estimate of the size of a gumball.

Jen knows that she needs the size of the gumballs to be expressed in terms of volume so that she can compare the volume of 25,000 gumballs to the volume of the glass case. She needs to find the volume of a single gumball and then multiply this value by 25,000. Using the formula for the volume of sphere, she finds that a gumball that is 1 inch in diameter (having a half-inch radius) has a volume of approximately 2.09 cubic inches. Similarly, she finds that a gumball that is 2 inches in diameter (having a one-inch radius) has a volume of 4.19 cubic inches. At this point, she feels confident that the volume of a single gumball is somewhere between these two estimated bounds. To find bounds for the total estimated volume of 25,000 gumballs, she multiplies the bounds for the volume of a single gumball by 25,000. She is reasonably certain that the volume of 25,000 gumballs is between 52,359 and 104,750 cubic inches. Since the upper bound for her estimate of the total volume of the gumballs is less than the volume of the glass case, Jen could decide at this point that she is convinced that 25,000 gumballs will in fact fit into the case. However, just before she turns her response in for evaluation, she becomes aware of a major flaw in her reasoning.

The total estimated volume of the gumballs definitely gives her a better feeling for this problem, but she quickly realizes that this will not yield conclusive results because she has not considered the air space in between all of the gumballs. What she has really figured out is that if she were to chew up 25,000 gumballs and press them into the glass case, the huge blob of gum would fit (especially if the gumballs are hollow). Little does Jen know that this is an example of a sphere-packing problem—a classic problem in mathematics for which there is no standard solution. Nevertheless, she is on the right track to making a fairly good estimate.

While thinking of a way to refine her estimation methods, she imagines having each gumball wrapped perfectly into a little box. She understands that this idea will lead to an overestimate of the amount of airspace in the glass case because the gumballs do not stack on top of each other perfectly. She will use the values found using this method as upper bounds for the total volume of the gumballs taking airspace into account.

Each gumball-wrapping box would have length, width, and height equal to the diameter of the gumball. For a gumball with a 1-inch diameter, the surrounding box would have a volume of 1 cubic inch. For a gumball with a 2-inch diameter, the box would have a volume of 8 cubic inches. The question now is whether or not 25,000 of these surrounding boxes will fit into the glass case. Multiplying by 25,000, she finds lower and upper bounds for the total volume of all of the wrapping boxes. The total volume of the boxes is between 25,000 and 200,000 cubic inches.

At this point, Jen stops to think about her progress so far. Since she can show that her largest estimate of 200,000 cubic inches—found by packing gumballs into boxes that overcompensate for airspace—is smaller than the volume of the glass case (221,184 cubic inches), she feels fairly certain that 25,000 gumballs fit into the glass case. (Note that if any of Jen's estimates were greater than 221,184 cubic inches, then she would have had to either refine her estimate of the diameter of a gumball or come up with a more accurate way to account for air space.)

POPULATION SAMPLING

Wildlife conservationists are often confronted with the task of estimating how many members of a certain species of animal are living in a given area. For example, suppose that a team of conservationists needs to estimate the number of fish in a small lake (without draining the lake and counting all of the fish). This may seem a daunting task because fish move around the lake, reproduce, and die. However, the team will be able to use population sampling techniques to find an estimate that is suitably accurate for their needs.

To begin, the team catches a sample of 300 fish. Each of these fish is tagged and returned to the lake. The team then makes a simplifying assumption that will be critical to the estimation process: over time, all of the fish in the lake move about at random. This is a reasonable assumption based on previous studies about these fish.

After waiting a week for the fish to redistribute themselves, the team again catches 300 fish and finds that 25 fish out of this sample are tagged. This time, the team members must do their best to select the fish at random from the total population of the lake. To ensure that they collect a random sample, they may collect the sample from various areas of the lake.

Next, the team uses the basic sampling principle as it applies to their situation: the proportion of tagged fish in the second sample should reflect the proportion of tagged fish in the entire lake population, as long as the sample size is reasonably large.

In the second sample, the team found that 25 out of 300 fish were tagged, so the proportion of tagged fish in the second sample is 25 divided by 300, or 1/12. The team also knows that there were 300 tagged fish in the lake (barring any fatalities among the first sample), so the proportion of tagged fish in the entire lake is 300 out of the total population of fish, the value that the team is attempting to estimate. In accordance with the basic sampling principle, the formula $1/12 \approx 300/N$, where N is the total number of fish in the lake helps team members find an estimate for the total population of fish in the lake.

After dividing both sides by 300 and simplifying, the team finds that $N \approx 3,600$. The team of conservationists now has a rough estimate of 3,600 fish living in the lake. Depending on the requirements of the study, the team may or may not need to take more samples and find an average value. The team would not replace the fish after each sample, so that the fish are not counted twice. A relatively consistent number of tagged fish in each sample would be a good indication that the estimations are sufficiently accurate. Using a larger sample will usually result in higher accuracy as well.

DIGITAL IMAGING

A digital image is an arrangement of tiny square regions called pixels. In the case of a gray-scale (black-and-white with shades of gray) image, the brightness of each pixel is determined by a numeric value. A typical gray-scale image contains values ranging from 0 to 255, with 0 representing black, 255 representing white, and intermediate values representing shades of gray.

A color image can be represented using different mixtures of red, green, and blue. The color of each pixel

164

in the digital image is usually determined by a set of three numbers, one representing red, one representing green, and one representing blue. These values each range from 0 to 255, where 0 indicates that none of that color is present in that pixel and 255 indicates a maximum amount of that color. When a digital image is magnified many times, the pixels can be seen clearly. If only part of a magnified image is visible, it may look like nothing more than different colored squares.

Estimation is a key concept in digital image compression processes. The goal of digital image compression is to reduce the size of the image file (so that it can be efficiently stored and shared) without losing so much quality that the human eye will easily notice the change. The main difference between image formats is the way that they compress images. The graphics interchange format (GIF) and joint photographic experts group (JPEG or JPG) formats—two of the most common digital image formats—are good examples of the effects of the various image compression techniques.

GIF images only support 256 colors—not much compared to the millions of colors found in most color photographs. If an image is converted to the GIF format, a compression technique called dithering is used to compensate for any loss of color. Image dithering involves repeating a pattern of two or more available colors in order to trick the eye into seeing a color very close to the color found in the original photograph. For example, to represent a solid area of a shade of red that is not included in the available 256 colors, the dithering process may alternate every other pixel with the two closest available shades of red. As the image is magnified, the pattern becomes more and more apparent. The color patterns that result from the dithering process are determined by mathematical functions that perform operations for estimating unavailable colors. The GIF format is best suited for illustrations and graphics with large regions of solid color. On the other hand, when a photograph containing millions of colors is converted to GIF, it usually appears grainy because there are too many colors to be adequately represented by just 256 colors.

JPEG images are much better suited for photographs. The JPEG format supports millions of colors and its compression method is intended to handle quick changes in color from pixel to pixel. However, graphics with relatively large areas of solid color converted to JPEG images tend to display messy spots around the areas where colors change. For example, if a company logo consisting only of a blue word on a solid red background is saved as a JPEG image, it will most likely have fuzzy areas all around the border of the letters. These fuzzy areas are called compression artifacts and, as implied by their name, are results of the compression process. As with the dithering process of the GIF format, the compression process is heavily dependent on mathematical functions that attempt to reduce the file size while retaining the image as seen through the human eye.

CARBON DATING

One of the most influential concepts in the field of archeology is that of carbon-14 dating, which allows archeologists to estimate the age of fossils and human artifacts. The basic idea behind carbon-14 dating is that all living things, from plants to humans, contain the same ratio of carbon-14 and carbon-12 atoms at all times (for every carbon-14 atom, there are a certain number of carbon-12 atoms). In a living organism, both types of atoms are constantly being created and destroyed, but the ratio between the two remains constant.

As soon as a living organism dies, it stops producing new carbon atoms. The carbon-14 decays and is no longer replaced, while the carbon-12 does not decay at all. By comparing the ratio of carbon-14 to carbon-12 in a formerly living organism to that in a living organism, it is possible to estimate how long the former has been dead. This concept has allowed archeologists to uncover many important milestones in the history of humankind.

Potential Applications

THE HUBBLE SPACE TELESCOPE

The Hubble Space Telescope (HST), a high-powered telescope attached to a spacecraft, has revolutionized astronomy by allowing astronomers to view celestial sights that are billions of light years away. Due to the fact that the images are captured from billions of light years away, astronomers know that the events depicted took place billions of years ago! Breathtaking images that have been constructed using data transmitted from the HST can be found on the Internet, in books, magazines, and newspapers.

However, these images are not exact representations of what is truly out there. The HST is capable of detecting different types of light and heat, including visible light (that humans can see), ultraviolet light, and infrared light. The raw data transmitted by the HST are electronic black-and-white images that reveal very little detail. Astronomers must combine the data from the various images (created from the different types of light and heat) and interpret the overall picture. These interpretations require advanced estimation methods, as well as some imagination and creativity. Since its launch in 1990, the HST has undergone many revisions, including updates to its image-capturing tools. As space telescope technology is refined, astronomers are able to construct increasingly accurate representations of celestial activity, providing valuable insight into the vastness of the universe.

SOFTWARE DEVELOPMENT

Software developers strive to estimate the amount of time that it will take to complete a software development cycle. A single development cycle often involves a vast number of steps that can take anywhere from a few weeks to a few years. All of these steps must be accounted for in the development plan to ensure that the software is completed, tested, and revised in a specified amount of time. If the product is not ready on time, the software company may lose clients and funding. Some of the major steps in the development cycle include conception (coming up with initial ideas), planning (organizing ideas and time), design (working out the look and feel of the on-screen display and general functionality), coding (using a programming language to write the software), and testing (checking to make sure that things look and work correctly). All of these steps are made up of multiple smaller tasks. For example, design might be split into visual design and functional design. Visual design might be split into window design, menu design, and so on. If any part of testing fails, issues must be listed and categorized by severity. The development team must then go back to the development cycle. How far back in the cycle the development team must go depends on the issues found by the testing team.

For decades, people have tried to conceive a universally accepted method for estimating the time it will take to complete a software product. However, software companies continue to run into unforeseen snags in the process, causing them to miss deadlines. Compensating for less tangible aspects of the development process proves to be a difficult task. For example, the complexity of the project (the number of pages, the types of tasks the software performs, how information is processed, etc.) is a consistent source of error.

In spite of past limitations to software development strategies, the desire to streamline the development processes continues to grow. This is due to a steady increase in the demand for software products, a trend that is not expected to change in the near future.

Where to Learn More

Books

Klette, R., and A. Rosenfeld. *Digital Geometry—Geometric Methods for Digital Picture Analysis.* San Francisco: Morgan Kaufmann, 2004.

Periodicals

Lewis, A. P. "Large Limits to Software Estimation." *ACM Software Engineering Notes.* 26, No. 4 (2001): 54–59.

Web sites

Calkins, Keith G. "The How and Why of Statistical Sampling." Andrews University. October 4, 2004. <http://www.andrews.edu/~calkins/math/webtexts/stattoc.htm> (March 8, 2005).

U.S. Census Bureau. "About Population Projections." August 2, 2002. <http://www.census.gov/population/www/projections/aboutproj.html> (March 9, 2005).

An exponent is a number placed just above and to the right of another number to say how many times the lower number should be multiplied by itself. For example, $2^3 = 2 \times 2 \times 2$, where the exponent is 3.

We can handle some very large and very small numbers easily using exponents. For example, instead of 100,000,000,000,000,000,000 can write 10^{20}. Equations that contain a variable as an exponent, such as $y = 5^x$, are known as exponential equations. They are used to describe the breakdown of radioactive atoms, the growth of living populations, the interest paid on loans, the cooling of planets and other objects, the spreading of epidemic diseases, and many other situations.

Exponents

Fundamental Mathematical Concepts and Terms

BASES AND EXPONENTS

The expression 2^3 is read as "two to the third power" or "two to the power of three." Here 3 is the exponent and 2 is the "base." In 5^6, the exponent is 6 and the base is 5.

INTEGER EXPONENTS

An integer is a whole number, like 3, 0, or -12. When a positive integer like 3 is used an exponent, it tells us to take the base and multiply it by itself. Thus, for example, $10^4 = 10 \times 10 \times 10 \times 10 = 10,000$. (Notice that when 10 is the base, the exponent gives the number of zeroes in the product.) For any number a and any positive integer n,

$$a^n = \overbrace{a \times a \times a \cdots a}^{n \text{ times}}$$

For any two positive integers, which we can call m and n, $a^m a^n = a^{m+n}$. For example, if the base is $a = 10$ and the exponents are $m = 2$ and $n = 3$, then

$$10^2 10^3 = \overbrace{10 \times 10}^{2 \text{ times}} \overbrace{10 \times 10 \times 10}^{3 \text{ times}}$$

$$= \overbrace{10 \times 10 \times 10 \times 10 \times 10}^{2+3 \text{ times}}$$

$$= 10^{2+3} = 10^5$$

Several other useful rules apply to integer exponents such as that $(a^m)^n = a^{an}$ or that $(ab)^m = (a)^m(b)^m$. Here are examples of these rules in action:

- $(a^m)^n = a^{mn}$ means that $(10^2)^3 = 10^2 10^2 10^2$
$$= 10^{2+2+2}$$
$$= 10^{2 \times 3} = 10^6$$

- $(ab)^m = a^m b^m$ means that $(3 \times 10)^2 = 3^2 10^2$

As for negative integer exponents, they also have a simple meaning:

$$a^{-n} = \frac{1}{a^n} = \frac{1}{\underbrace{a \times a \times a \cdots a}_{n \text{ times}}}$$

What about using 0, which is neither positive nor negative, as an exponent? By definition, $a^0 = 1$ for any number a other than 0 itself. For example, $1^0 = 1$, $-10^0 = 1$, and $1,000,000^0 = 1$. But this doesn't work for 0^0. Raising 0 to the power of 0, like dividing by 0, is what mathematicians call "undefined"—it has no meaning. You might want to try raising 0 to the power of 0 (or dividing anything by 0) on your calculator, and see what happens.

NON-INTEGER EXPONENTS

So much for integer exponents. But how do we handle an expression with a fractional exponent, like $2^{1/3}$? We can't multiply 2 by itself one-third of a time! Therefore, we expand our definition of exponent to include rational numbers, that is, all numbers that can be written as fractions, such as 1/3. The rational numbers include the integers, because we can always write an integer as a fraction by putting a 1 in the denominator: 56 = 56 / 1. Any number in decimal form, such as 5.34, can also be written as a fraction:

$$5.34 = 5 + \frac{3}{10} + \frac{4}{100} = \frac{534}{100}$$

Let's start with rational numbers of the form

$$\frac{1}{n}$$

where is a positive integer. For two positive numbers a and b, $b = a^{1/n}$ means that $b^n = a$. For example, $3 = 9^{1/2}$ means that $3^2 = 9$, and $5 = 25^{1/2}$ means that $5^2 = 25$. When $b = a^{1/2}$, as in these two examples, we say that b is the "square root" of a; so 3 is the square root of 9, and

5 is the square root of 25. Taking the "square" of b (raising b to the power of 2) gives a back again: $3^2 = 9$ and $5^2 = 25$.

When $b = a^{1/3}$ we say that b is the "cube root" of a, meaning that $b \times b \times b = a$. When $b = a^{1/n}$ we say that b is the "nth root" of a, meaning that $b \times b \times b \ldots \times b$ (n times) $= a$.

By combining this rule for $1/n$ exponents with the rule that $a^{mn} = (a^m)^n$, we can see what it means to use rational numbers (fractions) as exponents, as in $a^{m/n}$: namely, $a^{m/n} = (a^m)^{1/n}$. And we already know how to deal with exponents like m and $1/n$ separately. For example,

$$3^{3/2} = (3^3)^{1/2}$$
$$= (3 \times 3 \times 3)^{1/2}$$
$$= 27^{1/2}$$

$27^{1/2}$, the square root of 27, is approximately 5.1961524. To write it down exactly, we would have to write an infinitely long string of digits to the right of the decimal point.

We've been looking at the meaning of rational exponents—exponents that can be expressed as fractions with integers in their numerators and denominators. Any number that can't be represented as a ratio of integers, like π, is termed irrational. Since we can't express an irrational number as a fraction, our method for dealing with rational exponents won't work for irrational exponents. The irrational exponent must be approximated as a rational exponent before it can be evaluated.

EXPONENTIAL FUNCTIONS

A function is a rule that relates numbers to each other. For example, the function $f(x) = 2x$ ("f of x equals 2x") means that for every number x there is another number, $f(x)$, that is related to it by being twice as large.

The exponential function is $f(x) = b^x$, where b is any number other than 1. The function behaves differently depending on whether x is greater than 1 or between 0 and 1. If b is greater than 1—say, $f(x) = 2^x$—then the exponential function behaves as shown in Figure 1.

Figure 1 shows the plot of the exponential function $f(x) = 2^x$. All functions of the form $f(x) = b^x$ with $b > 1$ have this shape, and all equal 1 at $x = 0$. The curve in this figure looks like it touches the x axis at the far left, but the curve never quite gets there, no matter how negative x becomes.

The key features of $f(x) = 2^x$ are its slow decline to the left, like a plane coming in for a landing that never quite touches the runway, and its upward zoom to the right. The curve increases to the right because we are raising 2 to increasingly large exponents : for $x = 2$ we have $f(2) = 2^2 = 4$, for $x = 5$ we have $f(5) = 2^5 = 32$, and so on. The functions tails off toward 0 as x gets more negative because we are raising 2 to increasingly negative exponents:

$$f(-1) = 2^{-1} = \frac{1}{2}, \; f(-5) = 2^{-5} = \frac{1}{2^5} = \frac{1}{32}$$

and so on.

If b is between 0 and 1, the exponential function $f(x) = b^x$ behaves as shown in Figure 2.

All functions $f(x) = b^x$ with $0 < b < 1$ have a similar shape, and all equal 1 at $x = 0$.

What do we do with negative exponents in an exponential equation, $f(x) = b^{-x}$? This can be rewritten using the rule that $(a^m)^n = a^{mn}$ $f(x) = b^{-x} = b^{(-1)(x)} = (b^{-1})^x$. Since

$$b^{-1} = \frac{1}{b}$$

using a negative exponent is the same thing as using a positive exponent with the base flipped upside down:

$$b^{-x} = \left(\frac{1}{b}\right)^x$$

For example,

$$2^{-x} = \left(\frac{1}{2}\right)^x$$

The rule $(a^m)^n = a^{mn}$ also tells us how to deal with numbers that multiply the exponent, as in $f(x) = b^{ax}$. We can always rewrite the function so that the exponent is plain old x: $f(x) = b^{ax} = (b^a)^x$.

Table 1 presents a summary of the laws of exponents.

The concept of the exponent boils down to repeated multiplication: take a number b, multiply it by itself, multiply that result by b, multiply that result by b, and so on. People began to play with this concept—geometric progression, as it is also called—very early in history.

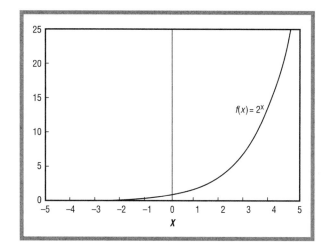

Figure 1: Plot of the exponential function $f(x) = 2^x$.

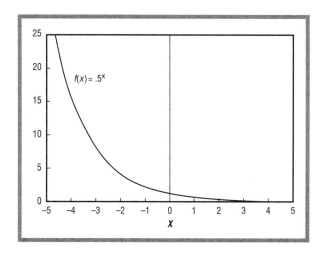

Figure 2: Plot of the exponential function $f(x) = (1/2)^x$.

Geometric progression was studied over 3,600 years ago by the ancient Egyptians and Sumerians and much later by the Greeks, including Euclid (c. 300 B.C.) and Archimedes (287?–212 B.C.).

A Brief History of Discovery and Development

Especially after the Middle Ages, scientists became aware of many real-life objects that behave in a geometric

Rule	Meaning	Example
$a^n a^m = a^{n+m}$	Multiplying two exponential terms having the same base is the same as raising that base to the product of the two exponents.	$2^2 2^3 = 2^5 = 32$
$\dfrac{a^n}{a^m} = a^{n-m}$	Dividing two exponential terms having the same base is the same as raising that base to the difference of the exponents.	$\dfrac{2^3}{2^2} = 2^{3-2} = 2^1 = 2$
$\left(a^m\right)^n = a^{mn}$	Applying an exponent to a base that is already raised to an exponent is the same as raising that base to the product of the two exponents.	$\left(2^3\right)^2 = 2^{3 \times 2} = 2^6 = 64$
$\left(ab\right)^n = a^n b^n$	Raising the product of two numbers to an exponent is the same as raising each number separately to that exponent and then multiplying.	$(3 \times 5)^2 = 3^2 5^2 = 9 \times 25 = 225$
$\left(\dfrac{a}{b}\right)^n = \dfrac{a^n}{b^n}$	Raising a fraction to an exponent is the same as raising the numerator and denominator separately to that exponent and then dividing.	$\left(\dfrac{3}{5}\right)^2 = \dfrac{3^2}{5^2} = \dfrac{9}{25}$

Table 1: Summary of the laws of exponents.

or exponential way, including the unrestrained breeding of animals; the cooling of hot objects; the shapes of natural spirals such as those found in pine cones, sunflowers, and ram's horns; the dimming of supernovae (exploding stars); the relationships between musical notes; and many others. Ancient records inscribed in clay show that in the Middle East, the Sumerians knew about the exponential properties of compound interest as early as 2000 B.C.

Our modern way of writing an exponent-placing a small number above and to the right of another number—was introduced in 1637 by the French philosopher and mathematician René Descartes (1596–1650). At abut that

time the relationship between the logarithm and the exponent (namely, that they are inverses of each other) was finally clarified.

In the eighteenth century, Swiss mathematician Leonhard Euler (1701–1783) first devised the complex exponential function, where a base is raised to the power of an "imaginary" number containing the square root of -1. The square root of -1 was a radical new idea because it seemed impossible: what number, when multiplied by itself, could give -1? The square root of $+1$ is simply itself (because $1^2 = 1$), but -1 cannot be its own square root (because $-1^2 = -1 \times -1 = +1$). Nevertheless, mathematicians have found numbers containing the square root of −1, called "complex" numbers, to be very useful. Euler also explored the use of the number $e = 2.7182818$ as a "natural" (that is, highly convenient) base for exponents and logarithms.

Today, exponents are applied throughout mathematics. They are used in physics and engineering to describe phenomena that fade with time, such as radioactivity, or periodic (repeating) phenomena like waves. They are used in biology to model populations of bacteria, animals, and people; in medicine to model the breakdown of drugs in the body; and in business and economics to describe interest and inflation.

Real-life Applications

SCIENTIFIC NOTATION

With scientific notation you can write down a number greater than the number of atoms in the universe with just a few strokes of your pen. This is how:

Recall that for powers of ten, the exponent gives the number of zeroes: $100 = 10^2$, $1000 = 10^3$, and so forth. Using this fact, an ugly number like 1,000,000,000,000, 000,000 becomes a user-friendly 10^{18}. We can also write a number like 1,414,000,000,000,000,000 as 1.414 times 10^{18}, namely, 1.414×10^{18}. A number written in this form is said to be in "scientific notation." Scientific notation makes large numbers much easier to handle.

The same trick also works for small numbers, because numbers like .1, .01, .001, and so forth can be written as tens raised to negative exponents; for example,

$$.01 = \frac{1}{100} = \frac{1}{10^2} = 10^{-2}$$

We can therefore write 10^{-20} instead of .0000000000 0000000001, and 1.675×10^{-24} instead of .0000000000 00000000000001675 (which happens to be the mass in grams of a single hydrogen atom).

Another useful feature of scientific notation is that changing the exponent is shorthand for moving the decimal point. Thus we write 4.5×10^{-2} for .045 and 4.5×10^{-4} for .00045.

To multiply two numbers written in scientific notation, all we have to do is multiply the numbers out front and add the exponents. So, 1.414×10^{18} times 1.675×10^{-24} is $1.414 \times 1.675 \times 10^{18-24} = 2.36845 \times 10^{-6}$. This is so easy that when multiplying simple numbers like 1×10^{12} and 5×10^9, it's actually easier to do the math in your head than to punch buttons on a calculator provided you can add 12 and 9 in your head without blowing a fuse. (The answer is $(1 \times 10^{12}) \times (5 \times 10^9) = (1 \times 5) \times 10^{12+9} = 5 \times 10^{21}$.) Division is equally easy, only you divide the numbers out front and subtract the exponents.

As for writing down a number larger than the number of atoms in the universe, this is it: Physicists estimate that there are fewer than 10^{100} atoms in the Universe, perhaps only about 10^{80}. If every atom in the Universe were inflated into a universe full of atoms, there would still be only $(10^{80})^2$ or 10^{160} atoms in existence. So writing 10^{160}, or 10^{200}, or 10^{300} gives a number much, much greater than the number of atoms in the Universe.

EXPONENTIAL GROWTH

A useful fact in science (and banking) is this: Any quantity that grows by a fixed percentage during each interval of time grows exponentially.

Consider a pair of rabbits. Say that this pair has two offspring by the end of one year. There are now four rabbits: the population has doubled in one year. Assume that both pairs will breed the following year, each producing two more offspring, and that their offspring will also breed, and so on, so that the total population keeps on doubling every year. This is the same as saying that the population increases by a fixed percentage every year, namely 100%: 2 rabbits plus 100% of 2 rabbits equals 4 rabbits (first year's population growth), 4 rabbits plus 100% of 4 rabbits equals 8 rabbits (second year's growth), and so forth. The growth of this rabbit population is described by an exponential equation. If we label the first year Year 0, then the number of rabbits at the beginning of each year, $R(t)$. is given by the following series of numbers: $R(0) = 2$, $R(1) = 4$, $R(2) = 8$, $R(3) = 16$, and so on. In general, $2R(t) = 2 \times 2^t$, where t is years.

Figure 3 shows the number of rabbits as an exponential function of time, starting with two rabbits and assuming a doubling time of 1 year. This curve is similar to the part of Figure 1 to the right of $x = 0$.

Do rabbits really do this? Sure—when they can; that is, if they have food to eat, room to live in, and air to breathe, and no enemies or diseases nasty enough to keep the population from growing. In reality, rabbits cannot have all these things, certainly not in infinite amounts. There are predators and diseases; there is only so much food, water, and room. So in one sense the exponential equation is not realistic—but in another sense, it is terribly realistic. It forces conflict between growing populations and their environments. If a population seeks to increase by any fixed percentage per year (that is, if it seeks to grow exponentially), then at some point

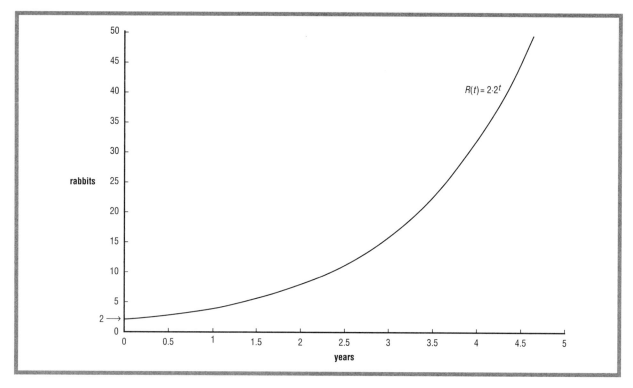

Figure 3: An exponential function.

deaths—whether from hunger or other causes—will inevitably outpace births. To see why this cannot be avoided, let's use the master rabbit equation, $R(t) = 2 \times 2^t$, to look at what would happen if our imaginary rabbit population was somehow, magically, able to keep on growing.

Let's calculate how long it takes to get a given number of rabbits, N. To do so, we find the "solution" to the equation $2 \times 2^t = N$, that is, that unique value of t for which the equation is true. Let's pick a nice, big value for N—say, enough rabbits to completely fill the Solar System. Pluto, usually regarded as the outermost planet, has an average distance from the Sun of 5.914×10^{12} km. Because the volume of a sphere of radius r is

$$\frac{4}{3}\pi r^3$$

(exponents again!), the volume of the Solar System is

$$\frac{4}{3}\pi(5.914 \times 10^{12})^3\,m^3 = \frac{4}{3}\pi 5.914 \times 10^{36}\,m^3$$

where 1 m^3 is a cubic meter (the amount of space in a cube 1 meter across). If we can pack 50 rabbits into each cubic meter of space, then the number of rabbits that can fit into the Solar System is

$$N = \underbrace{\left(\frac{4}{3}\pi 5.914 \times 10^{36}\,m^3\right)}_{\substack{\text{number of cubic meters} \\ \text{in Solar System}}} \times \underbrace{\left(\frac{50\,\text{rabbits}}{m^3}\right)}_{\substack{\text{number of rabbits} \\ \text{per cubic meter}}}$$

$$= \frac{4}{3}\pi 2.957 \times 10^{38}\,\text{rabbits to fill Solar System}$$

Because the number of rabbits at the start of year t is $N = 2 \times 2^t$, to find out the number of years till there are

$$N = \frac{4}{3}\pi 2.957 \times 10^{38}$$

rabbits we need to find t such that

$$N = \frac{4}{3}\pi 2.957 \times 10^{38}$$

To find the t that satisfies this equation, we must perform the mathematical operation known as "taking the logarithm" of both sides. Taking the logarithm undoes

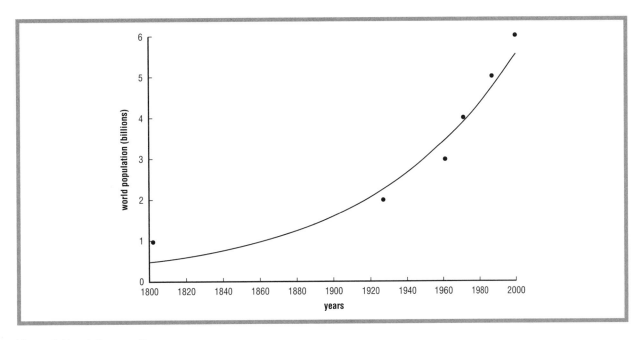

Figure 4: Population growth.

exponentiation (applying exponents) in much the same way that subtraction undoes addition or division undoes multiplication.

Solving using logarithms, we find that t equals approximately 129 years. This means that a rabbit population that doubled every year would fill the whole Solar System with long-eared rodents in only 129 years. It would take the first 128 of those years to fill half the Solar System, but just one more year to fill the other half!

A Solar System full of rabbits is, of course, physically impossible. The moral is that exponential population growth always runs up against physical limits, most often getting eaten or starving to death.

The equation $R(t) = 2 \times 2^t$ is an example of the general exponential equation $R(t) = R_0 b^t$, where b is some positive number and R_0 is the value of $R(t)$ at $t = 0$ (because $R(0) = R_0 \times b^0 = R_0 \times 1 = R_0$). We'll use this form of the exponential equation in the following application also.

ROTTING LEFTOVERS

Any quantity—say a population of rabbits, or of people—that grows by a fixed percentage each year, no matter how small, will double over some period of time. It will then double again after an equal period of time, and so on forever. Every exponentially growing quantity grows in this way, so every exponentially growing quantity is said to have a "doubling time."

Suppose you leave a dish of food out at midnight. The dish happens to have 10 bacteria sitting on it. Suppose also that this population of bacteria increases by 4% every minute. How long will it be before the number of bacteria in the dish doubles? And how many bacteria will you be consuming when you finish off the leftovers at noon the next day?

If the population is growing by 4% every minute, then it is growing exponentially, and can be described by an exponential equation of the form $R(t) = R_0 b^t$, just like the rabbit population in the previous example. We already know R_0, the number of bacteria at time $t = 0$; it's 10. But what is b?

Besides the fact that $R_0 = 10$, we also know that the bacterial population at the end of 1 minute, $R(1)$, is 4% greater than at $t = 0$, because we're told that the population is growing by 4% every minute. This fact can be written down as $R(1) = 10 + (.04 \times 10) = 10.4$.

We also know that $R(1)$ must be given by the exponential equation $R(t) = R_0 b^t$ with 10 plugged in for R_0 and 1 plugged in for t, namely, $R(1) = 10b^1 = 10b$. We can now set this expression for $R(1)$ equal to the number found in the previous paragraph: $10b = 10.04$.

Dividing both sides by 10 to solve for b, we find that $b = 1.004$.

We now have both R_0 and b, and so can write down the exact exponential equation that describes this bacterial population: $R(t) = R_0 b^t = 10 \times (10.004^t)$.

Population Growth

The human race has inhabited Earth for about 3 million years. For much of that time, the world's population was constant at about 10 million people. Life was difficult; most babies died, and people reached old age and usually died by 30. For food, people harvested wild plants and hunted animals.

With the invention of farming and cities about 10,000 years ago, larger local populations became possible. During the first century A.D., some 2,000 years ago, the world's population had grown to about 300 million people. Around the year 1600, as modern science and technology started to come into being, and population began to grow faster. By 1800, there were about 1 billion people on Earth. It took about 3 million years for the world to gain its first billion people, and only 130 years to gain its second billion. Today, there are over 6 billion people living on Earth.

The most common mathematical model for population growth is the exponential function, $Q(t) = {_0}k^t$ (see main text). As the population figure shows the approximate recent exponential growth of world population, the data becomes visible. Dots are actual world population at 1, 2, 3, 4, 5, and 6 billion; the smooth curve is exponential function $2.3236 \times 10^{-10} \times 1.0124^t$, where t is in years A.D.

The world's population will continue to grow for the near future. But it is physically impossible for the Earth's population to continue to grow exponentially, as there is a finite amount of space and potential for growing food.

We can now return to our first question: What is the doubling time? Let us call that unknown number of minutes T_D. Because we start out with 10 bacteria, the number of bacteria after the first doubling time will be 20 (double). So the population at time T_D is given by $R(T_D) = 20 = 10 \times 1.004_{TD}$. To solve for T_D, we must "take the logarithm" of both sides of this equation. Taking the logarithm undoes the exponential operation much the way that subtraction undoes addition or division undoes multiplication (see chapter on Logarithms). We find that $T_D = 173.63$ minutes (about 2 hrs 54 minutes).

The equation $R(t) = 10 \times (10.004^t)$ also tells us how many bacteria you'll be eating at noon the next day. All we need to know is the number of minutes between midnight and noon, which is 12 hours times 60 minutes per hour, or 720 minutes. Thus, $R(720) = 10 \times 1.004^{720}$, or about 177 bacteria.

Not bad, really. As you read this, your intestines contain trillions of bacteria. But remember the power of exponential growth. After 24 hours there will be 3,137 bacteria; after 48 hours, 984,205 bacteria; and after 3 days, 3.088×10^8 bacteria, about as many people as there are in the United States. Look out for a stomach upset—or worse.

EXPONENTS AND EVOLUTION

Predators and sickness often keep populations from growing exponentially, but if they don't there is one thing that certainly will: hunger. There can never be an infinite food supply—even if you could somehow turn the whole Earth into a ball of food, it would be limited. Therefore, sooner or later, any exponentially growing population must run out of food and either stop breeding or start starving.

This principle was first clearly explained English economist and minister Thomas Robert Malthus (1766–1834). In his 1798 book, *An Essay on the Principle of Population*, Malthus wrote: "Population, when unchecked, increases in a geometrical ratio [exponentially]. Subsistence [that is, food supply] increases only in an arithmetical ratio [like a straight line]. A slight acquaintance with numbers will show the immensity of the first power in comparison of the second. . . . This implies a strong and constantly operating check on population from the difficulty of subsistence."

By "a slight acquaintance with numbers" Malthus meant a knowledge of exponents. Population increases exponentially ("in a geometrical ratio") whenever it can, rising in a curve that gets ever steeper; but food supply increases (if it increases at all, say by the clearing and planting if more cropland) approximately as a straight line, that is, according to a "linear" function or "arithmetical ratio." And any exponential function will eventually outrun any linear function. Accordingly, any freely-breeding population must eventually outrun its food supply.

Malthus was talking about the human population, but his logic applies to any biological population. Two English biologists, Charles Darwin (1809–1882) and Alfred Russel Wallace (1823–1913)—realized this when they read Malthus's book in the mid-1800s. Both Darwin and Wallace were trying to think of a mechanism to explain biological evolution, the process whereby new species of animals and plants arise from older ones. People had been suggesting theories of evolution for years, but none of them could explain why evolution happened. However, Malthus's reasoning about population growth

Population Growth

Think of how a rumor spreads. Somebody starts a rumor by telling everyone they know, then those people tell everyone they know, then those people tell everybody they know, and so on.

In a finite environment, such as a high school (or the planet Earth, for that matter), there are a limited number of people to tell. Soon, people who have heard the rumor are having trouble finding people who haven't heard it yet. What happens then?

The function that describes the growing number of people who have heard a rumor is called the logistic curve.

The horizontal axis here is time, the vertical axis the fraction of the school that knows the rumor; let's call it the hip fraction. The curve starts out at time zero at some nonzero number, namely, the fraction of the school population that knows the rumor to begin with. As time goes on, the hip fraction approaches 1; everybody knows the rumor. Using calculus, we can show that the derivative of the logistic curve always has a maximum where

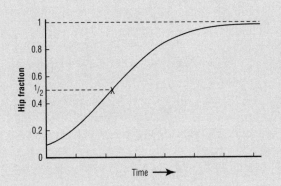

Logistics curve.

the hip fraction equals .5 (marked X on the curve). That is, the rate of change of the hip fraction decreases after that time. A rumor, therefore, spreads more slowly once it has been heard by half the people in a group.

Because germs, like rumors, spread by contact, mathematicians also use the logistic curve to describe the spread of a disease in a finite population.

triggered a fresh insight for both Wallace and Darwin. Working separately, they realized that the potential of every species for exponential population growth guaranteed struggle between organisms. In biology, "struggle" usually means not fighting, but competition to leave more offspring than one's rivals. In an article published jointly with Wallace in 1858, Darwin said, echoing Malthus: "[T]he amount of food for each species must, on an average, be constant, whereas the increase of all organisms tends to be geometrical, and in a vast majority of cases at an enormous ratio .... Now, can it be doubted, from the struggle each individual has to obtain subsistence, that any minute variation in structure, habits, or instincts, adapting that individual better to the new conditions, would tell upon its vigour and health? In the struggle it would have a better chance of surviving; and those of its offspring which inherited the variation, be it ever so slight, would also have a better chance.... Let this work of selection on the one hand, and death on the other, go on for a thousand generations, who will pretend to affirm that it would produce no effect ..."

Wallace and Darwin's insight was that competition (made inevitable by exponential population growth) is more than just a check or limit on population: it forces

nature to filter the chance changes that show up in every generation of creatures and so acts as a creative force, helping sculpt such marvels as the gull's wing, the eagle's eye, and the human brain.

RADIOACTIVE DECAY

In Nevada, about 90 miles (145 km) northwest of Las Vegas, stands an unremarkable-looking ridge of dry, brown rock, owned by the Federal government and known as Yucca Mountain. This is where the United States government hopes to bury 77,000 tons (69,853 tonnes) of highly radioactive nuclear waste from around the U.S. (about 60% of the total amount that had built up as of 2004).

This waste is what is left over when a nuclear power plant has finished using uranium fuel to produce electricity. Plans call for it to be mixed with molten glass and cooled to a solid ("vitrified"), sealed inside rust-resistant metal containers, and parked along 73 miles (117 km) of branching tunnels located 1,000 feet (305 m) below the surface of Yucca Mountain. When the tunnels are full, they will be sealed off and hopefully not entered again—especially by water, which might carry the waste back to the surface—for at least 10,000 years.

A common example of exponential decay is radioactive decay. A concrete sarcophagus covered the damaged nuclear reactor No. 4 at the power plant in Chernobyl, Ukraine, following the 1986 nuclear accident. A/P WIDE WORLD. REPRODUCED BY PERMISSION.

Why does the waste need to be put in such a special place at all? And why for as long as 10,000 years—or for only 10,000 years? Why not forever?

The Reason is Radioactive Decay A "radioactive" substance is one whose individual atoms break apart (fission) at random (chance) times, releasing energy. This energy takes the form of fast-moving particles or invisible rays that can cause cancer or other sickness. Some radioactive atoms are mixed naturally with the environment, whereas some are human-made. Regardless of where they come from, the fewer radioactive atoms we come in contact with, the better for our health. (Some medical tests and treatments do use radioactive substances, however, where the gain is thought to be larger than the risk.)

Radioactive substances disappear over time as their atoms change into other types of atoms. This change in atoms is called "radioactive decay." Like the curve in Figure 2, radioactive decay can be described by an exponential

equation with a base between 0 and 1. Someday, therefore, all the nuclear waste generated today will be harmless, but that date is tens or hundreds of thousands of years in the future.

Just as every exponentially increasing process has a doubling time, so every exponentially decaying process has a halving time. In the case of a radioactive substance, this halving time is called the substance's "half-life." Half the atoms in a lump of any radioactive substance will have changed into other substances after one half-life of that substance. Different radioactive substances have different half-lives. Half-lives vary from a tiny fraction of a second up to billions of years.

Consider the substance plutonium 238. Plutonium 238 (also written ^{238}Pu) is both poisonous and radioactive. It can be used as fuel for nuclear reactors or to make nuclear bombs. It is one of the ingredients in radioactive waste of the type that may someday be stored beneath Yucca Mountain (perhaps starting in 2010). ^{238}Pu has a half-life of about 25,000 years. If the amount of ^{238}Pu that we start out with at time $t = 0$ is Q_0 tons, then the amount at some later time t will be described by the exponential equation $Q(t) = Q_0 k^t$. Because we know that after the first 25,000 years there will be half as much ^{238}Pu as at time $t = 0$, we can write the following:

$$\frac{Q_0}{2} = Q_0 k^{25,000}$$

Because Q_0 appears on both sides of the equal sign, it cancels out. We can then solve for k using logarithms. We find that $k = .999972$. The radioactive decay of 1 ton of ^{238}Pu is shown in Figure 4.

In Figure 4, the radioactive decay of 1 ton of plutonium 238 (^{238}Pu). Time is shown in units of half-lives. Notice that at $t = 1$ (one half-life) the amount of ^{238}Pu is down to

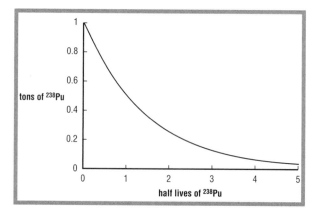

Figure 4.

1/2 ton. To read the time axis in units of years, multiply by 25,000. Note that this curve is exactly the same (except for vertical scale) as the part of Figure 2 to the right of $x = 0$.

As a rule of thumb, experts often say that we should wait at least 10 half-lives of a radioactive substance before considering that a chunk of it is harmless. In the case of plutonium, 10 half-lives are about 250,000 years. How much of an original quantity of any radioactive substance, say 1 ton, is left after 10 half lives? Since waiting one half-life cuts the amount in half, and waiting two half-lives cuts that amount in half, and so on, we can use exponents. A little thought shows that the fraction that is left after ten half-lives is

$$\left(\frac{1}{2}\right)^{10} = .0009766$$

That is, after 10 half-lives, only .0009766 tons (about one one-thousandth of the original ton) will be left.

The wastes intended for Yucca Mountain contain many radioactive substances besides plutonium. Many of these have much shorter half-lives than plutonium, so in about 300 years 99% of the radioactivity of the nuclear waste will have disappeared. After 10,000 years, the amount of time that the Yucca Mountain storage tunnels are supposed to be guaranteed for, these shorter-lived substances will be essentially gone. On the other hand, after 10,000 years only about one-fourth of the plutonium will be gone.

RADIOACTIVE DATING

"Radioactive dating" is an essential technique in geology and archaeology. By looking at the amounts of radioactive substances embedded in rocks or other objects and at the amounts of breakdown products (substances left over from radioactive decay) that are mixed with them, scientists can tell how much radioactive material must have been originally present in the object—and thus, by the exponential equation, how old the object is. For example, if a rock contains 1 gram of radioactive uranium 238 (^{238}U) and 1 gram of lead, which is a breakdown product, it is probable that the rock originally contained 2 grams of ^{238}U. Since ^{238}U has a half-life of about 4.5 billion years, it takes about 4.5 billion years for 2 grams to decay to 1 gram, so we deduce that this particular rock is about 4.5 billion years old.

In practice, breakdown products and radioactive dating are more complex than this. Scientists must look at many different samples of rock (or wood, or whatever material they are dating) and at a number of different radioactive substances and breakdown products, rather than just one. But by combining methods and measuring many different objects, error can be minimized. Through

radioactive dating, scientists have verified that the Earth, the Moon, and most meteorites are about 4.5 billion years old. That is, about half the ^{238}U that was present when the Solar System formed has turned into other elements. Some of it, in fact, has ended up in your car battery.

INTEREST AND INFLATION

Let's say that you earn $100 at a weekend job. Your parents insist that you put it in a nice, safe bank until you're 18. They try to comfort you with the idea that your money will earn interest, so you will end up with more money if you wait. How much more? And what exactly does it mean to "earn interest," anyway?

"Interest" is money that is paid to you by a bank in which you have deposited money. The bank invests the money in enterprises that it thinks will be worth more in the future. Banks make profit by taking in more money on their investments than they pay in interest to their depositors (that's you), but regardless of how well a bank's investments are doing, it is obliged to pay you the agreed-upon interest.

To pay interest, the bank looks at the money in your account at regular intervals, say every three months. It then calculates a fixed percentage of that amount (your interest) and adds this money to your account. (At the words "regular intervals" and "fixed percentage" your ears should prick up: "Regular intervals? Fixed percentage? My money will grow exponentially?" Yes, but wait.) The percentage used to calculate your interest is called the "interest rate."

After another three months (or whatever the interval happens to be), the bank calculates the fixed percentage again and adds it to your account. The interest from the previous time interval—also called a "conversion period"—earns interest during the next time interval, assuming that you haven't taken any money out. This arrangement, where interest earns interest, is called "compound" interest.

Let's go back to your $100. Assume the conversion period is three months (which is one quarter of a year, so it's also called a "quarter"). You get a quarterly interest rate of 1.5%, so the end of the first quarter, the bank adds 1.5% of $100 to your account, namely, $1.50. Your account now contains $101.50. At the end of the second quarter, the bank calculates 1.5% of $101.50, which is $1.52 (rounding down), and it adds that to your account. Your account now contains $103.02. Notice that the amount of interest you receive at the end of the second quarter is larger than the interest you receive at the end of the first quarter. The reason is that you've begun to earn interest on your interest.

Not surprisingly, this is an exponential process. Its equation is $S(n) = P(1 + r)^n$. Here $S(n)$ is the amount of money in your account after n quarters, P is your principal (the money you start off with, in this case $100), and r is the quarterly interest rate (1.5%, in this case). Since time, n, is in the exponent, this is an exponential function. Putting in our numbers for P, r, and n, we find that $S(n) = 100 (1 + .015)^n = 100 \times 1.015^n$.

For the end of the second quarter, $n = 2$, this gives the result already calculated: $S(2) = \$103.02$.

This equation for $S(n)$ should look familiar. It has the same form as the equation for a growing population, $R(t) = R_0 b^t$, with R_0 set equal to $100 and b set equal to 1.015.

If $100 is put in the bank when you're 14, then by the time you're 18, four years or 16 quarters later, it will have grown exponentially to $100 \times 1.015^{16} = \$126.90$ (rounded up). If you had invested $1,000, it will have grown to $1,268.99. That's lovely, but meanwhile there's inflation, which is exponentially making money worth less over time.

Inflation occurs when the value of money goes down, so that a dollar buys less. As long as we all get paid more dollars for our labor (higher wages), we can afford the higher prices, so inflation is not necessarily harmful. Inflation is approximately exponential. For the decade from 1992 to 2003, for example, inflation was usually around 2.5% per year. This is lower than the 6% per year interest rate we've assumed for your invested money, so your $100 of principal will actually gain buying power against 2.5% annual inflation, but not as quickly as the raw dollar figures seem to show: after four years, you'll have 26% more dollars than you started with ($126.90 versus $100), but prices will be 10.4% higher (i.e., something that cost $100 when you were 14 will cost about $110 when you are 18).

Furthermore, 6% is a rather high rate for a savings account: during the last decade or so, interest rates for savings accounts have actually tended to be lower than inflation, so that people who keep their money in interest-bearing savings accounts have actually been losing money! This is one reason why many people invest their money in the stock market, where it can keep ahead of inflation. The dark side of this solution is that the stock market is a form of gambling: money invested in stocks can shrink even faster than money in a savings account, or disappear completely. And sometimes it does.

CREDIT CARD MELTDOWN

When you deposit money in a bank, the bank is essentially borrowing your money, and pays you interest for the privilege of doing so. When you borrow money from a bank, you pay the bank interest, so if you don't pay off your debt, it can grow exponentially. Exponential interest

growth is why credit-card debt is dangerous. A credit-card interest rate, the percentage rate at which the amount you owe increases per unit time, is much higher than anything a bank will pay to you. (Fifteen percent would be typical, and if you make a late payment you can be slapped with a "penalty rate" as high as 29%.) So if you only make the minimum monthly payments, your debt climbs at an exponential rate that is faster than that of any investment you can make. This is why you can't make a living by borrowing money on a credit card and investing it in stocks. If you could, the economy would soon collapse, because everyone would start doing it, and an economy cannot run on money games; it needs real goods and services.

Those high credit-card interest rates are also the reason credit-card companies are so eager to give credit cards to young people. They count on younger borrowers to get carried away using their cards and end up owing lots of fat interest payments. And it seems like a good bet. In 2004, the average college undergraduate had over $1,800 in credit-card debt.

The good news is that to avoid high-interest credit-card debt, you need only pay off your credit card in full every month.

THE AMAZING EXPANDING UNIVERSE

The entire Universe is shaped by processes that are described by exponents.

All the stars and galaxies that now speckle our night sky, and all other mass and energy that exists today, were once compressed into a space much smaller than an atom. This super-tiny, super-dense, super-hot object began to expand rapidly, an event that scientists call the Big Bang. The Universe is still growing today, but at different times in its history it has expanded at different speeds. Many physicists believe that for a very short time right after the Big Bang, the size of the Universe grew exponentially, that is, following an equation approximately of the form $R(t) = Ka^t$, where $R(t)$ is the radius of the Universe as a function of time and t and K and a are constants (fixed numbers). This is called the "inflationary Big Bang" theory because the Universe inflated so rapidly during this exponential period. If the inflationary theory is correct, the Universe expanded by a factor of at least 10^{35} in only 10^{-32} seconds, going from much smaller than an electron to about the size of a grapefruit.

This period of exponential growth lasted only a brief time. For most of its 14-billion year history, the Universe's rate of expansion has been more or less proportional to time raised to the 2/3 power, that is, $R(t) = Kt^{2/3}$. Here $R(t)$ is the radius of the universe as a function of time, and K is a fixed number.

Most scientists argue that the Universe will go on expanding forever—and that it's expansion may even be accelerating slowly.

WHY ELEPHANTS DON'T HAVE SKINNY LEGS

The two most common exponents in the real world are 2 and 3. We even have special words to signify their use: raising a number to the power of 2 is called "squaring" it, while raising it to the power of 3 is called "cubing" it. These names reflect the reasons why these numbers are so important. The area of a square that is L meters on a side is given by $A = L^2$, that is, by "squaring" L, while the volume of a cube that is L meters on a side is given by $V = L^3$, that is, by "cubing" L.

These exponents—2 and 3—appear not only in the equations for the areas and volumes of squares and cubes, but for any flat shapes and any solid shapes. For example, the area of a circle with radius L is given by $A = \pi L^2$ and the volume of a sphere with radius L is given by $4/3\ \pi L^3$. The equations for even more complex shapes (say, for the area of the letter "M" or the volume of a Great Dane) would be even more complicated, but would always include these exponents somewhere—2 for area, 3 for volume. We say, therefore, that the area of an object is "proportional to" the square of its size, and that its volume is proportional to the cube of its size.

These facts influence almost everything in the physical world, from the shining of the stars to radio broadcasting to the shapes of animals' legs. The weight of an animal is determined by its volume, since all flesh has about the same density (similar to that of water). If there are two dogs shaped exactly alike, except that one is twice the size of the other, the larger dog is not two times as heavy as the smaller one but 2^3 (eight) times as heavy, because its volume is proportional to the cube of its size. Yet its bones will not be eight times as strong. The strength of a bone depends on its cross-sectional area, that is, the area exposed by a cut right through the bone. The bigger dog's bones will be twice as wide as the small dog's (because the whole dog is twice as big), and area is proportional to the square of size, so the big dog's bones will only be 2^2 (four) times as large in cross section, therefore only four times as strong. To be eight times as strong as the small dog's bones, the big dog's bones would have to be the square root of 8, or about 2.83 times wider.

You can probably see where this is leading. An elephant is much bigger than even a large dog (about ten times taller). Because volume goes by the cube of size, an elephant weights about $10^3 = 10 \times 10 \times 10 = 1000$ times as much as a dog. To have legs that are as strong relative to its weight as a dog's legs are, an elephant has to have leg bones that are the square root of 1,000, or about 31.62 times wider than the dog's. So even though the elephant is only 10 times taller, it needs legs that are almost 32 times thicker. If an elephant's legs were shaped like a dog's, they would snap.

Where to Learn More

Books

Durbin, John R. *College Algebra*. New York: John Wiley & Sons, 1985.

Morrison, Philip, and Phylis Morrison. *Powers of Ten: A Book About the Relative Size of Things in the Universe and the Effect of Adding Another Zero*. San Francisco: Scientific American Library, 1982.

Periodicals

Curtis, Lorenzo. "Concept of the exponential law prior to 1900," *American Journal of Physics* 46(9), Sep. 1978, pp. 896–906 (available at <http://www.physics.utoledo.edu /~ljc /explaw.pdf>.

Wilson, Jim. "Plutonium Peril: Nuclear Waste Storage at Yucca Mountain," *Popular Mechanics*, Jan. 1, 1999.

Web sites

Population Reference Bureau. "Human Population: Fundamentals of Growth: Population Growth and Distribution." <http://www.prb.org/Content/NavigationMenu/PRB/Educators/Human_Population/Population_Growth/Population _Growth.htm> (April 23, 2004).

Factoring

Factoring a number means representing the number as the product of prime numbers. Prime numbers are those numbers that cannot be divided by any smaller number to produce a whole number. For instance, 2, 3, 5, 7, 11, and 13 (among many others) cannot be divided without producing a remainder.

Factoring in its simplest form is the ability to recognize a common characteristic or trait in a group of individuals or numbers which can be used to make a general statement that applies to the group as a whole.

Another way to think of factoring is that every individual in the group shares something in particular. For example, whether someone is from France, Germany, or Austria is irrelevant in the statement that they are European, because all three of these countries share the geographic characteristic of being on the continent of Europe. The factor that can be applied to all three individuals in this particular group is that they are all European. The ability to recognize relationships between individual components is fundamental to mathematics. Factoring in mathematics is one of the most basic but important lessons to learn in preparation for further studies of math.

Fundamental Mathematical Concepts and Terms

A number which can be divided by smaller numbers is referred to as a composite number.

Composites can be written as the product of smaller primes. For example, 30 has smaller prime numbers which can be multiplied together to achieve the product of 30. These numbers are as follows: $2 \times 3 \times 5 = 30$. A number is considered to be factored when all of its prime factors are recognized. Factors are multiplied together to yield a specific product.

It is important to understand a few basic principals in factoring before further discussion can continue on how factoring can be applied to real life. One of the most important studies of mathematics is to study how individual entities relate to one another.

In multiplying factors which contain two terms, each term must be multiplied with each term of the second set of terms. For example, in $(a+b)\,(a+b)$, both the a and b in the first set must be multiplied by the a and b in the second set. The easiest way to accomplish this is by employing the FOIL method. FOIL refers to the order of multiplication: first, outer, inner, and last. First we multiply a by a to yield a^2, then the Outer terms of a and b to yield ab, then the Inner terms of b and a to yield another ab, finally we multiply the Last terms of b and b for b^2. Putting all of these together, we achieve $a^2 + 2ab + b^2$.

Greatest common factor (GCF) refers to two or more integers where the largest integer is a factor of both or all numbers. For example, in 4 and 16, both 2 and 4 are factors that are common to each. However, 4 is greater than 2, so therefore 4 is the greatest common factor. In order to find the greatest common factor, you must first determine whether or not there is a factor that is common to each number. Remember that common factors must divide the two numbers evenly with no remainders. Once a common factor is found, divide both numbers by the common factor and repeat until there are no more common factors. It is then necessary to multiply each common factor together to arrive with the greatest common factor.

Factoring perfect squares is one of the essentials of learning factoring. A perfect square is the square of any whole number. The difference between two perfect squares is the breaking of two perfect squares into their factors. For example $a^2 - b^2$ is referred to as the difference between two perfect squares. The variables a and b refer to any number which is a perfect square. In order to factor $a^2 - b^2$, we must see that the factors must contain both a and b. If we start with (a − b), and remove this expression from $a^2 - b^2$, we will have (a − b) remaining. This would yield a solution of (a − b) (a − b). Using the FOIL method, the product would be $a^2 - ab - ab + b^2$, which is a $a^2 - 2ab + b^2$ which is incorrect due to the presence of a middle term.

Alternatively, if we choose (a + b) and remove both a and b from the original equation, we have: (a + b) (a − b). Multiplying these factors back together yields $a^2 - ab + ab - b^2$ which simplifies to our original equation of $(a^2 - b^2)$. The difference between two perfect squares always has alternating + and − signs to eliminate the middle term.

Real-life Applications

Factoring is used to simplify situations in both math and in real life. They allow faster solutions to some problems. In the mathematical calculations used to model problems and derive solutions, factoring plays a key role in solving the mathematics that describe systems and events.

IDENTIFICATION OF PATTERNS AND BEHAVIORS

By learning the patterns and behaviors of factors in mathematical relationships, it is possible to identify similarities between multiple components. By being able

to quickly and accurately find similarities, a solution can usually be identified. The solution to any given problem is based on how each individual player or factor in the problem relates to one another for an effective solution. By being able to see these relationships, many times it is possible to see the solution in the relationship.

An example is commonly found in decision making. For example, a shopper enters an unfamiliar grocery store looking for Gouda cheese. The shopper could wander aimlessly, hoping to spot the cheese, but a smarter approach illustrates the intuitive process of factoring. Granted, with enough time, the shopper might eventually find the cheese, but a better approach is to search for a common factor to help narrow the search. What common factor does cheese have with other items in the store? The obvious choice would be to look for the dairy section and eliminate all other sections in the store. The shopper would then further factor the problem to locate the cheese section and eliminate the milk, eggs, etc. Finally one would only look at the cheese selections for the answer, the Gouda cheese. This is a fairly simple non-mathematical example, but it demonstrates the principle of mathematical factoring—a search for similarities among many individual numerical entities.

REDUCING EQUATIONS

In math, one of the most useful applications of factoring is in eliminating needless calculations and terms from complex equations. This is often referred to as "slimming down the equation." If you can find a factor common to every term in the equation, then it can be eliminated from all calculations. This is because the factor will eventually be eliminated through the calculation and simplification process anyway. An example of this is $(2+8)/4$ which can be slimmed down to $(1 + 4)/2$ by eliminating the common factor of 2. The value of the first expression was 10/4 and the value of the second one is 5/2, which is the same once 10/4 is simplified. As we can see, one advantage in eliminating factors is the answer is already simplified. Now let's take a look at a slightly more complicated example:

$$\frac{ax^3 + abx^2 - acx^2}{ax^2}$$

we can see that a common factor of ax^2 can be eliminated.

This expression then becomes:

$$\frac{ax^2 \, (x + b - c)}{ax^2} = (x + b - c)$$

This same technique can be employed in any mathematical equation in which there is a factor common to all parts of the equation.

DISTRIBUTION

Factoring is often used to solve distribution and ordering problems across a range of applications. For example, a simple factoring of 28 yields 4 and 7. In application, 28 units can be subdivided into 4 groups of 7 or 7 groups of 4, Again, by example, in application 28 players could be divided into 4 teams of 7 players or 7 teams of 4 players. This is intuitive factoring—something done every day without realizing that it is a math skill.

SKILL TRANSFER

In addition to factoring mathematical equations, the ability to mathematically factor has been demonstrated to transfer into stronger pattern recognition skills that allow rapid categorization of non-mathematical "factors." Essentially is it an ability to find and eliminate similarities and thus focus on essential difference.

When a defensive linebacker looks over an offensive set in football, he scans for patterns and similarities in numbers of players each side of the ball, in the backfield, in an effort to determine the type of play the opposing quarterback (or his coach) has called. This is not mathematical factoring, but psychology studies have shown that practice in mathematical factoring often leads to a general improvement in pattern recognition and problem solving.

CODES AND CODE BREAKING

Another example of mathematical factoring is in coding and decoding text. Humans have found clever ways of concealing the content of sensitive documents and messages for centuries. Early forms of coding involved the twisting of a piece of cloth over a rod of a certain length. On the cloth would be printed a confusing matrix of seemingly unrelated letters and symbols. When the cloth was twisted over a rod of the proper diameter and length, it would align letters to form messages. The concealed message would be determined by a mathematical factor of proper rod diameter and length that only the intended party would have in possession. Coding and decoding text today is far more complicated. In our new highly computerized age, coding and decoding text depends on an extremely complicated algorithm of mathematical factors.

GEOMETRY AND APPROXIMATION OF SIZE

While factoring is primarily taught and practiced in algebra courses, it is used in every aspect of mathematics. Geometry is no exception. In the field of geometry, there exists the rule of similar triangles. The rule of similar triangles shows that if two triangles have the same angles and the lengths of two legs on one triangle along with a corresponding leg on the other triangle is known, there exists a common factor that can be used to determine the lengths of the other legs. For example, if one wishes to determine the height of a flagpole, factoring through the use of similar triangles can be employed. This is accomplished by an individual of known height standing next to the flagpole. The shadows of both the individual and the flagpole will now be measured. Because the person in standing perpendicular to the ground, a 90-degree triangle is formed with the height of the person being one leg, the length of the shadow being the other leg, and the hypotenuse being the distance from the tip of the person's head to the tip of the head on the shadow. The flagpole forms a similar 90-degree triangle. Once the lengths of the shadows are known, divide the length of the flagpole's shadow by the length of the individual's shadow to determine the common factor. This factor is then multiplied by the height of the individual to find the height of the flagpole.

Potential Applications

In engineering, business, research, and even entertainment, factoring can become a valuable asset. Engineers must use factoring on a daily basis. The job of an engineer is either to design new innovations or to troubleshoot problems as arise in existing systems. Either way, engineers look for effective solutions to complex problems. In order to make their job easier, it is important for them to be able to identify the problem, the solution, and—with regard to the mathematics that describe the systems and events—the factors that systems and events share. Once equations describing systems and events are factored, the most essential elements (the elements that unite and separate systems) can often be more clearly identified. The relationship of each component in the problem will often lead to the solution.

In business, factoring can help identify fundamental factors of cost or expense that impact profits. In research applications, mathematical factoring can reduce complex molecular configurations to more simplified representations that allow researchers to more easily manipulate

Key Terms

Algorithm: A set of mathematical steps used as a group to solve a problem.

Hypotenuse: The longest leg of a right triangle, located opposite the right angle.

Whole number: Any positive number, including zero, with no fraction or decimal.

and design new molecular configurations that result in drugs with greater efficiency—or that can be produced at a lower cost. Factoring even plays a role in entertainment and movie making as complex mathematical patterns related to movement can be factored into simpler forms that allow artists to produce high quality animations in a fraction of the time it would take to actually draw each frame. Factoring of data gained from sensors worn by actors (e.g., sensors on the leg, arms, and head, etc.) provide massive amounts of data. Factoring allows for the simplified and faster manipulation of such data and also allow for mapping to pixels (units of image data) that together form high quality animation or special effects sequences.

Where to Learn More

Web sites

University of North Carolina. "Similar Triangles." <http://www.math.uncc.edu/~droyster/math3181/notes/hyprgeom/node46.html> (February 11, 2005).

AlgebraHelp. "Introduction to Factoring." <http://www.algebrahelp.com/lessons/factoring/> (February 11, 2005).

Financial Calculations, Personal

Overview

Unlike calculus, geometry, and many other types of math, basic financial calculations can be performed by almost anyone. These simple financial equations address practical questions such as how to get the most music for the money, where to invest for retirement, and how to avoid bouncing a check. Best of all, the math is real life and simple enough that anyone with a calculator can do it.

Fundamental Mathematical Concepts and Terms

Financial math covers a wide range of topics, broken into three major sections: Spending decisions deals with choices such as how to choose a car, how to load an MP3 player for the least amount of cash, and how to use credit cards without getting taken to the bank; Financial toolbox looks at the basics of using a budget, explains how income taxes work, and walks through the process of balancing a checkbook; Investing introduces the essentials of how to invest successfully, as well as sharing the bottom line on what it takes to retire as a millionaire (almost anyone can do it).

Real-life Applications

BUYING MUSIC

Today's music lover has more choices than ever before. Faced with hundreds of portable players, a dozen file formats, and millions of songs available for instant download, the choices can become a bit overwhelming. These choices do not just impact what people listen to, they can also impact the buyer's finances for years to come. Additionally, in many cases, comparing the different offers can be difficult.

One well-known music service ran commercials during the 2005 Super Bowl, urging music buyers to simply "Do the math" and touting its offer as an unparalleled bargain. The reasoning is that the top-selling music player in 2005 held up to 10,000 songs and allowed users to download songs for about a dollar apiece; buying that player along with 10,000 songs to fill it up would cost around $10,000. But the music service's ad offered a seemingly better deal: unlimited music downloads for just $14.95 per month. While this deal sounds much better, a little math is needed to uncover the real answer.

A good starting point is calculating the "break-even" point: how many monthly payments do we make before we actually spend the same $10,000 charged by the other firm. This calculation is simple: divide the $10,000 total by the $14.95 monthly fee to find out how many months it takes to spend $10,000. Not surprisingly, it takes quite a few: 668.9 months, to be exact, or about 56 years, which is the break-even point. This result means that if we plan to listen to our downloaded songs for fewer than 56 years, we will spend less with the monthly payment plan. For example, if we plan to use the music for 20 years, we will spend less than $3,600 during that time (20 years $\times$ $14.95 per month), a significant savings when compared to $10,000.

One question raised by this ad is, "How many songs does a typical listener really own?" Assuming the user actually does download 10,000 songs, the previous analysis is correct. But 10,000 songs may not be very realistic; in order to listen to all 10,000 songs just one time, a person would have to listen to music eight hours a day for two full months. In fact, most listeners actually listen to playlists much shorter than 10,000 tracks. So if a listener doesn't want all 10,000 tunes, is the $14.95 per month still the better buy?

Again, the calculations are fairly simple. Let's assume we want to listen to music four hours per day, seven days per week, with no repeats each week. By multiplying the hours times the days, we find that we need 28 hours of music. If a typical song is 3 minutes long, then we divide 60 minutes by 3 minutes to find that we need 20 songs per hour, and by multiplying 20 songs by the 28 hours we need to fill, we find that we need 560 songs to fill our musical week without any repeats. Using these new numbers, the break-even calculation lets us ask the original question again: how long, at $14.95 per month, will it take us to break-even compared to the cost of 560 songs purchased outright? In this case, we divide the $560 we spend to buy the music by the $14.95 monthly cost, and we come up with 37.5 months, or just over three years. In other words, at the end of three years, those low monthly payments have actually equaled the cost of buying the songs to start with, and as we move into the fourth and fifth year, the monthly payments begin to cost us more. Plus, for users whose music library includes only 200 or 300 songs, the break-even time becomes even shorter, making the decision even less obvious than before.

Several other important questions also impact the decision, including, "What happens to downloaded music if we miss a monthly payment?" Since subscription services typically require an ongoing membership in order to download and play music, their music files are designed to quit playing if a user quits paying. The result is generally a music player full of unplayable files. A second consideration is the wide array of file formats currently in

Today's music lover has more choices than ever before. Faced with hundreds of portable players, a dozen file formats, and millions of songs available for instant download, the choices can become a bit overwhelming. These choices don't just impact what people listen to, they can also impact the buyer's finances for years to come. KIM KULISH/CORBIS.

use. Some services dictate a specific brand of player hardware, while others work with multiple brands. Most users feel that the freedom to use multiple brands offers them better protection for their musical investment. Since some players will play songs stored in multiple formats, they offer users the potential to shop around for the best price at various online stores. A final question deals with musical taste and habits. For listeners whose libraries are small, or who expect their musical tastes to remain fairly constant, buying tracks outright is probably less expensive. For listeners who demand an enormous library full of the latest hits and who enjoy collecting music as a hobby, or for those whose music tastes change frequently, a subscription plan may provide greater value.

In the end, this decision is actually similar to other financial choices involving the question of whether to rent or buy (see sidebar "Rent or Buy?"), since the monthly subscription plan is somewhat like renting music. Math provides the tools to help users make the right choice.

CREDIT CARDS

Although the average American already carries eight credit cards, offers arrive in the mail almost every week encouraging us to apply for and use additional cards. Why are banks so eager to issue additional credit cards to consumers who already have them? Answering this question requires an examination of how credit cards work.

In its simplest possible form, a credit card agreement allows consumers to quickly and easily borrow money for daily purchases. Typically, we swipe our card at the store, sign the charge slip or screen, and leave with our goods. At this point in the process, we have our merchandise, paid for with a "loan" from the credit card issuer. The store has its money, less the fee it paid to the credit card company, and the credit card has paid our bill in exchange for a 2–3% fee and for a promise of payment in full at a later date. At the end of the month, we will receive a statement, pay the entire credit card bill on time to avoid interest or late charges, and this simplest type of transaction will be complete.

If this transaction were the norm, very few companies would enter the credit card business, as the 2–3% transaction fees would not offset their overhead costs. In reality, a minority of consumers actually pay their entire bills at the end of the month, and any unpaid balances begins accruing interest for the credit card issuer. These interest charges are where credit card companies actually earn their profits, as they are, in effect, making loans to thousands of consumers at rates that typically run from 9–14% for the very best customers, from 16–21% for average borrowers, and in the case of customers with poor credit histories, even higher rates. Countless individuals who would never consider financing a car loan or home mortgage at an interest of 16% routinely borrow at this and higher rates by charging various monthly expenses on credit cards, and consequently carrying a balance on their bill.

The average American household with at least one credit card in 2004 carried a credit card balance of $8,400 and as a result paid lenders more than $1,000 in interest and finance charges alone, making the credit card business the most profitable segment of the banking industry today. This fact alone answers the original question of why so many credit cards are issued each year: because they are highly profitable to the lenders. Card issuers mailed out three billion credit card offers in 2004 (an average of ten invitations for every man, woman, and child in the United States) because they know their math: half of all credit card users carry a balance and pay interest, so the more new cards the lenders issue, the greater their profits will be.

Loaning money in exchange for interest is an ancient practice, discussed in numerous historical documents, including the Jewish Torah and the Muslim Koran, which both discuss the practice of usury, or charging exorbitantly high interest rates. Modern U.S. law restricts excessive interest charges, and most states have usury laws on their books that limit the rate that an individual may charge another individual. These rates vary widely from state to state; as of 2005, the usury rate, defined as the highest simple interest rate one individual may legally charge another for a loan, is 9% in the state of Illinois. In contrast, Florida's rate is 18%, Colorado's rate is 45%, and Indiana has no stated usury rate at all. Ironically, these laws do not apply to entities such as pawn brokers, small loan companies, or auto finance companies, explaining why these firms frequently charge rates far in excess of the legal maximums for individuals. Credit card issuers, in particular, have long been allowed to charge interest rates above state limits, making them typically one of the most expensive avenues for consumer borrowing.

How much does it really cost to use credit cards for purchases? The answer depends on several factors, including how much is paid each month and what interest rate is being charged. For this example, we'll assume a credit card purchase of $400, an interest rate of 17%, and a minimum monthly payment of $10. After the purchase and making six months of minimum payments, the buyer has paid $60 (six months × $10 per month). But because more than half that amount, $33.06 has gone to pay the 17% interest, only $26.94 has been paid on the original $400 purchase. At this point, even though the buyer has paid out $60 of the original bill, in reality $373.06 is still owed ($400−$26.94).

This pattern will continue until the original purchase is completely paid off, including interest. If the buyer continues making only the required $10 monthly payment, it will take five full years, or 60 payments, to retire the original debt. Over the course of those five years, the buyer will pay a total of $194 in interest, swelling the total purchase price from $400 to almost $600. And if the item originally purchased was an airline ticket, a vacation, or a trendy piece of clothing, the buyer will still be paying for the item long after it's been used up and forgotten. While many factors influence the final cost of saying "charge it," a simple rule of thumb is this: Buyers who pay off their charges over the longest time allowed can expect to pay about 50% more in total cost when putting a purchase on the credit card, pushing a $10 meal to an actual cost of $15. Similarly, a $200 dress will actually cost $300, and a $1,000 trip will actually consume $1,500 in payments.

Credit cards are valuable financial tools for dealing with emergencies, safely carrying money while traveling, and in situations such as renting a car when required to do business. They can also be extremely convenient to use, and in most cases are free of fees for those customers who pay their balance in full each month. Only by doing the math and knowing one's personal spending habits can one know if credit cards are simply a convenient financial tool, or a potential financial time bomb.

CAR PURCHASING AND PAYMENTS

For most consumers, an automobile represents the second largest purchase they will ever make, which makes understanding the car buying process critically important. Several important questions should be considered before buying a new car. First, a potential buyer should calculate how much he can spend. Most experts recommend keeping car payments below 20% of take-home pay, so if a worker receives a check for $2,000 each month (after taxes and other withholding), then he should plan to keep his car payments below $400 (20% × $2,000). This figure is for all car payments, so if he already has a $150 payment for another car, he will be shopping in the $250 per month payment range.

Using this $250 monthly payment, the buyer can consult any of several online payment calculators to determine how much he can spend. For example, if the buyer is willing to spend five years (60 months) paying off his vehicle, this might mean he could afford to borrow about $13,000 for a vehicle (this number varies depending on the actual interest rate at the time of the loan). However this value must pay not just for the car, but also for additional fees such as sales tax, license fees, and registration, which vary from state to state and which can easily add hundreds or thousands of dollars to the price of a new vehicle. For this example, we will estimate sales tax at 6%, license fees at $200, and registration at $100; so a car priced at $12,000 will wind up costing a total of $13,020 (12,000 + .06 × 12,000 + $200 + $100), which is right at the target value of $13,000.

The second aspect of the buying equation is the down payment. A down payment is money paid at the time of sale, and reduces the amount that must be borrowed and financed. In the case of the previous example, a down payment of $2,000 would mean that instead of shopping in the $12,000 price range, the buyer could now shop with $14,000 as the top price.

Many buyers have a used car to sell when they are buying a new vehicle, and in many cases they sell this car to the dealer at the same time, a process known as "trading-in." A trade-in involves the dealer buying a car from the customer, usually at a wholesale price, with the intent to resell it later. A trade-in is a completely separate transaction from the car purchase itself, although dealers often try to bundle the two together. Here again, securing information such as the car's fair trade value will allow the savvy customer to receive a fair price for the trade.

Many consumers find the car-buying experience frustrating, and they worry that they are being taken advantage of. Automobile dealerships are among the only places in the United States where every piece of merchandise has a price tag clearly attached, but both the seller and the buyer know the price on the tag means very little. Most cars today are sold at a significant discount, meaning that a sticker price of $20,000 could easily translate to an actual sales price of $18,000. Incentives, commonly in the form of rebates (money paid back to the buyer by the manufacturer), can chop another $2,000-$5,000 off the actual price, depending on the model and how late in the season one shops. While dealers are willing to negotiate and offer lower prices when they must, they are also going to try to sell at a higher price whenever possible, which places the burden on the buyer to do the homework before shopping. Numerous websites and printed manuals provide actual dealer costs for every vehicle sold in the United States, as well as advice on how much to offer and when to walk away.

CHOOSING A WIRELESS PLAN

Comparing cellular service plans has become an annual ritual for most consumers, as they wrestle with whether to stay with their current cell phone and provider or make the jump to a new company. Beyond the questions of which service offers the best coverage area and which phone is the most futuristic-looking, some basic calculations can help determine the best value for the money.

There are normally three segments to wireless plans. The first segment consists of a set quantity of included minutes that can be used without incurring additional charges. These are typically described as "anytime" minutes, and are the most valuable minutes because they can be used during daytime hours. These minutes are typically offered on a use-it-or-lose-it basis, meaning that if a plan includes 400 minutes and the customer uses only 150, the other 250 minutes are simply lost. Some plans now offer rollover minutes, which means that in the previous example, the 250 minutes would roll to the next month and add to that month's original 400 minutes, providing a total of 650 minutes that could be used without additional charges.

Another segment is that many wireless plans include large blocks of so-called free time, during which calls can be made without using any of the plan's included minutes. These free periods are usually offered during times when the phone network is lightly used, such as late at night and on weekends when most businesses are closed. Users may talk non-stop during these free periods without paying any additional fees.

The third major component of a wireless plan is its treatment of any additional minutes used during non-free periods. In many cases, these additional minutes are billed at fairly high rates, and using additional minutes past those included in a plan's base contract can potentially double or triple the monthly bill.

Other features are sometimes offered, including perks such as free long-distance calling, premium features such as caller identification, and free voicemail. In other cases, providers allow free calls between their own members as part of so-called affinity plans. Cellular plans are typically sold in one- or two-year contracts.

Choosing a wireless plan can be challenging, since there are so many options, and choosing the wrong plan can be a costly choice. A few guidelines can help simplify this choice. First, users should estimate how many minutes will be needed during non-free periods, and then add 10–15% to this estimate in order to provide a margin of error. Next, users can consider whether an affinity plan or free long distance can impact their choices; in cases where most calls are made between family members, plans with these features can offer significant savings.

Finally, users can compare options among the several providers, paying careful attention to coverage areas. For most users, saving a few dollars per month by choosing a carrier with less coverage winds up being an unsatisfying choice. In addition, users should carefully weigh whether to sign a two-year contract, which may offer lower rates, or a one-year plan. One-year plans provide the most flexibility, since rates generally fall over time and a shorter contract allows one to reevaluate alternative plans more often. In addition, wireless providers are now required to let customers keep their cell numbers when they change providers (a feature called "portability"), simplifying the change-over process.

For users needing very few minutes each month, or those on extremely tight budgets, pay-as-you-go plans offer a thrifty alternative. These plans do not normally include free phones or bundles of minutes; instead, a user recharges the account by buying minutes in credit card form at a convenience store or similar outlet. For users who talk 30 minutes or less each month, these plans can be ideal.

When purchasing a wireless plan, add-ons will inevitably increase the final cost. A plan advertised at $39.95 per month will typically generate bills of $43.00 or more when all the taxes and fees are added in, so plan accordingly.

BUDGETS

Personal budgets fill two needs. First, they measure or report, allowing people to assess how much they are spending and what they are spending on. Second, budgets forecast or predict, allowing people to evaluate where their finances are headed and make changes, if necessary. A budget is much like an annual checkup for finances, and can be simple or complex. The simplest budget consists of two columns, labeled "In" and "Out."

The first step in the budgeting process consists of filling the in column with all sources of income, including wages, bonuses, interest, and miscellaneous income. In the case of income that is received more frequently, such as weekly paychecks, or less frequently, such as a quarterly bonus, one must convert the income to a monthly basis for budget purposes, with quarterly items being divided by three and weekly items being multiplied by four. In the case of semiannual items, such as auto insurance premiums, the amount is divided by six.

Next, in the out column, all identifiable outflows should be listed, such as mortgage/rent payments, utilities (electricity, gas, water), car payments and gasoline, interest expense (i.e., credit card charges), health care, charitable donations, groceries, and eating out. The details of this list will vary from person to person, but an effort should be made to include all expenditures, with particular attention paid to seemingly small purchases, such as soft drinks and snacks, cigarettes, and small items bought with cash. For accuracy, any purchase costing over $1 should be included.

The third step is to add up each column, and find the difference between them; in simplest terms, if the out column is larger than the in column, more money is flowing out than in, the budget is out of balance and the family's financial reserves are being depleted. If more money is flowing in than out, the family's budget is working, and attention should be paid to maintaining this state.

The fourth step in this process is evaluating each of the specific spending categories to determine whether it is consuming a reasonable proportion of the spendable income. For instance, each individual category can be divided by the total to determine the percentage spent; a family spending $700 of their monthly $2,000 on car payments, gas, and insurance should probably conclude that this expenditure (700/2000 = 35%) is excessive and needs to be adjusted. In many cases, families creating a first-time budget find that they are spending far more than they realized at restaurants, and that by cooking more of their own meals they can almost painlessly reduce their monthly deficits.

The previous four steps of this process ask "What is being spent?" The fifth and final step asks, "What should be spent?" or "What is the spending goal?" At a minimum, efforts should be made to bring the entire budget into balance by adjusting specific categories of spending. Ideally, goals can be set for each category and reevaluated at the end of each month. A budget provides a simple, inexpensive tool to begin taking control of one's personal finances. W. Edwards Deming, the genius who transformed the Japanese from makers of cheap trinkets into the worldwide experts on quality manufacturing, is often paraphrased as saying, "You can't change what you can't measure." A simple three-column budget provides the basic measurement tool to begin measuring one's financial health and changing one's financial future.

UNDERSTANDING INCOME TAXES

The United States Treasury Department collects around $1 trillion in individual income taxes each year from U.S. workers, most of it subtracted from paychecks. While income tax software has taken much of the agony out of tax preparation each April, most workers still have to interact with the Internal Revenue Service, or IRS, from time to time, especially in the area of filling out tax paperwork.

Employers are required by law to withhold money from employee paychecks to pay income taxes. But because each person's tax situation is different, the IRS has a specific form designed to tell employers how much to withhold from each employee. This form, the W-4, asks taxpayers a series of questions, such as how many children they have and whether they expect to file specific tax forms or not. By supplying this form to new employees, companies can ensure that they withhold the proper amount from each paycheck, as well as protect employees from penalties that apply if they do not have enough of their taxes withheld. In cases where family information changes, or where the previous year's withholding amount was too high or too low, a new form can be filed with the employer at any time during the year.

At the end of the calendar year, employers issue a report to each employee called a W-2. Form W-2 is a summary of an employee's earnings for the entire year, including the total amount earned, or gross pay and, amounts withheld for income tax, social security, unemployment insurance, and other deductions. The information from the W-2 is used by the employee when filing federal and state income returns each year. W-2 forms are required to be mailed to employees by January 31; if a W-2 is not received by the first week in February, the employee should contact the employer.

Other forms are used to report other types of income. The 1099 form is similar to W-2s and is sent to individuals who received various types of non-wage income during the year. For example, form 1099-INT is used by banks to provide account holders with a record of interest earned, form 1099-DIV is used to report dividend income,

and form 1099-MISC is used to report monetary winnings such as contest prizes, as well as other types of miscellaneous income. These forms should not be discarded, as the amounts on them are reported to the IRS, which matches these reported amounts with individual tax returns to make sure the income was reported and taxes were paid on it. Failure to report income and payroll taxes could lead to penalties and the possibility of a tax audit, in which the taxpayer is required to document all aspects of the tax return to an IRS official.

BALANCING A CHECKBOOK

Balancing a checkbook is an important chore that few people enjoy. A correctly balanced checkbook provides several distinct benefits, including the knowledge of where one's money is being spent, and the avoidance of embarrassing and costly bounced checks. A balanced account also allows one to catch any mistakes, made either by the bank or by the individual, before they create other problems. Balancing a checkbook is actually quite simple and can usually be accomplished in less than half an hour. Whether one uses software or the traditional paper-and-pencil method, the general approach is the same.

Balancing a checkbook begins with good recordkeeping, which means correctly writing down each transaction, including every paper check written, deposit made, ATM withdrawal taken, or check-card purchase made. Bad recordkeeping is a major cause of checkbook balancing problems.

Determining whether all of one's transactions have cleared the checking account is described as the process of a paper check winding its way through the financial system from the merchant to the bank, which can take several days. It also refers to deposits or withdrawals made after the statement date. The net effect of clearing delays is that most consumers will have records of transactions that are not in the latest bank statement, meaning this statement balance may appear either too high or too low. Determining whether all items have cleared involves a review of the records collected in the previous step. A checkmark is placed next to the item on the bank statement for each check, ATM receipt, or other record. Once this process is complete, and assuming good records have been kept, all the items in the bank statement will be checked, and several items that were not in the statement at all will remain. The process of adjusting for these uncleared items is called reconciling the statement.

To reconcile a check register with the bank statement, all the uncleared items must be accounted for, since these transactions appear in the personal check register but not in the statement. Specifically, deposits and other uncleared additions to the account must be subtracted, while withdrawals, check-card transactions, written checks, and other uncleared subtractions from the account must be added back in. The net effect of this process is to back the records up to the date of the bank statement, at which time the two totals, the check register and the bank statement, should match. Many banks include a simple form on the back of the printed bank statement to simplify this process.

For most customers, a day will arrive when the account simply does not balance. Since bank errors are fairly rare, the most common explanation is an error by the customer. A few simple steps to take include scanning for items entered twice, or not entered at all; data entry errors, such as a withdrawal mistakenly entered as a deposit; simple math errors; and forgetting to subtract monthly service charges or fees. Most balancing errors fall into one of these categories, and as before, good record-keeping will simplify the process of locating the mistake.

Balancing a checkbook is not difficult. The time invested in this simple exercise can often pay for itself in avoided embarrassment and expense.

SOCIAL SECURITY SYSTEM

The Social Security system was established by President Franklin Roosevelt in 1935, creating a national system to provide retirement income to American workers and to insure that they have adequate income to meet basic living expenses. Due largely to this program, nine in ten American senior citizens now live above the official poverty line.

But a Social Security number is important long before one retires. Because the United States does not have an official, government-issued identification program, Social Security numbers are frequently used as personal identification numbers by universities, employers, and banks. U.S. firms are also required by law to verify an applicant's Social Security number as part of the hiring process, making a Social Security card a necessity for anyone wanting to work. For this reason, most Americans apply for and receive a Social Security number and card while they are still minors.

Social Security numbers and cards are issued free of charge at all Social Security Administration offices. An applicant must present documents such as a birth certificate, passport, or school identification card in order to verify the person's identity. After these documents are verified, a number will be issued. A standard Social Security number is composed of three groups of digits, separated by dashes, such as 123-45-6789, and always contains a total of nine digits. Each person's number is unique, and

in some cases, the first three digits may indicate the region in which the card was issued. The simplest way for a child to receive a Social Security number is for the parents to apply at birth, at the same time they apply for a birth certificate. After age 12, a child applying for a card, in addition to providing documentation of age and citizenship, must also complete an in-person interview to explain why no card has been previously issued.

When a person begins working, the employer withholds part of the worker's earnings to be deposited into the Social Security system; as of 2005, these contributions are taken out of the first $90,000 in earned income each year at a rate of 7.65%. Starting at age 25, each worker receives an annual statement listing their income for the previous year; this information should be carefully checked for accuracy. While taking one's Social Security card to job interviews or loan applications is a good idea, the Social Security Administration recommends that cards be kept in a safe place, rather than carried on one's person. In the event that a Social Security card is lost or stolen, a new card can be requested at no charge by completing the proper form and submitting verification of identity. The new card will have the same number on it as the old card. In the case of a name change due to marriage, divorce, or similar events, a new card can be issued with the same number and the cardholder's new name. This process requires documentation showing both the previous name and the new name.

The Social Security system remains the largest single retirement plan in the country, is mandatory for most workers, and is expected to remain in place for the foreseeable future.

INVESTING

Investing simply means applying money in such a way that it grows, or increases, over time. In a certain sense, investing is somewhat like renting money to someone else, and in return, receiving a rental fee for the privilege. Investments come in an almost endless variety of forms, including stocks, bonds, real estate, commodities, precious metals, and treasuries. While this array of options may seem bewildering at first, all investment decisions are ultimately governed by a simple principle: "risk equals reward."

Risk is the potential for loss in any investment. The least risky investments are generally government-backed investments, such as Treasury bills and Treasury bonds issued by the United States government. These investments are considered extremely safe because they are backed by the U.S. Treasury and, barring the collapse of the government, will absolutely be repaid. For this reason, these investments are sometimes described as riskless. At the other end of the risk spectrum might be an investment in a company that is already bankrupt and is trying to pull itself out of insolvency. Because the risk of losing one's investment in such a firm is extremely high, this type of investment is often referred to as a junk bond, since its potential for loss is high. Between riskless and highly risky investments are a variety of other options that provide various levels of risk. Risk is generally considered higher when money is invested for longer periods of time, so short-term investments are inherently less risky than long-term ones.

Reward is the return investors hope to receive in exchange for the use of their money. Most investors are only willing to lend their money to someone for something in return. Investors who buy a rare coin or a piece of real estate are hoping that the value of the coin or house will rise, so they can reap a reward when they sell it. Likewise, investors who buy shares of a company's stock is betting that the company will make money, which it will then pass along to them as a dividend. Investors also hope that as the company grows, other investors will see its value and the stock price itself will rise, allowing them to profit a second time when they sell the stock. Investment rewards take many different forms, but financial returns are the main incentive for people to invest.

The principle "risk equals reward" states that investments with higher levels of risk will normally offer higher returns, while safer (less risky) investments will normally return smaller rewards. For this reason, the very safest investments pay very low rates. An insured deposit in a savings account at a typical U.S. bank earns about 1–2% per year, since these funds are insured and can be withdrawn at any time. Other safe investments, such as U.S. Treasury bills and U.S. savings bonds, pay low interest rates, typically 3–4% for a one-year investment.

Corporate bonds and stocks are two tools that allow public corporations to raise money. Bonds are considered a less risky investment than stocks, and hence pay lower returns, generally a few percentage points higher than Treasury bills. Historically, stocks in U.S. firms have returned an average of 9–10% per year over the long-term. However, this average return conceals considerable volatility, or swings, in value. This volatility means in a given year the stock market might rise by 30-40%, decline by the same amount, or experience little or no change. This variation in annual rates of return is one reason stocks are considered more risky than Treasuries, and hence pay a higher rate of return. Most financial experts recommend that those investing for periods longer than ten years place most of their funds in a variety of different kinds of stocks.

Among the riskiest investments are stock options and commodity futures. Because these types of investments are complex and can potentially lead to the loss of one's entire investment, they are generally appropriate only for experienced, professional investors. Other investments, such as rental real estate, can offer substantial returns in exchange for additional work required to maintain, repair, and manage the property.

A few tricks can help young investors take advantage of certain laws to invest their money. Because the government taxes most forms of income, any investment vehicle that allows the investor to defer (delay) paying taxes will generally produce higher returns with no increase in risk. As an example, consider a worker who begins investing $3,000 per year in a retirement account at age 29. If the worker deposits this money in a normal, taxable savings account or investment fund, each year he will have to pay income tax on the earnings, meaning that his net return will be lower. But if this same amount of money is invested in a tax-sheltered account, the money can grow tax-free, meaning the income each year is higher. Over the course of a career, this difference can become enormous. In this example, the worker's contributions to the taxable account will grow to $450,000 by age 65. But in a tax-sheltered account, those very same contributions would swell to more than $770,000, a 70% advantage gained simply by avoiding tax payments on each year's earnings.

One of the simplest ways to begin a tax-deferred retirement plan is with a Roth Individual Retirement Account (IRA). Available at most banks and investment firms, Roth accounts allow any person with income to open an account and begin saving tax-free. Beginning in 2005, the maximum annual contribution to a Roth IRA is $4,000, which will increase again in 2008 to $5,000. One notable feature of IRAs is the hefty 10% penalty paid on withdrawals made before retirement. While this may seem like a disadvantage, this penalty provides strong incentive to keep retirement funds invested, rather than withdrawing them for current needs.

Another outstanding investment option is a 401(k) plan, offered by many large employers under a variety of names. These plans not only allow earnings to grow tax-deferred like an IRA, they offer other advantages as well. For instance, most firms will automatically withdraw 401(k) contributions from an employee's paycheck, meaning he doesn't have to make the decision each month whether to invest or not. Also, some companies offer to match employee contributions with additional contributions. In a case where a company offers a 1:1 match on the first $2,000 an employee saves, the employee's $2,000 immediately becomes $4,000, equal to a 100% return on the investment the first year, with no added risk. In the case of a 50% match on the first $3,000, the firm would contribute $1,500. Company matches are among the best deals available and should always be taken advantage of.

Investing is a complex subject, and investing in an unfamiliar area is a chance for losses. By choosing a variety of investments, most investors can generate good returns without exposing themselves to excessive risk. And by taking time to learn more about investment options, most investors can increase their returns without unduly increasing their risk.

RETIRING COMFORTABLY BY INVESTING WISELY

Who wants to be a millionaire? More importantly, what chance does an average 18-year-old person have of actually reaching that lofty plateau? Surprisingly, almost anyone who sets that as a goal and makes a few smart choices and exercises self-discipline along the way can fully expect to be a millionaire by the time he retires. In fact, there are so many millionaires in the United States today that most people already know one or two, even though they are tough to pick out since few of them fit the common stereotype (see sidebar: Millionaire Myths).

Is a million dollars enough to retire comfortably on? Most people would scoff at the question, but the answer may not be as obvious as it first seems. Most members of the World War II generation clearly remember an era of $5,000 houses, $500 cars, and 5-cent soft drinks. What they may not recall so clearly is that in 1951, the average American worker earned only $56.00 per week, meaning that while prices are much higher today, wages have risen substantially as well.

This gradual rise in prices (and the corresponding fall in the purchasing power of a dollar) is called inflation. When inflation is low, and prices and wages increase 3–4% per year, most economists feel the economy is growing at a healthy pace. When inflation reaches higher levels, such as the double-digit rates experienced in the late 1970s, the national economy begins to collapse. And in rare situations, a disastrous phenomena known as hyperinflation takes over. In 1922, Germany experienced an inflation rate of 5,000%. This staggering rate meant that in a two-year period, a fortune of 20 billion German marks would have been reduced in value to the equivalent of one mark. One anecdotal account of hyperinflation in Germany tells of individuals buying a bottle of wine in the expectation that the following day the empty bottle could be sold for more than the full bottle originally cost. Hyperinflation has occurred more recently as well: Peru,

Brazil, and Ukraine all experienced hyperinflation during the 1990s; with prices rising quickly, sometimes several times each day, workers began demanding payment daily so they could rush out and spend their earnings before the money lost much of its value.

While hyperinflation can destroy a nation's economy, it is a rare event. A far more realistic concern for workers intent on retiring comfortably is the slow but steady erosion of their money's value by inflation. In the same way that the 5-cent sodas of the 1950s now cost more than a dollar, an increase of twenty-fold, one must assume that the one-dollar sodas of today may well cost $20 by the middle of the twenty-first century. And as costs continue to climb, the value of a dollar, or a million dollars, will correspondingly fall.

The million dollar question (will a million dollars be enough?) can be answered fairly simply using a mathematical approach and several steps. The first question: how much money will be needed in 50 years to equal the value of $1 million today? The first step of this process is determining how much buying power $1 million loses in one year. If the rate of inflation is 3%, a reasonable guess, then over the course of one year $1 million is reduced in buying power by 3%. At the end of the first year, it has buying power equal to $1,000,000 × 97%, or $970,000. This is still a fantastic sum of money to most people, but the true impact of inflation is not felt in the first year, but in the last.

These calculations could continue indefinitely, multiplying $970,000 × 97% to get the value at the end of the second year, and so forth. If this were done for 50 years, we could eventually produce an inflation "multiplier," a single value by which we multiply our starting value to find the predicted future buying power of that sum. In this example, the inflation multiplier is .22, which we multiply by our starting sum of $1 million to find that at retirement in 50 years the nest-egg will have the buying power of only $220,000 today. And while $220,000 is a nice sum of money, it may not be enough to support a comfortable retirement for very many years.

This raises another obvious question: how much will it take in 50 years to retain the buying power of $1 million today? This calculation is basically the inverse of the previous one. To determine how much is required one year hence to have the buying power of $1 million today, we simply multiply by 1.03 (based on our 3% inflation assumption), giving a need next year for $1,030,000. Again, we can carry this out for 50 years and produce a multiplier value, which in this case turns out to be 4.5. We then multiply that value times the base of $1 million to learn that in order to have the buying power of $1 million today will require one to have accumulated more than $4 million by retirement.

In summary, the answer to the question is simple: If a retirement fund of $220,000 would be adequate for today, then $1 million will be adequate in 50 years. But if it would take $1 million to meet one's retirement needs today, the goal will need to be quite a bit higher, since today's college students will likely retire in an era when a bottle of drinking water will set them back $20.

This example requires that we picture our bank account as a swimming pool and the money we save as water. The goal is to fill the pool completely by the time of retirement. Because the pool begins completely empty, the task may seem daunting. But like most challenging goals, this one can be achieved with the right approach.

In order to fill the pool, one must attach a pipe that allow water to flow in, and the first decision relates to the size of this pipe, since the larger the pipe, the more water it can carry and the faster the pool will fill. The size of the pipe equates to income level, or for this illustration, the total amount we expect to earn over an entire career. This first decision may be the single most important choice one makes on the road to millionaire status, since this first decision will largely determine the size of the pipe and the size of one's income.

Educational level and income are highly correlated, and not surprisingly, less education generally equates to less income. A report by the U.S. Census Office provides the details to support this claim, finding that students who leave high school before completion can expect to earn about $1 million over their careers. While this sounds like a hefty amount, it is far below what most families need to live, and almost certainly not enough to amass a million dollars in retirement savings. Just for comparison, this value equates to annual earnings of less than $24,000 per year. In our current illustration, this equates to a tiny pipe, and means the swimming pool will probably wind up empty.

The good news from the report is that each step along the educational path makes the pipe a little larger, and fills the pool a little faster. For high school students who stay enrolled until graduation, lifetime earnings climb by 20%, to $1.2 million, meaning that a high school junior who chooses to finish school rather than dropping out will earn almost a quarter of a million dollars for his or her efforts. And with each diploma comes additional earning power. An associate's degree raises average lifetime earnings to $1.5 million, while a bachelor's degree pushes average lifetime earnings to $2.1 million, more than double the amount earned by the high school dropout. Master's degrees, doctorates, and professional

degrees such as law and medical degrees each raise expected earnings as well, increasing the size of the pipe and filling the pool faster. Simple logic dictates that when the pipe is two to four times as large, the pool will fill far more quickly. For this reason, one of the best ways to predict an individual's retirement income level is simply to ask, "How long did you stay in school?"

Retirement savings are impacted by income level in multiple ways. First, since every household has to pay for basics such as food, housing, clothing, and transportation, total income level determines how much is left over after these expenses are paid each month, and therefore how much is available to be invested for retirement. Second, as detailed in the Social Security system section, Social Security pays retirement wages based on one's earnings while working, so those who earn more during their career will also receive larger Social Security payments after retirement. Third, employers frequently contribute to retirement plans for their workers, and the level of these contributions is also tied directly to how much the worker earns, with higher earnings equating to higher contributions and greater retirement income. Because each of these pieces of the retirement puzzle is tied to income level, each one adds to the size of the pipe, and helps fill the pool more quickly. Again, education is a primary predictor of income level.

Of course a few people do manage to strike it rich in Las Vegas or win the state lottery, which is roughly equivalent to backing a tanker truck full of water up to the pool and dumping it in. For these few people, the size of the income pipe turns out to be fairly unimportant, since they have beaten some of the longest odds around. To get some idea just how unlikely one is to actually win a lottery, consider other possibilities. For example, most people don't worry about being struck by lightning, and this is reasonable, since a person's odds of being struck by lightning in an entire lifetime are about one in 3,000, meaning that on average if he lived 3,000 lifetimes, he would probably be struck only once. And even though shark attacks make the news virtually every year, the odds of being attacked by a shark are even lower, around one in 12,000.

Since most people fully expect to live their entire lives without being attacked by a shark or being struck by lightning, it seems far-fetched that many would play the lottery each week, given that the odds of winning are astronomically worse. As an example, the Irish Lotto game, which offers some of the best odds of any national lottery on the planet, gives buyers a 1-in-5 million chance of winning, meaning a player is 1,600 times more likely to be struck by lightning than to win the jackpot. And the U.S. PowerBall game offers larger jackpots, but even lower

Millionaire Myths

Say the word "millionaire," and most Americans picture Donald Trump, fully decked out in expensive designer suits and heavy gold jewelry. To most Americans, yachts, mansions, lavish vacations, and fine wines are the sure signs that a person has made it big and has accumulated a seven-figure net worth. But recent research paints a very different picture: most millionaires live fairly frugal lives and tend to prefer saving over spending, even after they've made it big. In fact, the most surprising fact about real millionaires is this: they don't look or act at all like TV millionaires.

The average millionaire in the United States today buys clothes at J.C. Penney's, drives an American made car (or a pickup), and has never spent more than $250 on a wristwatch. He or she inherited little or nothing from parents and has built the fortune in such industries as rice farming, welding contracting, or carpet cleaning. This person is frugal, remains married to the first spouse, has been to college (but frequently was not an A student), and lives in a modest house bought 20 years ago.

In short, while most millionaires are gifted with vision and foresight, there is little they have done that cannot be duplicated by any hard-working, dedicated young person today. The basic principles of accumulating wealth are not hard to understand, but they require hard work and self-discipline to apply.

odds of winning: a player in this game is 16 times less likely to win than in the Irish Lotto, meaning the average PowerBall player should expect to be struck by lightning 26,000 times as often as he wins the jackpot. Of all the unlikely events that might occur, winning the lottery is among the most unlikely.

Once the pipe is turned on, which means we have begun making money, one may find the pool filling too slowly, which means assets and savings are accumulating too slowly. At this point it becomes necessary to notice that the pool includes numerous drains in the floor, some large and others small. Water is continually flowing out these drains, which represent financial obligations such as utility bills, tuition payments, mortgages, and grocery costs. In some cases, the water may flow out faster than the pipe can pump it in, causing the water level to drop

until the pool runs dry, meaning the employee runs out of money, and bankruptcy follows. In most families, the inflow and outflow of money roughly balance each other, and each month's bills are paid with a few dollars left, but the pool never really fills up. In either case, retirement will arrive with little or nothing saved, and retirement survival will depend largely on the generosity of the Social Security system.

A more pleasant alternative involves closing some of the drains in the pool, or reducing some expenditures. For most families, the largest drains in the pool will be monthly items such as mortgage and car loan payments that are set for periods of several years and may not be easily changed over the short-run. For these items, decisions can only be made periodically, such as when a new car or home is purchased.

However, some seemingly small items may create huge drains in the family financial pool. For most families, eating out consumes a majority of the food budget, even though eating at home is typically both cheaper. Numerous small bills such as cable, wireless, and internet access can add up to take quite a drink out of the pool, even though each one by itself seems small. Yet, while the total dollar value of such items may seem insignificant, their impact over time can be enormous. By removing just $50 from consumption and investing it at 8% each month during the 50 years of a career, this trivial amount will grow to almost $350,000. These types of choices are among the most difficult to make, but can be among the most significant, especially considering that $50 per month represents what many Americans spend on soft drinks or gourmet coffee. A good rule of thumb for this calculation is to multiply the monthly contribution times 7,000 to find its future value at retirement, assuming one begins at age 20 and retires at age 70.

The other major factor in retiring comfortably is time. To put it simply, the final value of one dollar invested at age 20 will be greater than the final value of four dollars invested at age 50. This means that $10,000 invested at age 20 will grow to $143,000 by age 75, while $40,000 invested at age 50 will be worth only $134,000 at the same time. In fact, a good general rule of thumb is for each eight years that pass, the final value of the retirement nest egg will be reduced by 50%. It is never too early to start saving for retirement.

CALCULATING A TIP

After the meal is over and everyone is stuffed, it's time to pay the bill and make one of the most common financial calculations: deciding how much to tip a server. Some diners believe that the term "tips" is an acronym for "to insure prompt service," hence they believe that the size of the tip should be tied to the level of service, with excellent service receiving a larger tip and poor service receiving less, or none. Others recognize that servers often make sub-minimum wage salaries (as of 2005, this could be as little as $2.13 per hour) and depend on tips for most of their income, hence they generally tip well regardless of the level of service. Another important consideration is that servers are often the victims of kitchen mistakes and delays, and therefore penalizing them for these problems seems unreasonable. A good general rule of thumb is to tip 20% for outstanding service, 15% for good service, and 10% or less for poor service. Regardless of which tipping philosophy one adopts, some basic math will help calculate the proper amount to leave.

For example, imagine that the bill for dinner is $56.97, which includes sales tax. By looking at the itemized bill, we determine that the pre-tax total is $52.75, since most people calculate the tip on the food and drink total, not including tax. Since the service was excellent we choose to tip 20%. Most tip calculations begin by figuring the simplest calculation, 10%, since this figure can be determined using no real math at all. Ten percent of any number can be found simply by moving the decimal point one place to the left. In the case of our bill of $52.75, we simply shift the decimal and wind up with 10% being $5.275, or five dollars twenty seven and one-half cents. Then to get to 20%, we simply double this figure and wind up with a tip of $10.55.

In real life, we are not concerned about making our tip come out to an exact percentage, so we generally round up or down in order to simplify the calculations. In this case, we would round the $5.275 to $5.25, which is then easily doubled to $10.50 for our 20% tip. Finding the amount of a 15% tip can be accomplished either of two ways. First, we can take the original 10% value and add half again to it. In this case, half of the original $5.27 is about $2.50, telling us that our final 15% tip is going to be around $7.75, which we might leave as-is or round up to $8.00 just to be generous. A second, less-obvious approach involves our two previous calculations of 10% and 20%. Since 15% is midway between these two values, we could take these two numbers and choose the midway point (a process that mathematicians call "interpolation"). In other words, 10% is $5.27 (or about $5.00) and 20% is $10.55 (or about $11.00), so the midway point would be somewhere in the $7.00–8.00 range. Either of these two methods will allow us to quickly find an approximate amount for a 15% tip.

CURRENCY EXCHANGE

Because most nations issue their own currency, traveling outside the United States often requires one to

A potential customer looks at exchange rates outside an exchange shop in Rome. AP/WIDE WORLD PHOTOS. REPRODUCED BY PERMISSION.

exchange U.S. dollars for the destination nation's currency. But this process is complicated by the fact that one unit of a foreign currency is not worth exactly one U.S dollar, meaning that one U.S. dollar may buy more or fewer units of the local currency. Currency can be exchanged at many banks and at most major airports, normally for a small fee. Banks generally offer better exchange rates than local merchants, so travelers who plan to stay for some time typically exchange larger amounts of money at a bank when they first arrive, rather than smaller amounts at various shops or hotels during their stay.

Consider a person who wishes to travel from the United States to Mexico and Canada. Before leaving the States, the traveler decides to convert $100 into Mexican currency and $100 into Canadian currency. At the currency exchange kiosk, there is a large board that displays various currencies and their exchange rates.

The official unit of currency in Mexico is the peso, and the listed exchange rate is 11.4, meaning that each

Currencies of the European Community. OWEN FRANKEN/CORBIS.

U.S. dollar is worth 11.4 pesos. Multiplying 100 × 11.4, the person learns that one is able to purchase 1,140 pesos with $100. Canadians also use dollars, but Canadian dollars have generally been worth less than U.S. dollars. On the day of the exchange, the rate is 1.3, meaning that each U.S. dollar will buy $1.3 Canadian dollars, so with $100 the person is able to purchase 130 Canadian dollars. At this point, the shopper might wonder about the exchange rate between Canadian dollars and pesos. Since it is known that 130 Canadian dollars equals the value of 1,140 pesos, the person can simply divide 1,140 by 130 to determine that on this date, the exchange rate is 8.77 pesos to one Canadian dollar.

Exchange rates fluctuate over time. On a business trip one year later, this same person might find that the $100 would now buy 2,000 pesos, meaning that the U.S. dollar has become stronger, or more valuable, when compared to the peso. Conversely, it might be that the dollar has weakened, and will now purchase only 800 pesos. These fluctuations in exchange rates can impact travelers, as the changing rates may make an overseas vacation more or less expensive, but they can be particularly troublesome

for large corporations that conduct business across the globe. In their situation, products made in one country are often exported for sale in another, and changing exchange rates may cause profits to rise or fall as the amount of local currency earned goes up or down.

In addition to U.S. dollars, other well-known national currencies (along with their exchange rates in early 2005) include the British pound (.52), the Japanese yen (105), the Chinese yuan (8.3), and the Russian ruble (27.7). Beginning in 2002, 12 European nations, including Germany, Spain, France, and Italy, merged their separate currencies to form a common European currency, the Euro (.76). Designed to simplify commerce and expand trade across the European continent, conversion to the Euro was the largest monetary changeover in world history.

Where to Learn More

Books

Stanley, Thomas, and William Danko. *The Millionaire Next Door*. Atlanta: Longstreet Press, 1996.

Key Terms

Balance: An amount left over, such as the portion of a credit card bill that remains unpaid and is carried over until the following billing period.

Bankruptcy: A legal declaration that one's debts are larger than one's assets; in common language, when one is unable to pay his bills and seeks relief from the legal system.

Bouncing a check: The result of writing a check without adequate funds in the checking account, in which the bank declines to pay the check. Fees and penalties are normally imposed on the check writer.

Inflation: A steady rise in prices, leading to reduced buying power for a given amount of currency.

Interest: Money paid for a loan, or for the privilege of using another's money.

Lottery: A contest in which entries are sold and a winner is randomly selected from the entries to receive a prize.

Mortgage: A loan made for the purpose of purchasing a house or other real property.

Reconcile: To make two accounts match; specifically, the process of making one's personal records match the latest records issued by a bank or financial institution.

Register: A record of spending, such as a check register, which is used to track checks written for later reconciliation.

Web sites

A Moment of Science Library. "Yael and Don Discuss Interpreting the Odds." <http://www.wfiu.indiana.edu/amos/library/scripts/odds.html> (March 6, 2005).

Balanced Scorecard Institute. "What is the Balanced Scorecard?" <http://www.balancedscorecard.org/basics/bsc1.html> (March 6, 2005).

Car-Accidents.com. "2004 Statistics." <http://www.car-accidents.com/pages/stats/2000_killed.html> (March 4, 2005).

CNN.com Science and Space. "Scared of Sharks?" <http://www.cnn.com/2003/TECH/science/12/01/coolsc.sharks.kellan/> (March 3, 2005).

Edmunds New Car Pricing, Used Car Buying, Auto Reviews. "New Car Buying Advice." <http://www.edmunds.com/advice/buying/articles/43091/article.html> (March 5, 2005).

Ends of the Earth Training Group. "W. Edwards Deming's Fourteen Points and Seven Deadly Diseases of Management." <http://www.endsoftheearth.com/Deming14Pts.htm> (March 7, 2005).

Euro Banknotes and Coins. "History of the Euro." <http://www.euro.ecb.int/en/what/history.html> (March 5, 2005).

Fidelity Personal Investments. "U.S. Treasury Securities." <http://personal.fidelity.com/products/fixedincome/potreasuries.shtml> (March 4, 2005).

First National Bank of St. Louis. "Roth IRA Calculator." <http://www.fnbstl.com/tools/calcs/tools.asp?Tool=RothIRA> (March 4, 2005).

Internal Revenue Service. "Form W-4 (2005)." <http://www.irs.gov/pub/irs-pdf/fw4.pdf> (March 5, 2005).

Internal Revenue Service. "Treasury Department Gross Tax Collections: Amount Collected by Quarter and Fiscal Year." <http://www.nlm.nih.gov/hmd/about/collectionhistory.html> (March 4, 2005).

Lectric Law Library. "State Interest Rates and Usury Limits." <http://www.lectlaw.com/files/ban02.htm> (March 7, 2005).

The Lottery Site. "Lottery Odds and Your Real Chance of Winning." <http://www.thelotterysite.com/lottery_odds.htm> (March 5, 2005).

Money Savvy: Yakima Valley Credit Union. "Keep Your Checkbook Up to Date." <http://hffo.cuna.org/story.html?doc_id=218&sub_id=tpempty> (March 7, 2005).

Snopes.com Tip Sheet. "Tip is an acronym for To Insure Promptness." <http://www.snopes.com/language/acronyms/tip.htm> (March 5, 2005).

Social Security Online. "Your Social Security Number and Card." <http://www.ssa.gov/pubs/10002.html> (March 5, 2005).

U.S.Info.State.Gov. "A Brief History of Social Security." <http://www.nlm.nih.gov/hmd/about/collectionhistory.html> (March 4, 2005).

William King Server, Drexel University. "Hyperinflation." <http://william-king.www.drexel.edu/top/prin/txt/probs/infl7.html> (March 5, 2005).

XE.com. "XE.com Quick Cross Rates." <http://www.nlm.nih.gov/hmd/about/collectionhistory.html> (March 7, 2005).

Fractals

A fractal is a kind of mathematical equation of which pictures are frequently made. A small unit of structural information structure forms the basis for the overall structure. The repeats do not have to be exact, but they are close to the original. For example, the leaves on a maple tree are not exactly alike, but they are similar.

The beauty principle in mathematics states that if a principle is elegant (arrives at the answer as quickly and directly as possible), then the probability is high that it is both true and useful. Fractal mathematics fulfills the beauty principle. Both in the natural world and in commerce, fractals are ever-present and useful.

A fractal has infinite detail. This means that the more one zooms in on a fractal the more detail will be revealed. An analogy to this is the coastline of a state like Maine. When viewed from a satellite, the ocean coastline of the state shows large bays and peninsulas. Nearer to the ground, such as at 40,000 feet (12,192 m) in a jet aircraft flying over the state, the convoluted nature of the coast looks similar, only the features are smaller. If the plane is much lower, then the convolutions become even smaller, with smaller bays and inlets visible but still have basically the same shape. A fractal is similar to the example of the Maine coastline. As the view becomes more and more magnified, the never-ending complexity of the fractal is revealed.

Fundamental Mathematical Concepts and Terms

Among the many features of fractals are their non-integer dimensions. Integer dimensions are the whole number dimensions that most people are familiar with. Examples include the two dimensions (width and length; this is also commonly referred to as 2-D) of a square and the three dimensions (width, length and height; commonly called 3-D) of a cube. It is odd to think that dimensions can be in between 2-D and 3-D, or even bigger than 3-D. But such is the world of fractals.

Dimensions of 1.8 or 4.12 are possible in the fractal world. Although the mathematics of fractals involves complex algorithms, the simplest way to consider fractal dimensions is to know that dimensions are based on the number of copies of a shape that can fit into the original shape. For example, if the lines of a cube are doubled in length, then it turns out that eight of the original-sized cubes can fit into the new and larger cube. Taking the log of 8 (the number of cubes) divided by the log of 2 (doubling in size) produces the number three. A cube, therefore, has three dimensions.

For fractals, where a pattern is repeated over and over again, the math gets more complicated, but is based on the same principle. When the numbers are crunched, the resulting number of dimensions can be amazing. For example, a well-known fractal is called Koch's curve. It is essentially a star in which each original point then has other stars introduced, with the points of the new stars becoming the site of another star, and on and on. Doing the calculation on a Koch's curve that results from just the addition of one set of new stars to the six points of the original star produces a dimension result of 1.2618595071429!

BUILDING FRACTALS

Fractals are geometric figures. They begin with a simple pattern, which repeats again and again according to the construction rules that are in effect (the mathematical equation supplies the rules).

A simple example of the construction of a fractal begins with a + shape. The next step is to add four + shapes to each of the end lines. Each new + is only half as big as the original +. In the next step, the + shapes that are reduced by half in size are added to each of the three end lines that were formed after the first step. When drawn on a piece of paper, it is readily apparent that the forming fractal, which consists of ever smaller + shapes, is the shape of a diamond. Even with this simple start, the fractal becomes complex in only a handful of steps. And this is a very simple fractal!

SIMILARITY

An underlying principle of many fractals is known as similarity. Put another way, the pattern of a fractal is the repetition of the same shaped bit. The following cartoons will help illustrate self-similarity.

In Figure 1, the two circles are alike in shape, but they do not conform to this concept of similarity. This is because multiple copies of the smaller circle cannot fit inside the larger circle.

In Figure 2, the two figures are definitely not similar, because they have different shapes.

The two triangles in Figure 3 are similar. This is because four of the smaller triangles can be stacked together to produce the larger triangle. This allows the smaller bits to be assembled to form a larger object.

A Brief History of Discovery and Development

Fractals are recognized as a way of modeling the behavior of complex natural systems like weather and

Figure 1.

Figure 2.

Figure 3.

animal population behavior. Such systems are described as being chaotic. The chaos theory is a way of trying to explain how the behavior of very complex phenomena can be predicted, based on patterns that occur in the midst of the complexity.

Looking at a fractals, one can get the sense of how fractals and chaos have grown up together. A fractal can look mind-bendingly complex on first glance. A closer inspection, however, will reveal order in the chaos; the repeated pattern of some bit of information or of an object. Thus, not surprisingly, the history of fractals is tied together with the search for order in the world and the universe.

In the nineteenth century, the French physicist Jules Henri Poincaré (1854–1912) proposed that even a miniscule change in a complex system that consisted of many relationships (such as an ecosystem like the Florida Everglades or the global climate) could produce a result to the system that is catastrophic. His idea came to be known as the "Butterfly effect" after a famous prediction concerning the theory that the fluttering of a butterfly's wings in China could produce a hurricane that would ravage Caribbean countries and the southern United States. The Butterfly effect relied on the existence of order in the midst of seemingly chaotic behavior.

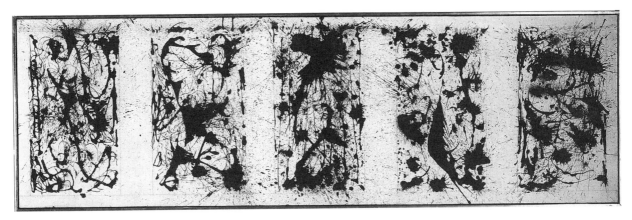

Polyptych, painting by Jackson Pollock. GIRAUDON/ART RESOURCE, NY. REPRODUCED BY PERMISSION.

In the same century, the Belgian mathematician P. F. Verhulst (1804–1849) devised a model that attempted to explain the increase in numbers of a population of creatures. The work had its beginning in the study of rabbit populations, which can explosively increase to a point where the space and food available cannot support their numbers. It turns out that the population increase occurs predictably to a certain point, at which time the growth in numbers becomes chaotic. Although he did not realize it at the time, Verhulst's attempt to understand this behavior touched on fractals.

Leaping ahead over 100 years, in 1963 a meteorologist from the Massachusetts Institute of Technology named Edward Lorenz made a discovery that Verhulst's model was also useful to describe the movement of complicated patterns of atmospheric gas and of fluids. This discovery spurred modern research and progress in the fractal field.

In the late 1970s, a scientist working at International Business Machines (IBM) named Benoit Mandelbrot was working on mathematical equations concerning certain properties of numbers. Mandelbrot printed out pictures of the solutions and observed that there were small marks scattered around the border of the large central object in the image. At first, he assumed that the marks were created by the unclean roller and ribbon of the now-primitive inkjet type printer. Upon a closer look, Mandelbrot discovered that the marks were actually miniatures of the central object, and that they were arranged in a definite order. Mandelbrot had visualized a fractal.

This initial accidental discovery led Mandelbrot to examine other mathematical equations, where he discovered a host of other fractals. Mandelbrot published a landmark book, *The Fractal Geometry of Nature,* which has been the jump-start for numerous fractal research in the passing years.

Real-life Applications

FRACTALS AND NATURE

Fractals are more than the foundation of interesting looking screensavers and posters. Fractals are part of our world. Taking a walk through a forest is to be surrounded by fractals. The smallest twigs that make up a tree look like miniature forms of the branches, which themselves are similar to the whole tree. So, a tree is a repeat of a similar (but not exact) pattern. The leaves on a softwood tree like a Douglass fir or the needles on a hardwood tree like a maple are almost endless repeats of the same pattern as well. So are the stalks of wheat that sway in the breeze in a farmer's field, as are the whitecaps on the ocean and the grains of sand on the beach. There are endless fractal patterns in the natural world.

In the art world, the popularity of the late painter Jackson Pollock's seemingly random splashes of color on his often immensely-sized canvasses relate to the fractal nature of the pattern. Pollock's paintings reflected the fractal world of nature, and so strike a deep chord in many people.

By studying fractals and how their step-by-step increase in complexity, scientists and others can use fractals to model (predict) many things. As we have seen above, the development of trees is one use of fractal modeling. The growth of other plants can be modeled as well. Other systems that are examples of natural fractals are weather (think of a satellite image of a hurricane and television footage of a swirling tornado), flow of fluids in a stream, river and even our bodies, geological activity like earthquakes, the orbit of a planet, music, behavior of groups of animals and even economic changes in a country.

The colorful image of the fractal can be used to model how living things survive in whatever environment

Fractals and Jackson Pollock

Early in his career as a painter, the American artist Jackson Pollock struggled to find a way to express his artistry on canvas. Ultimately, he unlocked his creativity by dripping house paint onto huge canvasses using a variety of objects including old and hardened paintbrushes and sticks. The result was a visual riot of swirling colors, drips, splotches, and cross-canvas streaks.

There was more to Pollock's magic than just the random flinging of paint onto the canvas. Typically, he would begin a painting by using fluid stokes to draw a series of looping shapes. When the paint dried, Pollock often connected the shapes by using a slashing motion above the canvas. Then, more and more layers of paint would be dripped, poured and hurled to create an amazing and colorful spider-web of trails all over the huge canvas.

Pollock's paintings are on display at several of the world's major museums of modern art, including the Museum of Modern Art in New York and the Guggenheim Museum in Venice, Italy, and continue to amaze many people. The patterns of paint actually traced Pollock's path back and forth and around the canvas as he constructed his images. One reason that these patterns have such appeal may be because of their fractal nature.

In 1997, physicist and artist Richard Taylor of the University of New South Wales in Australia photographed the Pollock painting *Blue Poles, Number 11, 1952*, scanned the image to convert the visual information to a digital form, and then analyzed the patterns in the painting. Taylor and his colleagues discovered that Pollock's artistry represented fractals. Shapes or patterns of different sizes repeated themselves throughout the painting. The researchers postulated that the fact that fractals are so prevalent in the natural world makes a fractal image pleasing to a person at a subconscious level.

Analysis of Pollock while he was painting and of paintings over a 12-year period from 1943–1952 showed that he refined his construction of fractals. Large fractal patterns were created as he moved around the edge of the canvas, while smaller fractal patterns were produced by the dripping of paint onto the canvas.

Pollock died in a high-speed car crash in 1956, long before the discovery of fractals that powered his genius.

---

they are in. The complexity of a fractal mirrors the complexity of nature. The rigid rules that govern fractal formation are also mirrored in the natural world, where the process of constant change that is evolution takes place in reasonable way. If a change is unreasonable, such as the sudden appearance of a strange mutation, the chance that the change will persist is remote. Fractals and unreasonable changes are not compatible.

Let us consider the fractal modeling of a natural situation. An example could be the fate of a species of squirrel in a wooded ecosystem that is undergoing a change, such as commercial development. The squirrel's survival depends on the presence of the woods. In the fractal model, the woods would be colored black and would be the central image of the developing fractal. Other environments that adversely affect the squirrel, such as smoggy air or the presence of acid rain, are represented by different colors. The colors indicate how long the squirrel can survive in the adverse condition. For example, a red color might indicate a shorter survival time than a blue color. When these conditions are put together in a particular mathematical equation, the pattern of colors in the resulting fractal, and the changing pattern of the fractal's shape, can be interpreted to help predict how environmental changes in the forest will affect the squirrel, especially at the border of the central black shape, where the black color meets the other colors in the image.

MODELING HURRICANES AND TORNADOES

Nonliving systems such as hurricanes and tornadoes can also be modeled this way. Indeed, anything whose survival depends on its surroundings is a candidate for fractal modeling. For example, a hurricane draws its sometimes-terrifying strength from the surrounding air and sea. If the calm atmosphere bordering a hurricane, and even the nice sunny weather thousands of miles away could be removed somehow, the hurricane would very soon disappear.

NONLIVING SYSTEMS

Other nonliving systems that can be modeled using fractals include soil erosion, the flicking of a flame and the tumbling or turbulent flow of a fluid like water. The movement of fluid through the tiny openings in rocks is another example. Indeed, oil companies use fractal modeling to try to unravel the movement of oil through rock formations to figure out where the best spot to drill might be to get the most oil with the least expense and danger.

ASTRONOMY

Fractals can be useful in understanding the behavior of events far from Earth. Evidence is mounting that the arrangement of galaxies in the inky vastness of space is fractal-like, in that the galaxies are somewhat similar in shape and are clustered together in a somewhat ordered way. "Clustered together" is relative; the galaxies are millions of light years apart. Still, in the infinity of space, the galaxies can be considered close neighbors. While the fractal nature of the universe is still controversial, it does make sense, because here on Earth the natural world beats to a fractal rhythm.

CELL PHONE AND RADIO ANTENNA

Fractals also have real-world applications in mechanical systems. One example is the design of the antennas that snag radio and other waves that pass through the air. A good antenna needs a lot of wave-trapping wire surface. Having a long and thin wire is not the best design. But, because some antennas need to fit into a narrow space (think of the retractable antenna on a car and in a cellular phone), there is not much room for the wire. The solution is fractals, whose mix of randomness (portions of the fractal) and order (the entire fractal) can pack a greater quantities of material into a smaller space.

By bending wires into the multi-star-shaped fractal that is the star-shaped Koch's curve, much more wire can be packed into the narrow confines of the antenna barrel. As an added benefit, the jagged shape of the snowflake-shaped fractal actually increases the electrical efficiency of the antenna, doing away with the need to have extra mechanical bits to boost the antenna's signal-grabbing power. Some companies use fractal antennas in cell phones. This innovation has proven to be more efficient than the traditional straight piece of wire antennas, they are cheaper to make, and they can be built right into the body of the phone, eliminating the pull-up antenna. The next time your cell phone chirps, the incoming connection might be due to a fractal.

COMPUTER SCIENCE

Another use of fractals has to do with computer science. Images are compressed for transmission as an email attachment in various ways such as in JPEG or GIF formats. A route of compression called fractal compression, however, enables the information in the image to be squeezed into a smaller, more easily transmitted bundle at one end, and to be greatly enlarged with a minimal loss of image quality.

There are many fractal equations that can be written, and so there are many images of fractals. The images are often beautiful; many sites on the Internet contain stunning fractal images available for download.

Where to Learn More

Books

Barnsley, M.F. *Fractals Everywhere*. San Francisco: Morgan Kaufmann, 2000.

Lesmoir-Gordon, N., W. Rood, R. Edney, and R. Appignanesi. *Introducing Fractal Geometry*. New York: National Book Network, 2000.

Mumford, D., C. Series, and D. Wright. *Indra's Pearls: The Vision of Felix Klein*. Cambridge: Cambridge University Press, 2002.

Web sites

Connors, M.A. "Exploring Fractals." *University of Massachusetts Amherst* <http://www.math.umass.edu/~mconnors/fractal/fractal.html> (September 8, 2004).

Lanius, Cynthia. "Why Study Fractals?" *Rice University School Math Project* <http://math.rice.edu/~lanius/fractals/WHY/> (September 8, 2004).

Overview

A fraction is a number written as two numbers with a horizontal or slanted line between them. The value of the fraction is found by dividing the number above the line by the number below the line. Not only are fractions a basic tool for handling numbers in mathematics, they are used in daily life to measure and price objects and materials that do not come in neatly countable, indivisible units. (We must often deal with a fraction of a pizza or a fraction of an inch, but we rarely have to deal with a fraction of an egg.) Fractions are closely related to percentages.

Fundamental Mathematical Concepts and Terms

WHAT IS A FRACTION?

Every fraction has three parts: a horizontal or slanted line, a number above the line, and a number below the line. The number above the line is the "numerator" and the number below the line is the "denominator." For example, in the fraction 3/4 (also written ¾), the numerator is 3 and the denominator is 4.

The fraction 3/4 is one way of writing "3 divided by 4." In general, a fraction with some number a in the numerator and some number b in the denominator, a/b, means simply "a divided by b." For example, writing 4/2 is the same as writing $4 \div 2$. Because division by 0 is never allowed, a fraction with 0 in the denominator has no meaning.

You can think of a fraction as a way to say how many portions. For example, if you slice 1 pizza into 8 equal-sized parts, each piece is an eighth of a pizza, 1/8 of a pizza. If you put 3 of these pieces on your plate, you have three eighths of the pizza, or 3/8.

TYPES OF FRACTIONS

There are different kinds of fractions. A proper fraction is a fraction whose value is less than 1, and an improper fraction is a fraction whose value is greater than or equal to 1. For example, 3/5 is a proper fraction, but 5/3 is an improper fraction. Despite the disapproving sound of the word "improper," there is nothing mathematically wrong with an improper fraction. The only difference is that an improper fraction can be written as the sum of a whole number and a proper fraction: 5/3, for example, can be written as $1 + 2/3$.

A unit fraction is any fraction with 1 in the numerator. This kind of fraction is so common that the English language has special words for the most familiar ones: 1/2 is a "half," 1/3 is a "third," and 1/4 is a "quarter."

Gas prices for (from top) plus, premium, and diesel, are typically shown with fractions denoting tenths of a cent. AP/WIDE WORLD PHOTOS. REPRODUCED BY PERMISSION.

Two or more fractions are called equivalent if they stand for the same number. For example, 4/2 and 8/4 are equivalent because they both equal 2.

A lowest-terms fraction is a fraction with all common terms canceled out of the numerator and denominator. A "common term" of two numbers is a number that divides evenly into both of them: 2 is a common term of 4 and 16 because it goes twice into 4, eight times into 16. For the fraction 2/16, therefore, 2 is a common term of both the numerator and denominator, and so the fraction 2/16 is not a lowest-terms fraction. We can make 2/16 into a lowest-terms fraction by dividing the numerator and the denominator by 2.

A mixed fraction is made up of an integer plus a fraction, like 1 + 1/2. In cooking and carpentry (but never in mathematics), a mixed fraction is written without the "+" sign: 1 1/2.

RULES FOR HANDLING FRACTIONS

To be useful, fractions must be added, subtracted, multiplied, and divided by other numbers. The rules for how to do each of these things are given in Table 1.

FRACTIONS AND DECIMALS

Fractions are closely related to another mathematical tool used in science, business, medicine, and everyday life,

namely decimal numbers. A number in decimal form, such as 3.1415, is shorthand for a sum of fractions: each of the numbers to the right of the decimal point (the "." in 3.1415) stands for a fraction with a multiple of 10 in its denominator. The first position to the right of the decimal point is a tenth, the second is a hundredth, the third is a thousandth, and so forth: .1 = 1/10, .01 = 1/100, .001 = 1/1,000, and so on. Therefore we can write any decimal number as a sum of fractions; for example, 3.1 = 3 + (1/10).

FRACTIONS AND PERCENTAGES

Fractions are also close cousins of percentages, which are fractions with 100 in the denominator. For example, to say "50 percent" is exactly the same as saying "fifty hundredths" (50/100). This fraction, 50/100, can be reduced to a least-terms fraction by dividing the numerator and the denominator by 50 to get 1/2. Accordingly, "50 percent" is the same as "half."

However, if percentages are just fractions, why use percentages? We do so because they give us a quick, useful way of relating one thing (a count or concept) to another. Say, for example, that we want to know how many people in a population of 150 million are unmarried. We conduct a survey and find out that the answer is 77 million. To describe this fact by reeling off the raw data—"77 million out of 150 million people in this population are unmarried"—would be truthful but clumsy. We can make things a little better by writing the two numbers as a fraction, 77,000,000/150,000,000, and then converting this into a least-terms fraction by dividing the numerator and denominator by 1,000,000. This gives us 77/150, which is more compact than 77,000,000/150,000,000, but is still hard to picture in the mind: how much is 77/150? Most of us have to do a little mental arithmetic to even say whether 77 is more than half of 150 or not. (It's a little more.) The handiest way to express our results would be to use a fraction with a familiar, easy-to-handle denominator like 100—a percentage. One way to do this is to divide 77 by 150 on a calculator, read off the answer in decimal form as .5133333 (the 3s actually go on forever, but the calculator cannot show this), and round off this number to the nearest hundredth. Then we can say, "51 percent of this population is married"—51/100.

By rounding off, however, we throw away a little information. (If all you keep from .5133333 is .51, the .0033333 is gone—lost.) In this case, however, as in great many real-life cases, the loss is not enough to matter. It is small because a hundredth is a small fraction. If we rounded off to thirds instead of hundredths, we would lose much more information: the closest we could come

Operation	Rule	Example
Multiply a fraction by an integer, n	$n\dfrac{a}{b} = \dfrac{na}{b}$	$4 \times \dfrac{3}{5} = \dfrac{4 \times 3}{5} = \dfrac{12}{5}$
Divide a fraction by an integer, n	$\dfrac{a}{b} \div n = \dfrac{a}{bn}$	$\dfrac{1}{2} \div 4 = \dfrac{1}{2 \times 4} = \dfrac{1}{8}$
Multiply fractions	$\dfrac{a}{b}\dfrac{c}{d} = \dfrac{ac}{bd}$	$\dfrac{1}{3} \times \dfrac{4}{5} = \dfrac{1 \times 4}{3 \times 5} = \dfrac{4}{15}$
Divide fractions	$\dfrac{a}{b} \div \dfrac{c}{d} = \dfrac{a}{b}\dfrac{d}{c} = \dfrac{ad}{bc}$	$\dfrac{1}{3} \div \dfrac{4}{5} = \dfrac{1}{3} \times \dfrac{5}{4} = \dfrac{5}{12}$
Add fractions	$\dfrac{a}{b} + \dfrac{c}{d} = \dfrac{ad + bc}{bd}$	$\dfrac{2}{9} + \dfrac{1}{7} = \dfrac{2 \times 7 + 1 \times 9}{9 \times 7} = \dfrac{23}{63}$
Subtract fractions	$\dfrac{a}{b} - \dfrac{c}{d} = \dfrac{ad - bc}{bd}$	$\dfrac{2}{9} + \dfrac{1}{7} = \dfrac{2 \times 7 - 1 \times 9}{9 \times 7} = \dfrac{5}{63}$

Table 1: Rules for handling fractions.

to 77/150 would be 2/3, which is 66%, which is much farther from the truth than 51% is. By expressing information as percentages (in numbers of hundredths), we get three advantages: (1) accuracy, because hundredths allow for pretty good resolution or detail; (2) compactness, because a percentage is usually easier to write down than the raw numbers; and (3) familiarity—because we are used to them.

ALGEBRA

All the rules that apply to adding, subtracting, multiplying, and dividing fractions are used constantly in algebra and higher mathematics. Simple fractions have only an integer in the numerator and an integer in the denominator, but there is no reason not to put more complicated mathematical expressions in the numerator and denominator—and we often do. For example, we can write expressions such as $x^2 / (9 + x^2)$ where x stands for an unknown number. Any material above the line in fraction-like expression such as 1/2 or to the left side of a fraction written in linear form such as 1/2), no matter how complicated it is, is the "numerator," and any material below the line is the "denominator." Such expressions are added, subtracted, multiplied, and divided using exactly the same rules that apply to ordinary number fractions like 3/4.

A Brief History of Discovery and Development

Fractions were invented about 4,000 years ago so that traders could keep track of how they were dividing goods and profits. For instance, if three traders own a grain-selling stall together, and 10 sacks of grain are sold at 1 denarius apiece, how much profit should the books record for each owner? Three does not go evenly into 10, so we have a fraction, 10/3. This is an improper fraction, and can be reduced to 3 + 1/3. Each trader should get 3 whole denarii and credit on the books for 1/3 more.

Many ancient civilizations developed some form of dealing with fractions—the Mayan, the Chinese, the Babylonian, the Egyptian, the Greeks, the Indians (in Asia), and the Romans. However, for centuries these systems of writing and dealing with fractions had severe limits. For example, the Egyptians wrote every fraction as a sum of unit fractions (fractions with a "1" in the numerator). Instead of writing 2/7, the Egyptians wrote their symbols for 1/4 + 1/28 (which, if you do the math, does add up to 2/7). The Romans did not write down fractions using numbers at all, but had a limited family of fractions that they referred to by name, just as we speak of a half, a third, or a quarter. All the other ancient systems had their own problems; in all of them it was very difficult to do calculations with fractions, like adding and subtracting and multiplying them. Finally, around 500 A.D., a system of number-writing was developed in India that was similar to the one we use now. In fact, our system is descended from that one through Arab mathematics.

Fraction theory advanced in the 1600s with the development of practical applications for continued fractions. Continued fractions are fractions that have fractions in their denominators, which have fractions in their denominators, and so on, forever if need be.

Continued fractions were originally used for designing gears for clocks and other mechanisms. Today they are used in the branch of mathematics called "number theory," which is used in cryptography (secret coding), computer design, and other fields.

Real-life Applications

COOKING AND BAKING

Fractions are basic to cooking and baking. Look at any set of cup measures or spoon measures: they are all marked in fractions of a cup (or, in Europe, fractions of a liter). A typical cup-measure set contains measures for 1 cup, 1/2 cup, 1/3 cup, 1/4 cup, and 1/8 cup; a typical spoon set contains measures for a tablespoon, a teaspoon, 1/2 teaspoon, 1/4 teaspoon, 1/8 teaspoon; some also include 1/2 tablespoon and 1 1/2 tablespoon.

Not only are measurements in cooking and baking done in fractions, a cook must often know at least how to add and multiply fractions in order to use a recipe. Recipes might only given for a single batch: if you want to make a half batch, or a double or triple batch, you must halve, double, or triple all the fractional measurements in the recipe. Say, for example, that a cookie recipe calls for 2 2/3 cups of flour and you want to make a triple batch. How much flour do you need to measure?

There are several ways to do the math, but all require a knowledge of fractions. One way is to write the mixed number 2 2/3 as a fraction by first noting that 2 = 6/3. Therefore, by the rule for adding fractions that have the same denominator, 2 2/3 = 6/3 + 2/3 = 8/3. To triple the amount of flour in the batch, then, you multiply 8/3 cups by 3:

$$\frac{8}{3} \times 3 = \frac{8 \times 3}{3} = \frac{24}{3}$$

At this point you can either get out your 1/3-cup measure and measure 24 times, which is a lot of work, or you can try reducing 24/3 to a mixed fraction. If you try reducing

the fraction, you will probably discover at once that 24 / 3 = 8. Therefore, you can measure eight times with your 1-cup measure and move on to the next ingredient on the list.

RADIOACTIVE WASTE

A continuing political issue in nuclear capable countries is the question of what to do with nuclear waste. Such waste is an unwanted by-product of making electricity from metals like uranium and plutonium. Nuclear waste gives off radiation, a mixture of fast-moving atomic particles and invisible, harmful kinds of light that at low levels may cause cancer and at high levels can kill living things. Only over very long periods will radioactive waste slowly become harmless as it breaks down naturally into other elements. This happens quickly for some substances in the waste mixture, slowly for others. How quickly a substance loses its radioactivity is expressed as a fraction, the "half-life" of the substance. The half-life of a substance is the time it takes for any fixed amount of the substance to lose 1/2 of its radioactivity. For the element plutonium, which is found in most nuclear waste, the half-life is about 24,000 years. That is, no matter how much plutonium you start out with at time zero, after 24,000 years you will have half as much plutonium left. (But not half as much radioactivity, exactly, since some of the elements that plutonium breaks down into are radioactive themselves, with half-lives of their own, and must break down further before they can become harmless.)

By multiplying fractions, it is possible to answer some questions about how much radioactive waste will remain after a certain time. For instance, after two half-lives, how much of 1 kilogram (kg) of plutonium will be left? This is the same as asking what is a half of a half, which is the same as multiplying 1/2 times itself: 1 kg × 1/2 × 1/2 = 1/4 kg.

This can be carried on for as many steps as we like. For example, how much of 1 kg of plutonium will be left after 10 half-lives (that is, after 240,000 years)? The answer is 1 kg × 1/2 × 1/2 × 1/2 × 1/2 × 1/2 × 1/2 × 1/2 × 1/2 × 1/2 × 1/2 = 1 / 1024 kg.

This shows that the plutonium will never completely disappear. The denominator gets larger and larger, which makes the value of the fraction smaller but cannot make it equal to zero.

MUSIC

Fractions and rhythm Fractions are used throughout music. In Western music notation, the time-values of notes are named after fractions: besides the whole note, which lasts one full beat, there is the half-note, which lasts only half a beat, and the quarter note, eighth note, six-teenth note, and so forth. Notice that these fractions—1/2, 1/4, 1/8, 1/16—all have multiples of 2 in the denominator. In fact, each fraction in the series is the previous fraction times 1/2. That is, each standard type of note lasts 1/2 as long as the next-longest type. Music notation also has "rest" symbols, marks that tell you how long to be silent. Just as there are notes with various values, there are rest symbols with various time values—whole, half, quarter, and eighth rests.

Nor are we limited to the beat fractions given above. Each of the standard notes can also be marked with a dot, which indicates that the duration of the note is to be increased by 1/2. This is the same as multiplying the time value of the original note by 3/2. So, for example, the time value of a dotted eighth note is given by $1/8 \times 3/2$, which is 3/16. And by tying three notes together with an arc-shaped mark (a "tie") and writing the number "3" by the arc, we can show that the musician should play a "triplet," a set of three notes in which each note lasts 1/3 of a beat.

Fractions and the musical scale A single guitar string can produce many different notes. The guitar player pushes the string down with a fingertip on a steel bar called a "fret," shortening the part of the string that vibrates freely. The shorter the freely vibrating part of the string, the higher the note. The Greeks also made music using stringed instruments, and they noticed several thousand years ago that the notes of their musical scale—the particular notes that just happen to be pleasing to the human ear—were produced by shortening a string to certain definite fractions of its full length. Sounding the open string produced the lowest note: the next-highest pleasing note was produced by shortening the string to 4/5 of its open length. Shortening the string to 3/4, 2/3, and 3/5 length—each fraction smaller than the last, each note higher—produced the three notes in the Greek 5-note scale. Shortening a string to 1/2 its original length produces a note twice as high as the open string does: musically, this is considered the same note, and the scale starts again.

Modern music systems have more than 5 notes; in the Western world we use 12 evenly-spaced notes called "semitones." Seven of these notes have letter names—A, B, C, D, E, F, and G—and five are named by adding the terms "sharp" or "flat" to the letters. These musical choices are built right into our instruments. If you look at the neck of a guitar, for example, you will see that the frets divide it up into 12 parts. Why 12? The ancients decided to see what would happen if they divided the fractional string lengths of the Greeks' 5-note scale into similar fractions. That is, if one pleasing note is produced by

shortening the string to 2/3 its open length, what note do we get if we shorten the string to 2/3 of that shorter length? The vibrating part of the string is then $2/3 \times 2/3 = 4/9$ the length of the open string. But this is less than half the length of the string, making the note an octave too high, so we double the fraction to lengthen the string and lower the note: $4/9 \times 2 = 8/9$. And indeed, the fret for playing a B on the A string of a guitar does shorten the string to 8/9 of its open length. By similarly multiplying the fractions that gave the other notes in the original Greek scale, people discovered 12 notes—the semitones we use today. Later, in the 1600s, people decided that they would space the notes slightly differently, based on multiples of the 12th root of 2 rather than on fractions. This makes it easier for instruments to be tuned to play together in groups, as the notes are spaced perfectly evenly, and as long two instruments match on one semitone they will match on all the others too. These modern notes are close to the fraction-based notes, but not exactly the same.

SIMPLE PROBABILITIES

Many U.S. states make money through lotteries, public games in which any adult can buy one or more tickets. The money spent on tickets is pooled, the state keeps a cut, and the rest is given to a single winning ticket-buyer who is chosen by chance. Some states have become dependent on the money they make from the lotteries, which now totals many billions of dollars every year. If you bought a lottery ticket, what would your chances of winning? Mathematically, we would ask: what is the probability that you will win?

A "probability" is always a number between 0 and 1. Zero is the probability of an event that can't possibly happen; 1 is the probability of an event that is sure to happen; and any number between 0 and 1 can stand for the probability of an event that might happen. If you buy one lottery ticket in which, say, 10 million other people have bought a ticket, then the probability that you will win is a unit fraction with 10 million in the denominator: a 1-ticket chance of winning = 1/10,000,000.

If you buy two tickets, your chance of winning is this fraction multiplied by 2: a 2-ticket chance of winning = 2/10,000,000. This fraction can be reduced to a lowest-terms fraction by dividing both the numerator and denominator by 2 to yield 1/500,000. Accordingly, buying two tickets doubles your chances of winning. On the other hand, double a very small chance is still a very small chance.

Lottery chances are typical of a certain kind of probability encountered often in everyday life, namely, when some number of events is possible (say, N events), but only one of these N events can actually happen. If all

N events are equally likely, then the chance or probability of any one of them happening is simply the fraction 1/*N*.

The math of probabilities gets much more complicated than this, but simple fractional probabilities can be important in daily life. Consider, for instance, two people who are considering having a baby. Many serious diseases are inherited through defective genes. Each baby has two copies of every gene (the molecular code for producing a certain protein in the body), one from each parent, and there are two kinds of defective genes, "dominant" and "recessive." For a disease controlled by a dominant defect, if the baby has just 1 copy of the defective gene from either parent, it will have the disease. If one parent carries one copy of the dominant defective gene in each of their cells, the probability that the baby will have the dominant gene (and therefore the disease) is 1/2; if both parents have one copy of the dominant gene, the probability that the baby will have the dominant gene is 3/4. Parents who are aware that they carry defective genes cannot make informed choices about whether to have children or not unless than can understand these fractions (and similar ones).

OVERTIME PAY

In many jobs, workers who put in hours over a certain agreed-only weekly limit—"overtime"—get paid "time and a half." This means that they are paid at 3/2 or 1 + 1/2 times their usual hourly rate.

Multiplying this fraction by your usual hourly rate gives the amount of money your employer owes you for your overtime.

TOOLS AND CONSTRUCTION

Most of us have to use tools at some time or another, and millions of people make a living using tools. An understanding of fractions is necessary to do any tool-work much more complicated than hammering in a nail. To begin with, all measurements, both in metal and wood, are done using rulers or measuring tapes that are divided into fractions of an inch (in the United States) or of a meter (in Europe). The fractions used are based on halving: if the basic unit of measure is an inch, and then the ruler or tape is marked at 1/2 inch, 1/4 inch, 1/8 inch, and 1/16 inch, each fraction of being half as large as the next-largest one. So to read a ruler or a tape measure it is necessary to at least be able to read off the fractions. Further, in making anything complex—framing a house, for example—it is necessary to be able to add and subtract fractions. For example, you are framing a wall that is 6 feet (72 inches) wide. You have laid down "two-by-fours" at the ends of the wall, at right angles to it where

the other walls meet, like the upright arms of a square "U." (A two-by-four was 2 inches thick and 4 inches wide many years ago, but today is 1 1/2 inches thick and 3 1/2 inches wide. Notice that in carpentry, as in cooking, it is acceptable to write "1 1/2" for 1 + 1/2.) Now you want to cut a two-by-four to lay down along the base of the 6-foot wall, in the space that is left by the two two-by-fours that are already down at right angles: you want to put in the bottom of the square "U." How long must it be?

You could, in this case, just measure the distance with a tape measure. But there are many occasions, in building a house, when it is simply not possible to measure a distance directly, and we'll pretend that this is one of them (because it's relatively simple). Each of the two-by-fours uses up 3 1/2 inches of the 72 inches of wall. There are several ways to do the problem: one is to add 3 1/2 + 3 1/2 to find that the two two-by-fours use up 7 inches of space. Since 72 − 7 = 65, you want to cut a board 65 inches long.

Another place where fractions pop up in the world of tools and construction is in dimensions of common tools. United States drill bits, for instance, typically come in widths of the following fractions of an inch: 1/4, 3/16, 5/32, 1/8, 7/64, 3/32, 5/64, and 1/16.

FRACTIONS AND VOTING

Simple fractions like and 1/4, 1/3, and 2/3 have a common-sense appeal that leads us to use them again and again in everyday life. They appear often in politics, for example. The United States Constitution states that the President (or anyone else who could be impeached) can only be convicted if 2/3 of the members of the Senate who are present agree. Some other fraction could have been used—4/7, say—but would not have been as simple.

One of the most famous fractions in political history, 3/5, appears in the United States Constitution, Article I, Section 2, which reads as follows: "Representatives and direct Taxes shall be apportioned among the several States which may be included within this Union, according to their respective Numbers, which shall be determined by adding to the whole Number of free Persons, including those bound to Service for a Term of Years, and excluding Indians not taxed, three fifths of all other Persons." Translated into plain speech, this means that the more people live in a state, the more congresspeople would be needed to represent it in the House of Representatives, giving it more voting power. The phrase "all other Persons" was an indirect reference to "slaves." Because of this clause in the Constitution, slaves, though they had no human rights, would count toward allotting congresspeople (and thus political power) to Southern states. The Southern states wanted the Constitution to count slaves as equal to free persons for the purposes of allotting state power in Congress, and the Northern states wanted slaves counted as a smaller fraction or not at all; James Madison proposed the fraction 3/5 as a compromise. The rule was ultimately canceled by the Fourteenth Amendment after the Civil War, but it did play an important part in U.S. history: Thomas Jefferson was elected to the Presidency in 1800 by Electoral College votes of Southern states derived from the three-fifths rule. (By the way, neither the original Constitution or the 14th

Amendment counted women at all: you might say that they were counted at 0/5 until the 19th Amendment gave them the legal right to vote in 1920.)

In 2004, a bill was proposed to give teenagers fractional voting rights in California. If the bill had passed, 14- and 15-year-olds would have been given votes worth 1/4 as much as those of adults and 16- and 17-year-olds would have been given votes worth 1/2 as much as those of adults. (All people 18 years and older already have the right to vote, each counted as one full vote.) The intent was to teach teenagers to take the idea of participating in democracy seriously from a younger age. Some European countries such as the United Kingdom have seriously considered lowering the voting age to 16—with no fractions involved.

Where to Learn More

Web sites

"Egyptian Fractions." Ron Knott, Dept. of Mathematics and Statistics, University of Surrey, Guildford, UK, October 13, 2004. <http://www.mcs.surrey.ac.uk/Personal/R.Knott/Fractions/egyptian.html#ef> (April 6, 2005).

"Causes of Sea Level Rise." Columbia University, 2005. <http://www.columbia.edu/~epg40/elissa/webpages/Causes_of_Sea_Level_Rise.html> (April 4, 2005).

"Fractions." Math League Multimedia, 2001. <http://www.mathleague.com/help/fractions/fractions.htm#whatisafraction> (April 6, 2005).

"Fraction." Mathworld, Wolfram Research, 2005. <http://mathworld.wolfram.com/Fraction.html> (April 14, 2005).

Functions

A function is a rule for relating two or more sets of numbers. The word "function" is from the Latin for perform or execute, and was first used by the German mathematician Gottfried Wilhelm von Leibniz, one of the inventors of calculus, in the late 1600s. A function is often written as an equation, that is, as two groups of mathematical symbols separated by an equals sign. A function can also be expressed as a list of numbers on paper or in computer memory.

A function in mathematics is, more or less, what the sentence is in English: it expresses a complete mathematical thought. Much of mathematics consists of functions and the rules that relate them to each other. To use mathematics in the real world means looking for functions to describe how things behave, and then using those functions to design, understand, predict, or control the things they describe.

Fundamental Mathematical Concepts and Terms

Mathematicians divide all known kinds of function into types. Each type of function has different properties and different uses. Some of the most common types of function are polynomials, exponentials, and trigonometric functions. Lists of more complicated, unusual functions are published as lists of "special functions."

FUNCTIONS AND RELATIONS

A function is a particular type of "relation." A relation is a set of ordered pairs of numbers. For example, the three pairs (2,1), (4,2), and (3,1) give a relation. It is usual in mathematics to refer to each left-hand number in a pair as an x and to each right-hand number as a y. In this relation, x can be 2, 4, or 3, and y can be 1 or 2.

A relation is also a function if each y goes with only one x. The relation (2,1), (4,2), and (3,1) is not a function because a y value of 1 goes with two values of x, namely 2 and 3. For a relation to be a function, each x must be paired with one and only one x.

In practice, the word "function" is often used to include equations that assign more than one y value to each x value. Such equations are sometimes called "multiple-valued functions" and are also useful in dealing with the real world.

HOW FUNCTIONS ARE DESCRIBED

Functions are usually written as an equations. This is done because most functions relate not only a few pairs of numbers, as in the example already given, but many pairs of numbers—too many to write down. For example, if we decide to use x and y to stand for any of the positive counting numbers $(1, 2, 3, \ldots)$, then the equation $y = 2x$ describes a function that relates an infinite number of x's to an infinite number of y's as shown in Table 1.

The only practical way to write down such a function is in the form of an equation.

Can such a simple function as $y = 2x$ describe something in the real world? Certainly. If you are climbing a steep mountain that goes up two feet for every foot you go forward, then the vertical (upward) distance you travel is twice as great as the horizontal (sideways) distance you travel. If we use y to stand for the distance you have traveled vertically and x to stand for the distance you have traveled horizontally, then $y = 2x$.

The letters x and y are called "variables" because they can vary, that is, take on different values. The numbers that x can stand for are called the function's "domain" and the numbers that y can stand for are called its "range." In the example above, the domain and the range both consist of all the positive counting numbers, 1, 2, 3, and so forth. Many symbols other than x and y are used to stand for variables; these two are merely the most common, by tradition.

A function can be thought of as a sort of number machine that takes in an x value and puts out a y value. The y value can thus be thought of as depending on the x value, so x is sometimes called the "independent variable" and y the "dependent variable." If we are using a function to describe cause-and-effect in the real world, we describe the cause as an independent variable and the effect as a dependent variable. For example, if a rocket is pushing on a spacecraft with a certain force, and we want to write a function that relates this force to the acceleration (increase of speed) of the spacecraft, we write the force as the input or independent variable and the acceleration of the spacecraft as the output or dependent variable. It is the acceleration that depends on the force, not the other way around.

There are also functions that have more than one independent variable. We put two (or more) numbers into the function rather than one, and the function produces a single number as output.

y's		x's
2	$= 2 \times$	**1**
4	$= 2 \times$	**2**
6	$= 2 \times$	**3** ... and so on, forever

Table 1. The function $y = 2x$, where x can be any positive counting number.

moved from one form to another. For example, a fluid (gas or liquid) can store energy in several ways: as heat, as motion, or as pressure. A function called Bernoulli's equation, named after its discoverer, Swiss mathematician Daniel Bernoulli (1700–1782), describes how a moving fluid can move its energy between pressure and motion. According to Bernoulli's equation, the faster a fluid flows, the lower its pressure.

Bernoulli's equation—or, rather, the physical effect described by it—is what keeps airplanes up. An airplane wing is shaped so that as it slices through the air, the air that flows over the rounded top of the wing has to travel farther than air under the flat bottom of the wing. The air on top and on bottom must make the trip from the leading edge of the wing to the trailing edge in the same amount of time, so the wing on the upper surface, which has farther to go, is forced to flow faster. But this means, by Bernoulli's equation, that its pressure is decreased. As a result, there is more pressure—more force—per square foot on the wing's bottom than on its top. This difference in pressure is what holds the plane up. In the case of a Boeing 747, the pressure difference between the top and bottom sides of the wing is over 100 pounds per square foot.

You can test the Bernoulli principle by holding a sheet of paper by one edge so that it droops away from you, and blowing on the top of it. The paper will rise into the jet of air from your mouth because the pressure in the moving air is lower.

Bernoulli's equation only holds true for flight slower than the speed of sound (about 741 mph [1,193 km/h], at sea level). At speeds faster than the speed of sound—supersonic flight—the behavior of air is so different that functions other than Bernoulli's equation must be used to design aircraft wings.

Real-life Applications

MAKING AIRPLANES FLY

One of the basic laws of physics is that energy can neither be created nor destroyed. It can, however, be

GUILLOCHÉ PATTERNS

Look at any piece of paper money from almost anywhere in the world. Somewhere on the bill—on United States currency, it is around the border—you will see a dense, complex pattern of curving, intersecting lines. This design is known as a Guilloché pattern and they have been used since the 1850s on paper money, stocks, bonds, and other official documents in order to make them more difficult to counterfeit. Today they are also used on laminated plastic cards such as some identification cards and driver's licenses.

Guilloché patterns were originally produced by mechanical means. Today they are produced mathematically on computers, using the functions called sinusoids. Sinusoids are functions that look, when graphed, like smooth, wavy lines. A Guilloché pattern is produced by multiplying sinusoids, shifting them, adding them, and placing one on top of another. (Straight lines may also be added to make the pattern even more complex.)

Modern money is protected from counterfeiting more by invisible inks, hidden plastic threads, and other hard-to-copy features than by its Guilloché patterns, but a finely-printed Guilloché pattern still makes it more difficult to produce a convincing fake of a piece of money or an identification card. A number of software packages are for sale that make Guilloché patterns.

THE MILLION-DOLLAR HYPOTHESIS

In 1859, the same year that saw the publication of Charles Darwin's theory of evolution in *The Origin of Species*, the German mathematician Georg Friedrich Bernhard Riemann (1882–1866) made a guess about a special function, the "zeta function." His guess, or hypothesis, was that all the zeroes of this function—all the values of the independent variable for which the function equals 0—lie on a certain line.

Mathematicians have been trying to prove the Riemann hypothesis for well over century. So important is the Riemann hypothesis that in 2000, the Clay Mathematics Institute of Cambridge, Massachusetts, announced that it would give $1,000,000 to the first person to prove it. Computers have been used to test millions of zeroes of the zeta function and found that all of them, so far, do lie on the line that Riemann described. But the zeta function has an infinite number of zeroes, so no computer study can prove that *all* the zeroes lie on the line.

The Riemann hypothesis matters to real-world mathematics because it is one of the most important ideas in the study of prime numbers. A prime number is any whole number that cannot be evenly divided by any number smaller than itself except 1. (The first 10 prime numbers are 2, 3, 5, 7, 11, 13, 17, 19, 23, and 29.) Large prime numbers (primes dozens or hundreds of digits long) are at the heart of modern cryptography, the science of sending secret messages—and cryptography is at the heart of our modern economy. Cryptography allows

banks and Internet users to send money, credit-card numbers, and other private information as coded messages without fear that thieves will be able to read or fake those messages. If the Riemann hypothesis concerning the zeta function is proved, it will become easier to discover new large primes.

In June 2004, a French mathematician working at Purdue University, Louis de Branges de Bourcia (1932–) claimed to have discovered a proof of the Riemann hypothesis. His proof was long and complicated, and mathematicians could not quickly agree whether he had solved the puzzle or not. As of late 2005, de Bourcia's claim had not been proved or disproved by other mathematicians.

FINITE-ELEMENT MODELS

A computer can predict the weather—more or less—by "modeling" it, that is, by working with a three-dimensional mesh or net of numbers. Each separate hole in the mesh or net—also called an element of the mesh—stands for a small piece of the atmosphere. Numbers attached to each element represent the pressure, temperature, motion, and other properties of that piece of air. Next, functions are programmed into the computer to describe the physical laws that the element must obey. Foremost among these functions are the Navier-Stokes equations, which relate air pressure to speed and acceleration. This set of equations is named after French physicist Claude-Louis Navier (1785–1836) and Irish mathematician Gabriel Stokes (1819–1903), the men who discovered them.

The computer feeds the numbers on the mesh into the functions that describe what should happen to them next. As output, the functions produce a new set of numbers that is, in effect, a prediction of what the weather will be a little bit into the future. This computation is done repeatedly, until eventually a prediction for the weather hours, days, or weeks into the future is produced. Will the hurricane head for shore? If so, where will the hurricane strike? How bad will it be?

This method of modeling or computing the behavior of physical systems is called "finite-element modeling" because the size of each piece or element of the model mesh is finite (limited) in size. Finite-element modeling was developed in the 1950s for aircraft design and is used today in a many real-life applications, including machine design, ocean current modeling, medical imaging, climate change prediction, and others.

Real weather is not really made up of separate elements or pieces, but just as a picture made of dots can get more realistic as the dots are made smaller and more numerous, a finite-element model can get more realistic as its elements are made smaller and more numerous. However, there is always a practical limit, because even the most powerful computers can only handle a limited number of calculations. In addition, some results, like hurricane forecasts, are needed in hours, not days or months. For this reason—and also because complex systems such as the weather are often "chaotic" or unpredictable over the long term by their very nature—weather forecasting will always remain imperfect. For example, when Hurricane Ivan approached the Gulf Coast of the United States in September 2004, computer models were unable to predict exactly where it would come ashore or how severe it would be by the time it did so. Repeated calculations starting from slightly different assumptions produced significantly different results.

SYNTHS AND DRUMS

Many of the characteristic sounds of modern pop music depend on math. From the deep "thump" of a driving hip-hop beat to the high-pitched "ting" or clashing sounds of high-hat cymbals, and lots of the sounds in between—snare drums, piano, strings, horns, hand claps, or sounds never heard before: are all produced mathematically, using functions.

Sounds are rapidly repeating pressure changes in air. If the pressure changes that we hear as a pure musical note are graphed, they look like a sinusoid (a type of function shaped like a wavy line). But music made only of pure sinusoids would sound flat and dull. More interesting sounds—sounds that growl, snap, clap, sing, and make our feet want to move—can be made by adding many functions together.

One of the ideas most commonly used in the design of sounds and sound systems is the fact, first discovered by French mathematician Jean Baptiste Fourier (1768–1830), that every possible sound can be built up by adding together the functions called sinusoids. Electronic sound synthesizers ("synths") allow a musician or sound designer to build sounds on this principle. In analog synths, an "oscillator bank" produces a number of sinusoids that can be shifted, amplified, and added to make new sounds.

This is not the only way that sounds can be created. Another method is called "physical modeling synthesis." In this approach, mathematical functions that describe the air vibrations made by a drum or other instrument are based the physical laws that describe the vibration of a drum head, string, soundboard, or other object. These functions are then evaluated by a computer to produce the desired sound.

NUCLEAR WASTE

Radioactive materials are made of atoms that break apart at random or unpredictable times. When an atom breaks apart ("decays"), it releases fast-moving particles and high-energy light rays. These particles and rays ("radiation") can kill or damage living cells. Only when most of the atoms in a lump of radioactive material have decayed is it no longer dangerous.

Radioactive waste is produced by the manufacture of nuclear weapons and by nuclear power plants (which make electricity) and are stored above the surface of the ground at about 130 locations around the United States. This is not a good long-term solution because accidents or terrorism might allow them to mix with the air and water. A possible solution is to bury the wastes deep underground in a part of the country that gets little rain, so that they can remain isolated until most of their radioactive atoms have decayed.

The time needed for radioactive material to become harmless is given by a mathematical function called an exponential function. This function states that after a certain amount of time—different for each radioactive substance—about half the atoms in a lump of that substance will have decayed. This amount of time is called a "half-life" of that substance. After another half-life has gone by, half of the atoms left after the first half-life will also have decayed, and so on, half-life after half-life, until the very last atom is gone. The amount of radiation given off by a quantity of radioactive waste thus decreases over time. The half-life of some radioactive substances is a fraction of a second; for others it is tens of thousands of years, or even millions of years.

If the amount of radioactivity at the beginning is R, then the amount after one half-life will be $1/2 \times R$. After another half-life it will be half of that, or $1/2 \times (1/2 \times R)$. If N is the number of half-lives that have passed, then the amount of radioactivity left will be R times $1/2$ multiplied by itself N times, which can be written as $(1/2)^N R$. As a general rule, physicists say that a sample of radioactive substance can be considered be safe after 10 half-lives. By that time, the radiation will be down to $(1/2)^{10} R$, about $1/1,000$ as much as there was at the start.

In the case of plutonium, a radioactive element found in nuclear waste, the half-life is about 20,000 years, so by this rule of thumb plutonium should be isolated from the environment for at least 200,000 years (10 half-lives).

These numbers are involved in a political dispute. Since 1978, the United States Department of Energy has been studying Yucca Mountain, in Nevada, as a place to bury about 60% of the nuclear waste that has accumulated around the country. Seventy-three miles (117 km)

Key Terms

Domain: The domain of a relation is the set that contains all the first elements, x, from the ordered pairs (x,y) that make up the relation. In mathematics, a relation is defined as a set of ordered pairs (x,y) for which each y depends on x in a predetermined way. If x represents an element from the set X, and y represents an element from the set Y, the Cartesian product of X and Y is the set of all possible ordered pairs (x,y) that can be formed.

Function: A mathematical relationship between two sets of real numbers. These sets of numbers are related to each other by a rule which assigns each value from one set to exactly one value in the other set. The standard notation for a function y = f(x), developed in the eighteenth century, is read "y equals f of x." Other representations of functions include graphs and tables. Functions are classified by the types of rules which govern their relationships.

Prime number: Any number greater than 1 that can only be divided by 1 and itself.

---

of huge tunnels to put the waste in have been dug 1,000 feet (305 m) below the surface at a cost of over $9 billion. However, no waste has yet been stored there. The site is still being studied, and government has decided that Yucca Mountain must be able to keep all its waste from leaking out for 10,000 years. Opponents of the plan—including the State of Nevada, which objects strongly to receiving the whole country's nuclear waste—argue that this is not long enough. In July 2004, a Federal Court of Appeals decided against the Department of Energy (which manages Yucca Mountain) in a lawsuit brought by the State of Nevada and environmental groups. The court ruled that 10,000 years was not long enough, and that the Department of Energy must come up with a tougher standard.

BODY MASS INDEX

Government experts said in 2000 that 11% of the United States population between the ages of 12 and 19 was overweight. That's more than double the rate measured in 1984; experts are talking of an "epidemic of obesity." But what is it, exactly, to be "overweight"?

Doctors decide whether a child or teen is overweight using something called the "body mass index (BMI) for children," also known as "BMI-for-age." Your BMI is your weight in kilograms divided by the square of your height in meters. If you are 1.7 meters (5 feet, 7 inches) tall and weigh 59 kilograms (130 pounds), then your BMI is

$$\frac{59 \text{ kg}}{(1.7 \text{ m}) \times (1.7 \text{ m})} = 20.4 \text{ kg/m}^2$$

If you are heavy for your height, you will have a high BMI.

So is 20.4 a good BMI or a bad BMI? It's not necessarily either. Based on measurements of many thousands of young people, the Centers for Disease Control of the United States government have graphed average BMI as a function of age. That is, age is graphed as the independent variable (horizontal or x axis) and BMI as the dependent variable (vertical or y axis) to produce a curve. BMI is a good example of a function that is described not by an equation, but by a collection of number pairs. (There is one chart for boys and another for girls.)

When you go for a checkup, the doctor may compare your BMI to a chart to see how many people your age have BMIs less than yours. If 95% of people your age have a lower BMI than yours, you are considered overweight. If only 5% of people your age have a lower BMI than yours, you are considered underweight.

Where to Learn More

Books

Benice, Daniel D. *Finite Mathematics with Algebra*. Philadelphia, PA: W. B. Saunders Co., 1975.

Wheeler, Ruric E., and W.D. Peeples, Jr. *Finite Mathematics: An Introduction to Mathematical Models*. Monterey, CA: Brooks/Cole Publishing Co., 1974.

Web sites

National Centers for Disease Control. "BMI for Children and Teens." April 8, 2003. <http://www.cdc.gov/nccdphp/dnpa/bmi/bmi-for-age.htm> (September 16, 2004).

Overview

Mathematics has played a role in games for centuries. Some games are purely mathematical in form, such as numbers theory, where solving a tricky math problem in of itself becomes the game. Other games, such as logic problems and puzzles, rely heavily on math to reach the solution. There are games where knowing a bit of math, such as probability, can help you determine your playing strategy, and still others where math helps you keep score.

Fundamental Mathematical Concepts and Terms

Game math covers a wide variety of entertainments, including board games such as Monopoly or chess, card games such as blackjack or poker, casino games such as roulette, logic puzzles, and number games. Number theory, a very specific sort of game math, deals with the make up of numbers themselves, and involves puzzling through the relationship between different numbers in order to find patterns in the ways that they relate. Often, number theory requires complex equations using algebra or calculus to come up with solutions, and in many cases there are mathematical questions that have yet to be answered—puzzles still unsolved. Games that involve the rolling of dice or the drawing of cards use addition, probability, and odds to determine a player's next move. Any game that involves keeping score requires someone to add up the points, and games that involve money require basic bookkeeping on the parts of the players. Computer game designers program their games to take into account the probability of each player's actions and the various reactions that the game might offer, working through every possible permutation in order to provide a realistic experience.

Probability and odds are two terms often used in relation to games that involve math, particularly games of chance. However, the terms are not interchangeable. Probability involves the outcome of a trial of chance in relation to the number of different outcomes that were possible. For instance, if you were to flip a coin, there would be two potential outcomes: heads or tails. Flip the coin once, and you will get one of those two choices. Therefore, if you flip a coin a single time, the probability that it will land heads up is one in two or 1/2. Likewise, the probability that it will land tails up is also one in two.

In the event that you have more potential outcomes, such as when you roll a single die, the principle remains the same. If you roll a die with six sides, numbered consecutively, there are six possible outcomes. The probability that you will roll a two is one out of six, or 1/6.

Odds refer to the chance that you will achieve a specific outcome compared to the chance that you will achieve any of the other potential outcomes. For instance, when tossing a coin a single time, you have only two potential outcomes, so the odds are that you will either have the coin come out heads up or tails up, one to one. These are also knows as even odds. But when rolling a single die, you have six different sides that might turn up. Therefore there is one chance of rolling any given number, such as two, while there are five chances that you will not roll a two, but one of the other potential outcomes. In this instance, the odds are one to five that you will roll a two, or 1/5.

In order to determine the odds against something happening, you take the mathematical reciprocal of the odds that it will happen. For example, in the instance of rolling the single die, if there is a 1/5 chance that you will roll a two, the odds are five to one or 5/1 against rolling a two in a single throw.

Permutations are the various choices at each stage of certain games, such as checkers, chess, or many computer games. Each time a player moves, they make a decision, and these decisions add up eventually to become their route through the game on that specific occasion. However, what would have happened if the player made a different move the first time? Or the second? In some instances there may have only been one choice, because an opponent blocked other routes or other moves led in the wrong direction, but ultimately, there were still a variety of options, and all of those put together are the potential permutations of the game. Mathematically, there are hundreds—sometimes thousands—of potential play scenarios available to each player, depending on how they mix and match their moves.

A Brief History of Discovery and Development

Game math has a long history, with some forms dating back to early human history. The throwing of dice originated in an ancient game that involved rolling bones, potentially man's earliest game of chance. Although it is unlikely that early gamblers were aware of the probability behind the game, it survived and eventually evolved into its more modern equivalent. Nor were dice games limited to specific nations. During his travels to China, Marco Polo reportedly encountered the dice rolling as both a means of divination and simple entertainment. Native Americans, the Aztecs and Mayans, Africans, and Eskimos all have evidence of dice in their cultures.

Early dice were fashioned out of animal teeth and bones, stones, and sticks, and used by witches or a tribal shaman to foretell the future. As they evolved and began to be used for diversion, the dice were shaped to match their uses. The modern look is believed to have originated in Korea, as part of a Buddhist game called Promotion. Dice as a means of gaming spread rapidly, particularly through the Roman Empire, where there is evidence not only of the rolling of dice, but of cheating. By the tenth century, dice games appear to have been a part of most cultures.

Chess is another mathematical game that has developed through the centuries. The earliest game resembling chess is Shaturanga, a board game between four armies that was designed by an Indian philosopher in the sixth century. The original version was played on a board made up of sixty four squares, and each of the four players had an army controlled by a rajah or king. Other pieces included infantry or pawns, calvary or knights, and an elephant that moved like the modern rook. In the early history of the game, dice were rolled to determine each player's moves. Only later, when Hindu law forbade gambling, were the dice eliminated. At that point, the game became a two-player contest, and the pieces were combined, with two of the kings demoted to prime ministers among other changes. The new version of the game was called Shatranj, and its first mention is in approximately A.D. 600 in a Persian text. As of A.D. 650, there was evidence of Shatranj being played in the Arab kingdoms, the Byzantine court, Greece, Mecca, and Medina.

Shatranj made its way to Europe some time during the eighth century, but there are several theories as to how. One theory is that the Saracens brought the game to Andalusia after conquering North Africa and it then traveled on to the court of Charlemagne in approximately A.D. 760. Another theory has Charlemagne engaged to the Empress Irene of the Byzantine Court. A Shatranj set was given to Charlemagne during one of their meetings, but instead of the standard prime ministers there were two queens mistakenly included. They were considered the most powerful pieces on the board and Charlemagne took it as a bad omen and cancelled the engagement. The third, and most likely explanation is that knights returning from the Crusades brought the game back to Europe with them.

The game continued to evolve across Asia and Europe, eventually breaking out into several versions. The European game most closely resembling modern chess became popular at the end of the fifteenth century. Certain pieces gained more power at that point, and several new moves were added to the game.

The common element through all of these versions of the game were the multiple permutations possible depending on the strategy of the players. This became increasingly evident in modern times, when computer programs were written to simulate a chess game in the 1960s. The early programs were easy to beat, but as the programs became more complex, taking into account the possible mathematical permutations for each move and the likely responses of the opponent, the games became more sophisticated and more closely resembled the playing style of a live player. In 1997, Gary Kasparov, considered the best chess player in the world, was beaten by a computer chess program.

Magic squares are blocks of cells—three by three, four by four, etc.,—where each cell contains an integer and the integers in each row, column, and diagonal add up to the same number. Coming up with working combinations of numbers is a pastime that dates back to as early as 2200 B.C. The first known square of this sort was a third-order magic square, three cells across and three cells down, recorded in a Chinese manuscript. Early squares have been discovered engraved into metal or stone in both China and India. There is a legend surrounding the first magic square that says that Chinese Emperor Yu discovered it while walking by the Lo River. He spotted a turtle on the river's bank, and that turtle had a series of dots on its shell in the formation of a magic square, with each row, column, or diagonal of the dots adding up to fifteen. The Emperor took the turtle home to his palace to study it, and the turtle continued to live at court, with famous mathematicians traveling to examine it. The pattern of dots on the turtle's back became known as the Lo-shu.

Nearly every civilization since has made a study of magic squares, often attributing them with mystical properties and using them in rituals and as the foundation of prophecies and horoscopes. They appeared in Europe during the first millennium A.D., with the Greek writer Emanuel Moschopoulus being the first to write about them. The numbers have been equated with planets, elements, and religious symbols, and have appeared in works of art and on the backs of coins. As an entertainment, they have provided generations of mathematicians and laymen alike with a diverting puzzle, as people continue to attempt to discover new combinations of numbers that work in the magic square format.

Card games of all types also use a certain amount of math. It is unknown precisely when the first card games appeared, but there are references to card games in Europe starting in the thirteenth century, and the first actual playing card can be traced back to Chinese Turkestan in the eleventh century. It is suggested that the Chinese might have invented the cards, as they were the first to create paper.

Chess is essentially a game of mathematical options. WILLIAM WHITEHURST/CORBIS.

Specific card games, such as blackjack, are not referenced until more recently. Blackjack originated in France as twenty-one, or vingt et un, in the early eighteenth century. The game traveled to the United States in the nineteenth century and became popular in the West, where gambling was a growing pastime. Las Vegas legalized gambling in 1931, and blackjack became a regular attraction in the new casinos. By the 1950s, books were being written on how to count cards and predict the odds for any given hand of cards. Blackjack games began using more decks of cards, making it nearly impossible for an individual to keep track of the permutations, and in the 1970s, with the advent of mini-computers and calculators, cheating at cards rose to a new level.

Slot machines, which combine probability, statistics, and a variety of potential permutations and corresponding payouts, first appeared during the early part of the twentieth century. They were invented in 1895 by Charles Rey but not manufactured until approximately 1907 when Rey and the Mills Novelty Company joined forces to create the first machine, the Liberty Bell, which had a cast iron case and reels with pictures of playing cards. Different themes became popular through the years and other pictures were added, such as fruit, castles, and eagles. The modern devices are electronic and provide even more potential for different combinations. Most illustrate the odds of achieving any given payout based upon the number of coins played. An understanding of probability helps a player realize the rate at which he will probably lose his money, and how slim his chances of hitting one of the jackpots.

Real-life Applications

CARD GAMES

Math is an important skill for various card games, both in order to strategize against one's opponents and as an actual part of certain games as well. In blackjack, basic addition is required in order to determine how to play each hand, in addition to an understanding of the value of each card in the deck. Numeric cards two through nine are worth the same number of points as their face value, so that a four is worth four points and so on. This is true of all four suits. The ten, jack, queen, and king are all worth ten points apiece. The value of the ace depends on the combination of cards in the hand, as it adapts itself accordingly, worth either one point or eleven, depending on which is more advantageous to the player. In order to get blackjack, the player's had must add up to exactly twenty-one points. This can happen as a natural blackjack—consisting of an ace and either a face card or a ten—or through another combination of cards.

There is more to blackjack than scoring a perfect twenty one, however. The game is usually played at a table with a dealer and one or more players. It is the dealer's duty to distribute cards both to the players and to himself—the house. In order to win, a player needs to come as close to twenty one as possible without going over, known as going bust, and yet still have either the same number of points or more points than the house. As a result, the game requires more math skills than simply adding up the cards in one's hand. Once the dealer distributes two cards to each player, the player must decide whether the cards he has are sufficiently close to twenty-one, or whether he wants to risk taking an additional card. Because the dealer only reveals one of his own cards at the onset of the round, the player also has to guess what the chances are that the dealer has a better hand. The player's only advantage is knowing that the dealer must continue to take additional cards until the house's hand reaches at least a total of sixteen. The house must also stand, or take no additional cards, once the dealer has reached a total of seventeen points in the hand.

In order to play blackjack effectively, a player must determine the mathematical likelihood that they will have a hand equal to or better than the dealer's hand, without going bust. There are some basic rules of probability that come into play at this point. For instance, looking strictly at the player's hand, if the two cards dealt total seventeen or higher, it is advisable for the player to stand and refuse any further cards. It is easy to see why when you look at the possible permutations. If a player already has a total of seventeen and chooses to draw an additional card, the only cards that would give him a new total of twenty one and lower are the ace—acting as a one and not eleven—or the two through the four. Five or higher will result in the player going bust. This means that out of a potential thirteen cards in a suit, four of them would result in a win, while the remaining nine would result in a loss. Not very good odds.

The math becomes more complicated when the player's original hand totals less than seventeen, and it becomes necessary to also consider the value of the single card the dealer has revealed for the house. The higher the dealer's standing card, the more likely it is that when the second card is revealed the total will be above seventeen. Therefore, the higher the dealer card, the more likely it is that the player will need to choose an additional card for his own hand. This becomes doubly true if the two original cards dealt to the player are substantially lower than sixteen. Once the various permutations are taken into account, it becomes clear that if the player's hand totals twelve or less and the dealer holds anything other than a four through a sex, the odds are better if the player draws an additional card. If the player holds thirteen to sixteen points, and the dealer holds a seven or higher, the player should also draw another card. It is important to remember when totaling a hand that an ace can count as either a ten or a one, so an ace and a six can be considered either a sixteen point hand or a seven point hand. Also, the probability of drawing any given hand changes depending on how many decks the dealer is using and how many hands have already been played.

Counting cards is a skill that can help a player get a better idea of his chances of winning over the course of a game. The system involves assigning each card a value other than their point value in the game. The ace and any card with a point value of ten is equal to negative one, cards two through six are equal to positive one, and the others are counted as zero. For each hand, the player keeps a running tally based on the cards dealt and the number of decks being used. There is a higher chance of winning when the card count total is above a positive two.

In other card games, counting cards merely refers to keeping track of what has already been played. If a single deck of fifty two cards is divided among four players, each player guesses what the other players are holding based on their knowledge of their own hand and their knowledge of the cards in the deck. Depending on the game, as cards are thrown down on the table, a player remembers what they have seen because that means those particular cards have been eliminated from play. The player, in a sense, subtracts those cards from the total set available.

Probability is also used to determine what sort of hand constitutes a winning hand in many card games. For instance, in Poker, a hand that contains a pair of like cards may have some value, but it will still be worth less than three of a kind or a flush, and far less than a royal flush. The probability of getting a pair is fairly high, given that there are more than one million potential combinations in a standard deck of cards that would qualify. A royal flush, however, which requires you to not only attain the major face cards but for all of them to be in the same suit, is limited to only four potential combinations—the face cards for hearts, diamonds, spades, and clubs. It is easy to see why a royal flush is such a difficult hand to obtain, and why it will beat out any of the other potential combinations.

OTHER CASINO GAMES

Game math in the form of probability plays a huge role in most other casino games as well. Roulette is an obvious example. A player chooses a number on the board and places their chips on that square. Once all of the bets are made, the dealer spins the wheel, setting loose a small ball that circles the wheel until the wheel comes to a halt, then drops down into one of the slots that indicate the winning number. It is similar to playing the lottery, but on a smaller scale. In order to win, a player must have bet on the same number as the one on which the ball lands.

Other bets are possible in roulette, but as with anything, the greater the odds of winning, the less money you lose—or win. Half of the numbers in Roulette are black while the remaining half are red. If a player decides to simply bet on black in general, they have a fifty/fifty chance of winning and therefore can do no more than double their bet with each spin. At the other end of the spectrum is a straight up bet, where a player chooses a number and hopes that is the one to win. Because there are thirty six numbers on the board, the odds of winning become one in thirty five or 1/35. In between these two bets are other combinations that provide various different odds, such as a trio bet in which three individual bets are placed on separate numbers and the odds of winning become three in 36 or one in 12.

It is important to understand the odds of a game before playing because the mathematics behind the action can help a player from making foolish mistakes. It is possible to place a bet in Roulette thirty six times in a row and never win, even though there are thirty six numbers, because the win spins independently each time and each time the odds of a given number appearing remain virtually unchanged. Common sense tells us that eventually different numbers will result, but experience teaches us that it is unlikely that each of the thirty six numbers will come up once. However, if a player takes those same thirty six bets and places them all at once, with one bet on each number, the outlay of chips is identical but there is a guarantee of winning because all of the potential outcomes have been covered.

Slot machines offer players the chance to play with minimal interaction. All one needs to do is put money into the machine and either pull the lever or push a button. If the machine offers the opportunity to bet more than one coin at a time, the player can also make that decision. However, there is no other human interaction—the rest of the game is up to the machine. Players win at various levels based on whether any matched sets appear on the reels when they stop spinning, but these wheels have been programmed and are encased within the machine, so a player has to trust to the payout odds listed on the front of the device to determine whether the probability of winning is worth the risk of losing. As with many casino games, the odds are in favor of the casino.

The payout percentage is listed on most slot machines. This is the amount of money that particular machine is required to pay back out in winnings over the course of its lifetime on the casino floor. However, a high percentage, while favorable, is no guarantee, as the machine could pay that entire amount in one huge jackpot every so often and then not provide any smaller winnings in the interim, or conversely, could pay the money in small, unimpressive amounts on a steady basis. There is no way to know the machine's history, or whether it might have paid out a fairly large amount of money to another player just an hour earlier. The machine should, however, illustrate some of the typical combinations that might appear when the reels stop spinning, and what the odds are of that type of payout occurring. It will also show if the machine accepts multiple coins for each spin, and what the difference is in the payout if you play one, two, or more coins.

Multiple line payout machines seem to offer multiple chances to win by paying for lines other than the one straight across the middle. For instance, a three line machine might offer payout for a line across the top, middle, and bottom of the reels. However, often these lines only payout if you are making the corresponding bet, so

Nigel Downing poses with his board game, "Enterprise Profit Ability." The board game teaches players how cash flow and profit affect business. AP/WIDE WORLD PHOTOS. REPRODUCED BY PERMISSION.

any non-jackpot winnings tend to be extremely small. Also, many progressive machines require players to bet the maximum number of coins in order to win the progressive, so if the jackpot comes up and a player has only bet a single coin out of a possible three coins, there won't be a payout.

BASIC BOARD GAMES

Board games rely on several mathematical principles to enable the players to advance around or across the board. Simple games for children, such as Chutes and Ladders and Candyland, provide the players with an arrow to spin, and whatever number the arrow lands on is the number of spaces the player can advance. This basic system enables younger children to grow accustomed to counting and adding as they work their way through the game.

More advanced games for older children and adults often use dice to determine the number of spaces a player advances. Again, they use simple addition, but often take other factors into consideration as well. It is common for a game to require players to roll a specific number at some point during the game, whether it is double sixes for bonus points or an exact number that allows the player to reach the final square of the game in the precise number of moves.

Monopoly provides players with numerous chances to apply their math skills. Not only does this game use dice to travel around the board, but it doles out money to each player based on various events over the course of the game and also requires players to pay out from their accumulated earnings in order to advance. For instance, players receive a two hundred dollar salary each time they pass the initial starting point on the board, the "Go" square. They can also land on squares that allow them to draw cards that occasionally reward other cash prizes, such as a lottery win or an inheritance. Expenses for the player fall into two categories: those that are unavoidable and those that serve as an investment. If a player lands on a square—or property—owned by one of his opponents, he must pay rent to the owner of that property. Players can also lose funds by landing on squares that require them to pay taxes or alimony or to go to jail and forgo passing "Go" and collecting that round's salary. However, if a player lands on a property that is not owned by another player, he has the option of buying it himself. This costs money from his own fund, but provides him with the potential to earn additional cash if and when his opponents land on that space. Rents are further escalated when players manage to buy several properties of a set, and when houses or hotels are purchased to add to the value of the square.

that one coin activates one line, two coins activate two lines, and so on. A player has to take a close look at the probability of their winning and determine if it is worth playing several coins at a time.

Progressive machines are a relatively new type of slot machine. A progressive machine reserves a small amount of money from each bet made and adds it to a grand jackpot total that continues to grow until someone finally wins. What makes this impressive is that multiple machines are linked together, with all of them feeding into the one prize total. They can be part of a local network, with machines all over that particular casino or even several nearby casinos under the same management, or they can be part of a wide area network, which could include slot machines over a fairly large region. This means that the jackpot has the potential to be a far larger amount than if it were just based on a single machine. However, because money is constantly reserved to add to that total, the odds of winning are greatly reduced, and

Chess Mathematics

Chess is a highly mathematical game, as the only way in which to excel is to study the board and determine the various potential moves for each piece in play—both your own and those of your opponent—for several moves into the future. This understanding of probability and permutations allows a player to strategize for the best outcome. The chess moves themselves, when written out, use a form of algebraic notation to explain what piece has been moved to which square on the board, and the board is referenced as a grid.

However, there is more to chess than the traditional game of white against black. Within the game itself is another puzzle called the Knight's Tour.

The Knight's Tour involves traveling a knight using only the traditional combination move of two squares either horizontally or vertically and one square at a right angle to that move so that the knight lands once and only once on each square of the board. This puzzle has likely provided an intriguing challenge nearly as long as Chess itself has existed, and is referenced as early as one thousand years ago. The first in-depth study of the math behind the puzzle was published by the mathematician Euler in 1759.

There is no rule that a Knight's Tour must take up the entire traditional board comprising sixty-four squares.

Smaller puzzles exist, and in some cases the smaller blocks are solved and then combined in an effort to find a new way of traversing the larger board.

Math puzzles can be combined, so that a Knight's Tour forms a quasi-magic square. In order to do this, each move the knight makes is numbered consecutively, with the number written into the landing square. Once the tour is complete, the numbers are then added across and down to make a partial magic square.

As difficult as the Knight's Tour is, there is an even more challenging version, known as a closed Knight's Tour, in which the knight finishes the tour of the board on the same square from which it began. This is also referred to as re-entrant. Other ways to make the tour more difficult include turning the flat chess board into a cube, where each face is a separate board of sixty-four squares and the knight must tour each face in such a way as to ensure that the last move allows it to jump to the next side of the cube.

Because the knight has the most complex way of moving around a chess board, it is the subject of a variety of these types of mathematical puzzles. Another poses the question of how many knights would need to rest on a board in order that every single square would be reached in a single move. On a standard chess board, the fewest number of knights required would be twelve.

MAGIC SQUARES

Magic squares are a form of math puzzle that dates back thousands of years. A magic square is made up of a number of cells, an equal number per row as per column, with an integer in each cell. The integers used are the numbers $1, 2, 3, \ldots N^2$, where N is the number of rows or columns. When added together, the sum of each row is equal to the sum of each individual column, as well as the sum of each of the diagonals. Magic squares are essentially puzzles with very little in the way of practical value. However, the patterns are often used in art and geometric designs.

The smallest possible magic square consists of nine cells—three across by three down. A magic square of two cells by two cells would only work if every cell held the number one, and so is considered too simplistic and therefore not a true magic square. The number of cells in a row or column, N, determines the order of the magic square. Therefore, a square consisting of three cells by three cells

is a third order square. The order of the square also determines what the sum of each row and column will be. Every third order magic square consists of rows, columns, and diagonals that add up to the number fifteen. A fourth order magic square—four cells by four cells—consists of rows, columns, and diagonals that total thirty four.

In order to determine what the sum is for a particular order of magic squares, all you have to do is plug the order number into a simple equation. For instance, assume you are interested in knowing the sum of each line for a fifth order magic square. N = the order of the square, so in this case N = 5.

$S = (N/2) (N^2 + 1)$ where S is the sum for any row or column in the magic square. In this example, for a fifth order magic square, the sum is sixty five. The sum is often referred to as the magic sum or magic constant.

But how do you determine what numbers go on which lines so that each of the lines adds up properly?

With small orders of magic squares, it is easy to see how to distribute the numbers because there are only a few potential equations to use. For example, in a third order magic square, where each line totals fifteen and each cell holds a single digit number, one simply examines the various different equations that meet those terms. Each equation needs to consist of three numbers. So, for a third order magic square, here are the possible equations:

$$6 + 5 + 4 = 15, 7 + 5 + 3 = 15, 7 + 6 + 2 = 15, 8 + 4 + 3 = 15,$$

$$8 + 5 + 2 = 15, 8 + 6 + 1 = 15, 9 + 4 + 2 = 15, 9 + 5 + 1 = 15$$

In this instance, the first thing that becomes apparent is that the number five appears in four of the eight potential equations for the third order magic square. This means that the number five should be placed in the center cell of the square, since four separate equations are required to use the center cell as one of their integers: the middle column, the middle row, and each of the two diagonals. Once those equations are in place, it is simply a matter of determining how the other cells need to be filled to match the equations as listed above.

The number of potential magic squares increases with each increase in the order. There is only one true third order magic square, although it is possible for it to appear as if there are more, depending on which way one flips the square itself. However, there are 880 fourth order magic squares, and over thirteen million fifth order magic squares. Obviously, one cannot simply list potential equations in order to determine how to fill the cells when the possibilities become so numerous. However, there is a way to generate a new magic square based on the layout of an existing one. For each cell in the magic square, subtract the integer from $N2 + 1$, then insert the new number into the cell. The new magic square is sometimes called the complement of the original square.

There are different types of magic squares. Some refer to the classic magic square described above, where the numbers in the cells consist of the integers from 1 through N^2, as a normal magic square. A magic square that allows other integers in the cells as long as the sums of the rows, columns, and diagonals work is then referred to as simply a magic square. Creating normal magic squares, however, is by far the more challenging puzzle.

In a normal magic square in which N is equal to an odd number, there are several systems for filling in the rest of the integers. Perhaps the simplest is called de la Loubere's algorithm. This method has you start by placing the number one in the cell that is at the top center of the magic square. Then working in numerical order, you add each number to the square by working in an upward diagonal direction to the right. So, after the number one, you move one cell up and one cell to the right and place the

number two in that cell. Of course, because you have commenced the square at the top, this movement takes you out of the square entirely. In order to determine where the number two should go, you must imagine that the magic square repeats in all directions. Once you've filled the cell in with the number two, and have determined where in the square it goes, you can transfer it to the same position in the original square. In the case of a third order magic square, the two would therefore appear in the bottom right hand cell.

De la Loubere's algorithm enables you to continue filling in the cells of the magic square until all of them have been filled. In some cases, the system of moving one cell up and one cell to the right will lead to a cell that has already been filled. When this happens, you simply drop straight down by one cell from the most recently filled cell and insert the next integer there instead. Then resume the standard movement of up one and over one to the right with the next consecutive number. In the third order magic square, this leads to the number four appearing below the three, and the seven beneath the six. This system also works if you spread the move out to resemble a knight's move in chess, in which you move the piece two cells either horizontally or vertically and then one cell at a right angle to the original direction. For the purpose of the magic square, the movement must always consist of two cells up and one cell to the right.

There is no algorithm to create magic squares where N is equal to an even integer, however. Despite this, many have been discovered. Perhaps the most famous even integer magic square is the fourth order square that appears in artist Albrecht Dürer's etching, "Melencolia I." The German Renaissance artist drew the magic square into the top right hand column of the work. Astrologers of that period linked fourth order magic squares to the planet Jupiter, and they were thought to battle melancholy. As the etching depicts a woman thinking while surrounded by uncompleted chores, it is possible that this was Dürer's purpose in including the magic square.

The square itself, however, has certain intriguing properties. The cells read across as 16, 3, 2, 13 for the first row, 5, 10, 11, 8 in the second, 9, 6, 7, 12, in the third row, and 4, 15, 14, 1 along the bottom row. As expected with a fourth order magic square, each row, column, and diagonal adds up to thirty four. In addition, however, the four corner numbers (16, 13, 4, 1) also add up to thirty four, as do the four numbers that comprise the inner square of two cells by two cells (10, 11, 6, 7). The sum of the remaining numbers is equal to sixty eight, which is twice the standard sum for any given line. As an added detail, the two central bottom cells of the square read 15 and 14; 1514 was the year in which Dürer created the etching.

MATH PUZZLES

Math puzzles or logic puzzles frequently consist of a short story in which a problem is given and the aim is to solve the puzzle through math. In a school setting, they are often called word problems, but they have a history of being offered up as challenges to see who can find the conclusion first—or in some cases, at all. Often these problems combine math with common sense to see if the person attempting to solve it is paying attention to all of the details.

Here's an example: an injured mouse is trapped at the bottom of a hole that is ten feet deep. Each day, the mouse is able to climb up three feet, but each night he slides two feet back down. How many days will it take the mouse to get free of the hole?

Well, at first glance, it seems that his is a simple math problem. The hole is ten feet deep. Every twenty-four hours, the mouse travels three feet forward and two feet back, which means he makes precisely one foot of progress per day. Therefore, the automatic response would be that, at one foot per day, it would take ten days for the mouse to travel ten feet and thereby climb out of the hole.

Except, it's not quite that straightforward because the mouse does make that three feet of progress every day, even if it does then lose a good portion of it during the night. Day one sees the mouse crawl to the three-foot mark, then slide down, so he starts day two at the one foot mark. He then crawls to four feet and slides down to two. And so on. On day eight, however, the mouse is starting at the seven-foot mark, because that is where he slid to the previous night. But when the mouse crawls three feet up, he has reached the ten foot mark, thereby reaching his goal and climbing out of the hole.

Other traditional math puzzles such as this include problems where the ages of several people are given but only in relation to each other, and the aim is to determine how old each of them are; time and distance puzzles, where one must try and determine which train will reach the city first based on when each leaves in relation to each other and how far they are traveling at a particular speed; and puzzles where an unstated number people choose items from a box and based on which items and how many each takes, one must determine how many people there actually are. These games are enjoyable for sheer entertainment value, but also help develop problem-solving skills that can be used in the real world.

Potential Applications

Game math serves purely as entertainment if you look at it as a form of recreation, but it can also mean big business if you are on the other end of the spectrum. The casino industry relies heavily on a solid knowledge of probability and the ways in which they can keep the odds in their favor, as that is how casinos make their money. A casino needs to keep players entertained and interested, or they will not continue to gamble. Therefore, the players must be able to win often enough that they are having a good time, but not so frequently that the casino ceases to make money. The casino's goal is to earn as much of a profit as possible while still keeping their customers happy. In order to keep everything in balance, they analyze the odds on every game on their floor and adjust them whenever necessary. Mechanized games, in particular, can be manipulated to pay out at a different percentage rate to keep players entertained and coming back for more.

Probability comes into play in any game that enables you to place a bet, including horse racing and other professional sports. Odds are determined based on numerous factors including previous performance results and other analysis. As with any form of gambling, the higher the odds one of the participants might win, the greater the potential payout, and the greater the chance of losing one's money.

Computer games are another enormous industry that relies heavily on various types of game math. Interactive games that allow you to choose a character's moves based on a set scenario are programmed with thousands of different permutations to take each possible choice into account. Each individual choice leads down numerous different paths, and each one results in a different ending. The best game designers understand that a variety of options makes the game more entertaining and insures that the players will be able to play over and over instead of simply solving the puzzle once and having to move on to something else.

Where to Learn More

Books

Falkner, Edward. *Games Ancient and Oriental, and How to Play Them.* New York, NY: Dover Publications, Inc., 1961.

Hunter, J.A.H. and Joseph S. Madachy. *Mathematical Diversions.* Princeton, NJ: D. Van Nostrand Company, Inc., 1963

Pickover, Clifford A. *The Zen of Magic Squares, Circles, and Stars.* Princeton, NJ.: Princeton University Press, 2002.

Web sites

Casinos and Gambling History. <http://www.worlds-best-online-casinos.com/Articles/History.html> (May 6, 2005).

Key Terms

Odds: A shorthand method for expressing probabilities of particular events. The probability of one particular event occurring out of six possible events would be 1 in 6, also expressed as 1:6 or in fractional form as 1/6.

Permutations: All of the potential choices or outcomes available from any given point.

Probability: The likelihood that a particular event will occur within a specified period of time. A branch of mathematics used to predict future events.

Delphi for Fun. "The Knight's Tour." <httphttp://www.delphi forfun.org/Programs/knight's_tour.htm> (May 7, 2005).

E-How. "How to Count Cards." <http://www.ehow.com/ how_4369_count-cards.html> (May 6, 2005).

Interactive Mathematics Miscellany and Puzzles. <http://www .cut-the-knot.org/games.shtml> (May 7, 2005).

Magic Squares. "Vignette 20." <http://www.jcu.edu/math/ vignettes/magicsquares.htm> (May 8, 2005).

Math Forum. "Magic Squares." <http://mathforum.org/alejandre/ magic.square.html> (May 7, 2005).

Math World. "The Knight's Tour." < http://mathworld.wolfram .com/KnightsTour.html> (May 7, 2005).

Math World. "Magic Squares." <http://mathworld.wolfram .com/MagicSquare.html> (May 7, 2005).

Mathematica Gallery. "Distribution of the Knight." <http://www .tri.org.au/knightframe.html> (May 7, 2005).

Online Guide to Traditional Games. <http://www.tradgames .org.uk/games/Chess.htm> (May 7, 2005).

The Knight's Tour. <http://www.borderschess.org/Knight Tour.htm> (May 7, 2005).

Wizard of Odds. <http://www.wizardofodds.com> (May 7, 2005).

Overview

Game theory is an approach to the way humans interact and behave that is rooted in mathematics. Game theory attempts to understand, explain, and even to predict the "give and take" between people and other organisms (even bacteria follow game theory) that allows an outcome to be reached.

Game theory does not involve entertaining games in the manner of tag or cards. The game in game theory, however, resembles tag or poker in that decisions have to be made to reach an outcome, and there are winners and losers based on the decisions that are made. Also, like tag and poker, game theory operates on the premise that the two or more people who are involved have an interest in the outcome. In most cases, the participants want to win. It is the strategies used in trying to achieve a winning outcome that game theory helps explore.

Some aspects of human behavior lead to individual gain. By performing in a certain manner, an athlete could attract the attention of the basketball coach, and so increase his chances of making the starting squad. One person making the team, however, comes at the expense of someone else not making the team. This area is part of non-cooperative game theory. The subject acts on his own, for his own benefit.

At other times, chances of success come when people get together and pool resources with others. This is called mutual gain. For example, in the basketball analogy, after a person makes the team, chances of success (winning games and being a star) happen only if that person works together with his teammates. By cooperating, the team wins and everyone benefits. The part of game theory that looks at this sort of behavior is known as cooperative game theory.

Game theory was first developed to help those who study the economy learn how decisions are made in the face of conflict. Whether in a small business deal or in economic relationships between two nations, conflict is one of the driving forces of the proceedings. By understanding how rational decisions that benefit the person or organization making them can be made in such an atmosphere, the chances of making a good decision improve.

As the power of game theory became more recognized over time, the theory was applied to other areas such as sociology, psychology, biology, and evolution. Game theory is also important in sports, mathematics, social science, and psychology (psychologists often refer to an aspect of game theory as the theory of social situations). It even appears that evolution operates according

Game Theory

to aspects of game theory. A genetic change is tried out in the face of a survival pressure. If there is a benefit to the change, it is conserved and the change continues.

Game Theory Issues

The following issues, posed as questions, illustrate the relevance of game theory. Some might seem familiar:

- A strategy is chosen to achieve a desirable outcome. How can this strategy be chosen rationally when the outcome depends on the strategies that are selected by others and when you do not have all the available information on the topic when you select your strategy?
- When a situation permits all those involved to gain (or to lose), is it rational to cooperate with the others to realize the mutual gain or mutual loss, or is it more rational to act independently and aggressively in order to claim a larger share, even at an increased risk of loss?
- If it sometimes beneficial to cooperate with others and sometimes beneficial to act independently, how are you to know which decision is the rational choice at a particular time?
- Do all of the questions above apply in the same manner to an ongoing relationship as compared to a one-time relationship?

These questions help to point how the important features of game theory. It is about how people get along with one another as much as it is about how we compete with each other. If we were completely competitive, such a predatory environment probably would have ended humans as a species long ago. We thrived because we learned to cooperate.

Game theory, therefore, goes to the heart of human nature, illustrating cooperation and independent behavior at different times. To understand which course is best at a particular time is what game theory is all about.

Fundamental Mathematical Concepts and Terms

The mathematics that govern game theory attempt to model and predict the outcome of interactions between people, or, in a field like evolutionary game theory, between an organism (such as a bacterial cell) and something else (such as an antibiotic). As summarized above, there are two main branches to this interaction: the cooperative game theory and the non-cooperative game theory.

The thing that drives game theory is the outcome. Most of us want the outcome of a decision to be something that is good for us, makes us feel happy, and which benefits us. In the jargon of the game theory world, this good stuff is called the "payoff."

In game theory, there are decision makers (the players) who each have at least two choices or a defined series of choices (the plays). All the different combinations of plays leads to some end (win, loss, or draw), and this ends the 'game.' Some games end when one player wins and the other loses at the same time. This is called a zero-sum game. Capturing your opponent's Queen in the game of chess is a perfect example of a zero-sum game.

Chess also provides another example of an outcome, namely a draw. Both players may realize that winning is not to be. So, they cooperate to end the game by deciding that there will not be a winner or a loser.

In game theory it is assumed that each decision maker has all the necessary and important information needed for the decision. Everybody knows as much as everybody else. As we saw above, this is not always the reality. However, it makes for a starting assumption. Furthermore, it is assumed that all players make their decisions rationally (for an interesting twist on this assumption, see "The Prisoner's Dilemma").

A Brief History of Discovery and Development

The popular root of game theory dates back to 1944. It was then that the renowned mathematician John von Neumann, in collaboration with the mathematical economist Oskar Morgenstern, published a book called *The Theory of Games and Economic Behavior*. This book laid out the framework of game theory upon which others have added to over the years.

However, the roots of game theory go back much further than the middle of the twentieth century. For example, the Babylonian Talmud gathered all the then-existing laws and traditions as a basis of Jewish religion, criminal law, and social interactions. One recommendation (Mishna) concerning the division of marital property upon the death of the husband among his wives (more than one being the norm at that time) has various options depending on the size of the estate. In the case, three wives whose marriage contracts specify that in the case of this death they receive proportions of his estate of 100, 200, and 300, respectively (there being some sort of seniority in place). If the estate was only 100 (a relative figure for the purposes of the example), the Talmud

The Prisoner's Dilemma

This classic example of game theory was first devised in the 1950s by two researchers of the RAND Corporation, when the "Cold War" threat of nuclear weapons was part of everyday life. The game was an attempt to understand decision making in times of stress.

The Prisoner's Dilemma relates the story of two prisoners (we will call them Fred and John). In the story, Fred and John team up to rob a store. Because they both planned the heist, they both know all the details. Later they are both picked up by the police under suspicion of being the men who committed the crime. They are carted downtown to the police station, and are put in holding cells that are some distance away from one another. There is no way they can communicate with each other.

A prosecutor offers Fred and then John a break on the length of their jail sentence if they give him information of the other man's participation in the holdup. He tells Fred that he has offered the deal to John, and tells John that he has offered the deal to Fred.

The deal offered to Fred and John is this:

- If one of the prisoners confesses that the two of them committed the crime and the other prisoner denies that he had anything to do with the heist, then the one who confessed will be set free and the one who denied any wrong doing will get a 5-year prison sentence.
- If both Fred and John deny doing anything wrong, they will probably be convicted and each receive a 2-year jail sentence.
- If both Fred and John confess that they committed the crime as a team, they both get a 4-year sentence.

The Prisoner's Dilemma now shifts to consider one prisoner. Let us consider the case of Fred. If Fred implicates John and he denies having anything to do with the convenience store holdup, he is free. However, if Fred says that John was involved and John tells of Fred's involvement, both get a jail sentence of four years. Finally, if Fred does not mention John's involvement but John says that Fred was a part of the robbery, Fred is off to jail for five years.

It is to Fred's advantage to implicate John, and hope that John tells the police that Fred was also a part of the crime, and so go to jail for a shorter time than if Fred denies involvement hoping that John also denies involvement. This is because Fred runs the risk of John implicating Fred while Fred denied doing anything, and thus Fred would be up for a 5-year prison stay.

The action		Jail sentence (the payoff)	
Fred	John	Fred	John
cooperate	cooperate	2 years (R)	2 years (R)
cooperate	deny	5 years (S)	0 years (T)
deny	deny	0 years (T)	5 years (S)
deny	cooperate	4 years (P)	4 years (P)

Table 1.

The rational thing to do is implicate John. Of course, John arrives at the same decision. They both end up confessing and so both go to the state penitentiary for four years. However, if both denied the crime, they both would be facing jail time of only two years.

By acting rationally, both have come out worse! It turns out that Fred and John's best strategy would have been to act irrationally and admit that they did do the crime. In this game, cooperation is both the best thing to do and an irrational thing to do!

The Prisoner's Dilemma explains how this has come about in terms of what is called a payoff (the consequence of the action). There are four payoff categories:

- R = reward for mutually cooperating
- S = sucker (admitting guilt thinking that the other person will do the same)
- T = temptation to deny
- P = punishment for mutual denial.

The dilemma is set up so that the rank of these preferences, from the most desirable to the least desirable, is T, R, P, and S. As well, the reward payoff R can be greater than the average of T and S payoffs.

Table 1 presents a chart that can help make things clearer to understand.

Another fascinating aspect of the Prisoner's Dilemma is the outcome if the same participants 'play' again. The next time Fred and John wind up in the jail together and get a offer of time off for spilling the beans on each other, Fred could well select a different option, based on your memory of John's response the first time around. So, the game becomes more complex as the same people play it over again. Likewise, if more than two people are involved, there are more interactions that can occur, which also increases the complexity.

The Prisoner's Dilemma model has been applied to many real-life situations, from the way business is conducted to political relations between countries, and even to how bacteria deal with the presence of an antibiotic.

John Forbes Nash PHOTOGRAPH BY ROBERT P. MATTHEWS. © REUTERS NEWMEDIA INC./CORBIS. REPRODUCED BY PERMISSION.

recommends that the estate be divided in three equal portions among the three wives. However, if the estate totaled 300, then the wives divide up the property in a different way, in a ratio of 50, 100, and 150 (the numbers in each example do not add up). Finally, while for an estate that totals 200, the Talmud recommends a proportional division of 50, 75, and 75. The reasons for the different divisions, and how the ratios were arrived at was, for a long time, a confusing mystery. But, in the 1980s scholars recognized that the options were based on what we now know as cooperative game theory. Each of the solutions was a logical response to the given situation (or 'game'). Thus, even in A.D. 500, game theory guided decisions.

Another milestone in games theory made it to the Hollywood screen. The movie *A Beautiful Mind* the Oscar-winning performance by Russell Crowe chronicled the troubled genius of Princeton University researcher John Forbes Nash. Between 1950 and 1953 Nash produced four papers that proved to be of major significance to game theory and made game theory very useful to non-cooperative situations like the bargaining process that goes on between workers and management in seeking a new work contract. Subsequently, Nash went through decades of torment, due to the development of schizophrenia. Ultimately, he was able to deal with his

mental illness. The brilliance of Nash's insights culminated with his being awarded a Nobel Prize in economic sciences in 1994.

In the mid-1950s, game theory was applied to the political arena. Two researchers (L.S. Shapley and M. Shubik) developed and used a calculation (the Shapley value) to determine who in the United Nations Security Council wielded the power. At about the same time the link between game theory and philosophy was recognized.

The 1960s also saw game theory applied to automobile insurance, where the rates that are set are influenced by the degree of risk.

Just several decades later, in 1982, a book entitled *Evolution and the Theory of Games* was written by John Maynard Smith. In the book, Smith applied game theory to evolutionary biology; the inherited biological changes that are driven by evolutionary pressures.

The intervening years have seen the applications of game theory expanded still further, and the tools used to apply the theory become more refined.

Real-life Applications

ECONOMICS AND GAME THEORY

A major application of game theory is to economics; the generation of wealth, creation of jobs, and the flow of money and goods that keeps societies from collapsing. Economic game theory also has three other related areas to consider. These are known as decision theory, general equilibrium, and mechanism design theory.

DECISION THEORY

You are golfing. You are getting ready to hit your drive on a particularly hard hole. The reason that this hole is so challenging is a pond that cuts across the fairway about 250 yards (228.6 m) away. If you hit a good shot, your ball will clear the pond. However, your shot could just as easily get wet. You could have a go at bashing your ball over the pond and leaving yourself a really easy second shot to the green. But that requires a pretty good shot. A not-so-good shot will be at the bottom of the pond, and you will have an added penalty stroke. Maybe you should play it safe and hit a shorter shot that does not make it the pond. That leaves you with a longer second shot to the green. How do you feel about your golf skills today? What are you going to do?

This example, which involves one person thinking about the particular situation, acquiring information, and using the information to make a decision that determines his or her outcome, is what decision theory is all about.

GENERAL EQUILIBRIUM

General equilibrium is not on the personal scale, like our previous golfing example. It is much larger in scope. The concept is suited to making or getting products to a large number of people. This can be literal, as in the manufacture of something and the distribution of the item to stores nationwide. However, general equilibrium can also be used to consider things like the stock market, and even politics.

NASH EQUILIBRIUM

The Nash equilibrium is named after John Nash. The premise that he developed (when he was still a graduate student, and before mental illness claimed him for several decades) is that participants in an activity have a number of options. The equilibrium exists when no one has any reason to change their selected option, since by doing so they earn less (Nash was thinking about economics) than if they hold their course. The outcome, however, does not have to be about money.

So, when someone contemplates a new strategy as a way of earning more or maximizing their payoff, he/she needs to consider what the others are doing and how they might change what they are doing. In a Nash equilibrium, the best individual response is to cooperate.

ECONOMICS

Economics and game theory go hand-in-hand. Indeed, game theory came about as a way of getting a handle on economic activities.

The link between economics and game theory is rational behavior. Although economists can disagree in the nuts and bolts of economic theory and which economic plan is best, the general consensus (often referred to as neoclassical economics) is that people are rational in their economic choices. We consider our options and make the choice that we feel gives the best opportunity for the best result. When the aim is to get maximum return on an investment, or to make the most money we can make out of an opportunity, a rational approach is certainly the wise approach.

If we were operating in a vacuum, with no outside forces affecting our economic decisions, then the decisions would be easier to make. However, life does not operate so simply. A person's decision is influenced by, and the outcome of a decision affected by, factors like political relationships between countries, the stock market, company fortunes, and the changing currency rates in your country and around the globe.

Game theory arose to enable economists to predict the outcome of an individual's decision in the face of all these unpredictable influences. In game theory, as in a real-life game like poker, a person chooses a strategy. The outcome of this choice depends on the strategies that are chosen by other participants.

Game theory applies to economics in more ways than just making money. The process of pondering a purchase can be guided by game theory. For example, a student decides to buy a new computer to help with homework. In pondering choices, the student considers the advantages of wireless Internet access knowing that the local board of education is considering installing wireless capability in schools. If the student purchases a computer with the extra expense of a wireless connection and the board comes through on its intent, both 'players' can be better off. The student gains the advantage of the Internet connection and the board benefits from having more capable and world-knowledgeable students. However, if the student spends the additional money to buy a wireless enabled computer and the Board decides not to proceed, the student may have wasted money on an accessory that does not help with homework (the original intent of purchase).

EVOLUTION AND ANIMAL BEHAVIOR

Game theory has been very useful when applied to evolution in the traditional sense; that is, the way living things change over time and with environmental pressure. Indeed, in the preface to his 1982 book *Evolution and the Theory of Games*, John Maynard Smith wrote that "[p]aradoxically, it has turned out that game theory is more readily applied to biology than to the field of economic behaviour for which it was originally designed."

The role of game theory in evolution was first developed in 1930 by Ronald Fisher to try to explain the observations that the ratio of males to females in many animals species (he specifically studied mice) are equal, even though the majority of males never mate. However, by using game theory, Fisher deduced that the seemingly 'excess baggage' non-mating males in fact help to raise and protect their grandchildren. Thus, it is in the best interest of the species to maintain fit males to take a role in childcare when females might be aging or in poorer health.

Another aspect of game theory concerns animal behavior. An example of this is the so-called 'Hawk-Dove game.' Animal (yes, even human) behavior can take the role of a hawk; initiating aggression and not backing down until injured or until the opponent backs down, or of a dove; immediate retreat when danger threatens. When a hawk meets a dove, the hawk will gain the territory, food, nest, etc. When two hawks meet, the feathers will literally fly. When two doves meet, there will be a sharing of the resources or territory. So, depending on who meets whom, the payoff for a hawk and a dove can be rather good or exceedingly bad (e.g., death).

Evolution will seek the path that is the most stable and which carries the most advantage for the species. To return to the hawk-dove example, a dove-dove relationship in a population does not make evolutionary sense, since the presence of even a single hawk throws the system into disarray. In contrast, the aggressive hawk strategy is evolutionarily stable—but if, and only if, the value of the resource in dispute is greater than the cost of being injured in a fight.

Of course, there can be variations on this all-aggression or all-retreat stance. Some animals may chose to fight or cut-and-run, depending on the circumstance. Over time, a dominant strategy (the one that produces the best payoff for the species) will emerge.

Evolutionary game theory need not just be concerned with the Darwinian kind of evolution. Social evolution—the way the beliefs and what is considered to be normal and acceptable behavior a society changes over time—can also be approached using game theory.

Potential Applications

INFECTIOUS DISEASE THERAPY

This application has some overlap to the evolutionary biology area. However, in an era of rising antibiotic resistance by bacteria, and the increasing frequency of infectious disease, the fight against certain infectious bacteria is a stand-alone category.

Game theory has a place in a bacterial population. When a population of millions of bacteria is exposed to an antibiotic, a large proportion of the cells are usually killed. But some will survive, either because they have acquired some genetic or other means of eluding or destroying the effects of the antibiotics, or just because the antibiotic was used up by the time it encountered that individual bacterium.

Game theory has potential in modeling the 'choices' faced by bacteria, and helping guide a strategy that attacks the bacteria after they have made the most probable choice. For example, bacteria exposed to an antibiotic may choose to seek shelter at a surface. By making the surface itself antibacterial, the cells are killed as they colonize the surface. This strategy exploits the bacterial choice as a weapon against infection.

In another tact, researchers at the Center for Genomic Sciences in Pittsburgh, Pennsylvania, and the Center for Interfacial Microbial Engineering at the University of Montana in Bozeman are testing the hypothesis that an individual bacterial cell is part of a larger community, and that the total number of genes in this community can be shared among individual members as needed. Thus, a single cell need not carry every single gene, since it can acquire that gene from its neighbor in time of need.

These sorts of strategies are currently being tested.

EBAY AND THE ONLINE AUCTION WORLD

This potential is real but will surge in coming years. Auction sites such as eBay have a fixed closing time for the auction. Experienced bidders will zoom in with their bid just before the close, often securing the sale. Such "sniping" requires a person to observe the auction, at least periodically, to time their bid. However, there are Internet sites that will automatically assume this function, for a price.

As has been explained by Harvard University economist, Al Roth, sniping web sites and the experienced auction snipers are examples of game theory in action. Late bidders are less influenced by the decisions of others. As well, those who recognize the value of an item will often reserve their bid for the end of the auction so as not to tip their hand about the object's value, thus hopefully keeping the price down.

Knowledge of game theory could make for more successful Internet auction hunting.

ARTIFICIAL INTELLIGENCE

A computer is not yet the equal of a human. Humans can make an independent decision, but computers must be pre-programmed with decisions that are based on the meeting of a number of conditions. Computers cannot make decisions they have not been programmed to make.

In the future world of artificial intelligence, it is anticipated that computers will be able to make decisions free of the constraints of meeting pre-set conditions. Game theory can play a role in this new world. Artificial intelligence programs would be able to make new decisions that produced the best payoff, based on the incoming information and on the previous experiences of the computer. In other words, a computer would be capable of "learning."

Key Terms

Nash equilibrium: A set of strategies, named after John Nash, that results in the maximum benefit of each player.

Player: In game theory, a decision maker.

Plays: In game theory, choices that can be made.

Zero-sum game: An outcome of a game where players choices have produced neither a win or a draw for all of the players.

Where to Learn More

Books

Camerer, C.F. *Behavioral Game Theory: Experiments in Strategic Interaction.* Princeton: Princeton University Press, February 2003.

Fisher, T.C.G., and R.G. Waschik. *Managerial Economics: A Game Theoretic Approach.* New York: Routledge, August 2002.

Miller J.D. *Game Theory at Work" How to Use Game Theory to Outthink and Outmaneuver Your Competition."* New York: McGraw-Hill Trade, March 2003.

Osborne M.J. *An Introduction to Game Theory.* Oxford: Oxford University Press, August 2003.

Periodicals

Greig D., and M. Travisano. "The Prisoner's Dilemma and polymorphism in yeast SUC genes," *Proc R Soc Lond B Biol Sci.* (February 2004) 271: S25–S26.

Holt C.A., and A.E. Roth. "The Nash equilibrium: a perspective," *Proc Natl Acad Sci USA* (March 2004) 101: 3999–4002.

Kirkup B.C., and M.A. Riley. "Antibiotic-mediated antagonism leads to a bacterial game of rock-paper-scissors in vivo," *Nature* (March 2004) 428: 412–414.

McNamara J.M., Z. Barta, and A.I. Houston. "Variation in behaviour promotes cooperation in the Prisoner's Dilemma game," *Nature* (April 2004) 428: 745–748.

Taylor P.D., and T. Day. "Behavioural evolution: cooperate with thy neighbour?" *Nature* (April 2004) 428: 611–612.

Web sites

Alexander, J.M. "Evolutionary Game Theory." <http://plato.stanford.edu/emtries/game-evolutionary> (May 27, 2004).

King W. "Game Theory: An Introductory Sketch." <http://william-king.drexel.edu/top/eco/game/intro.html> (May 27, 2004).

Levine DK. "What is Game Theory?" <http://levine.sscnet.ucla.edu/general/whatis.htm> (May 27, 2004).

McCain, Roger. "Essential Principles of Economics: A Hypermedia Text" <http://william-king.www.drexel.edu/top/prin/txt/EcoToC.html> (Oct 23, 2004).

Geometry

Overview

Geometry is defined as the branch of mathematics concerned with the properties and relations of points, lines, surfaces, and solids, and can be found in absolutely anything, from the shape of the tires on a car, to the movement of light throughout the Universe. The forms of buildings, the paths of flying objects, the patterns in strands of DNA, the characteristics of leaves on a tree, the markings on a sports field, and the behavior of atoms can all be investigated using some degree of geometric reasoning. Even someone who has never heard of the mathematical study of geometry cannot help but use its concepts in every action and observation. In its entirety, geometry relates all objects and phenomena to the mathematical language common to all human communities on Earth, existing at the heart of all indisputable logic.

Fundamental Mathematical Concepts and Terms

In order to understand how geometry can be used to expound upon just about anything in the Universe, it is important to fully grasp the concept of dimensions. Although all objects in the Universe have three dimensions, they can all be related to lesser dimensions.

Theoretically, points have no dimension. A point is a set of coordinates, a location, not something tangible. When a point is represented as a drawing on graph paper, a dot is used to represent the location, but the mark that is used to indicate the location of a point is always three-dimensional. For example, a dot made by a pencil on a piece of paper is roughly circular, having a crude diameter, perimeter, and area existing in two dimensions, and a measurable thickness, or height protruding from the surface of the paper to account for the third dimension.

To represent one dimension, a pencil can be dragged across a piece of paper to draw a curve (where a line is considered to be a curve with no curvature). This drawing also exists in three dimensions, having thickness, width, and length. However, the drawing is intended to represent only one of these dimensions, length. Lines are considered to continue forever along their path, while line segments include only the section of the line between two endpoints. When two straight lines or line segments meet, they create an angle. Two lines are said to be parallel if they will never meet. Perpendicular lines create a 90° angle, or right angle, like those found in rectangles. Angles smaller than 90° are called acute angles; larger angles are called obtuse angles.

The geometric alignment of the stones in Stonehenge served as an ancient astronomical calculator and calendar. PHOTOGRAPH BY JASON HAWKES. CORBIS. REPRODUCED BY PERMISSION.

When one-dimensional curves create an enclosed shape, the realm of two dimensions is reached. Two-dimensional figures exist in planes, which can be thought of as pieces of paper for drawing shapes, although planes have no thickness and exist only in the imagination. Polygons and ellipses are fundamental examples of two-dimensional figures. Any enclosed planar figure (existing in a plane) bound by straight lines is defined as a polygon. Basic polygons include triangles, quadrilaterals such as parallelograms and rectangles, pentagons, hexagons, and octagons. Because of their many interesting properties regarding lengths and angles, triangles create a large amount of useful concepts. The study of triangles is referred to as trigonometry and is most often taught in a separate academic course.

Ellipses include ovals and circles, and enclose two-dimensional space without creating any corners. The definition of an ellipse involves two points, but the concept is easier to understand by considering the definition of a circle. A circle is a special ellipse that has a boundary defined as the set of points that are equidistance (the same distance) from one point. That distance is called the radius and is usually represented by a line segment stretched from the center to the perimeter. The perimeter can be viewed as the path that the end of the radial line segment traces out when is rotated around the center like the hands of a clock. In any other ellipse, there are two points that determine the boundary, causing the figure to resemble a circle that has been stretched. Ellipses are always symmetrical, meaning that if they are cut in half at some angle, each side looks like a mirror image of the other side.

Vectors present another important geometric concept, representing both a number, known as a vector's magnitude or length, and an angle that determines its direction. Vectors are essential to depicting phenomena such as wind or the movement of a sailboat in water. When investigating the factors that determine the movement of a sailboat, the vectors representing the speeds and

With its five-sided geometric shape still intact in the inner rings, the west face of the Pentagon is shown in this aerial photo taken before the attack on September 11, 2001. UPI/CORBIS-BETTMANN. REPRODUCED BY PERMISSION.

directions of the wind and water currents can be added to the vector representing the motion of the vessel in order to model the true motion of the boat relative the seabed.

Any geometric figure can theoretically exist in three dimensions. For example, the point 5 feet (1.5 m) above the intersection of the borders between Utah, Colorado, Arizona, and New Mexico can be described in reference to other locations on Earth, but that point is not a tangible object and has no dimension. The border between Utah and Arizona can be described in reference to other geographic landmarks, but it has only one dimension that describes the imaginary path of the border. The border around Colorado can also be defined as a rectangle with a definite length and width, but this two-dimensional figure cannot be seen or touched. Someone standing at the intersection of the four states can imagine all of these figures, and even though they do not actually exist, their underlying concepts are vital to our perception and use of space.

In reality, the form of every object has three dimensions. The notions of lesser dimensions, however, provide helpful tools for understanding all that is real. Humans can simplify any physical problem by investigating the properties of points, curves, and areas. For example, the area of a circle is defined by $A = \pi r^2$, where A represents the area and r represents the radius; and the formula for the volume of a cylinder is $V = h(\pi r^2)$, where V represents the volume and h represents the height. It is easier to understand the formula for volume by thinking of a cylinder as a circular base extended upward. The volume is equal to the area of the two-dimensional base multiplied by height, the third dimension of the cylinder.

In the mind of anyone equipped with basic geometric tools, anything in the physical universe can be transformed into combinations of geometric figures spanning multiple dimensions. Whether too small to see, plainly visible, or located in a different solar system, any object seems less intimidating when represented by elegant geometric figures that have been systematically scrutinized for thousands of years.

A Brief History of Discovery and Development

Like most fields of mathematics, geometry found its beginning due to a necessity to understand and predict phenomena in the natural world. Geometry provided a tool to help humans understand their surroundings long before generalized mathematical formulas were conceived. As long as 6,000 years ago, people began using geometric reasoning as a visual aid to explain and predict phenomena in the world around them. Pure mathematical reasoning (and proofs that the reasoning was sound) would not appear until geometry was coupled with algebraic rules thousands of years later.

Possibly the first use of abstract geometric reasoning was invoked in early human settlements as result of the inception of monetary calculations. In Egypt, for example, nomadic peoples led simple yet inconsistent and insecure lives, roaming the land and setting up temporary shelters wherever the forces of nature (most notably weather and the availability of food) guided them. After thousands of years spent roaming the sparse desert, these people eventually discovered that the Nile River provided ample water, and hence vegetation, which in turn attracted animals, essential survival resources that are consistently sparse in the middle of the desert. The activities of their lives depended on the ebb and flow of the river. Even their first calendar was based on the cycles that governed their annual cultivating and harvesting periods. After generations of cultivation and expansion along the river, towns were set up and a system of leadership was established.

The concept of money was implemented in these towns in order to facilitate trades between the citizens, and eventually visitors. Similar to the taxation of modern cultures that enables important communal facilities, including schools and healthcare, the Egyptian leaders set up a system of taxation to support the needs of the growing culture. Most of the taxes implemented under the Pharaohs were used to build large, intricate structures to support their images as divine beings that should be feared and followed. Because the Nile held the key to the flourishing agricultural settlements, the River was believed to possess a godlike persona, and each year the taxes were based on the amount of flooding that took place. Taxes were vital to the success of the Pharaohs, and were therefore taken very seriously.

Arithmetic and algebra (the essential tools for analyzing numbers) had yet to be discovered, so Egyptian accountants used visual aids, now recognized as geometric figures, to determine the amount that each citizen should be taxed. The circle, for one, was an important figure in the calculation of taxes. This use of circles led Egyptians to approximate the value now commonly symbolized by the Greek letter π, which defines the relationship between radius and circumference. They estimated pi (π) to be 3.16, but it is closer to 3.1415927. The fact that Egyptian officials were happy to collect slightly too much tax from each citizen is probably the reason that they used an overestimate of the value for π, when an underestimate could surely be calculated just as easily. In a culture that called for the execution of mildly rebellious individuals, a percentage of income was not something to be quibbled over.

In ancient Egyptian culture, spatial measurements were eventually given a name that translates loosely to "Earth measure." The Egyptians and Babylonians used these ideas to describe the physical world around them, but made no advances in using mathematics to explain seemingly inexplicable events or reveal fundamental truths. The building of the pyramids is the most notable use of geometric reasoning by ancient Egyptians. A pyramid has a square base and four triangular sides that come to a point at the top. If a measurement is even slightly inaccurate, the top of the pyramid will not come to a point, but rather a flat grouping of stones. This would subtract from the pyramid's magnificence and almost certainly result in the execution of the designer at the order of the Pharaohs who demanded divine perfection.

To ensure that the base was square and that the triangular faces met at a point, the Egyptians conceived an ingenious method of measurement. Systems of ropes were stretched by a handful of workers to map out triangles. Knots tied in the ropes at equal distances enabled the workers to create triangles with proportional lengths, in turn controlling the resulting angles. For example, in a triangle with a side created by three knots, a side created by four knots, and a side created by five knots, the angle opposite the longest side will be a right angle.

With all of their advances in measurement, the Egyptians did not view the concepts behind their quantitative reasoning as mathematical notions that could be extended to explain other phenomena. They did not truly understand the theoretical points represented by the knots, or the lines represented by the ropes.

During approximately the same time period that gave rise the Egyptian pyramids, the Babylonians were making similar advancements in the perception of geometric logic.

In general, the civilizations that existed prior to the Greeks noticed and utilized many interesting and helpful properties about their physical surroundings, and were able to use their findings to accomplish relatively incredible feats. What separates these earlier cultures from the Greeks is that they did not possess, and perhaps did not desire, a deeper understanding of the measurements that they recorded.

About 2,500 years ago, a Greek merchant, engineer, and philosopher named Thales at long last expanded geometric reasoning beyond the measurements of taxes and sand, sparking a new wave of logic and abstraction that would later motivate the work of Pythagoras. Thales invested most of his life to the evolution of all types of knowledge, including astronomy. For part of his life, Thales lived and studied in Egypt. It is widely accepted that he was first to determine the true height of one of the pyramids, employing the triangles used in the precise measurement of the base in a completely new way. When he returned to Greece, he brought a multitude of new ideas regarding spatial measurements. These ideas and the resulting field of logical reasoning that followed were soon thereafter referred to as geometry, the term that is still used today. Like the term used in Egypt at the time, the word geometry translates loosely to "Earth measurements."

Ancient sources claim that, shortly before death, Thales received a visit from a young philosopher and mathematician named Pythagoras. During their meeting, Thales is believed to have suggested that Pythagoras travel to Egypt in order to further the advancement of Egyptian practical geometric concepts. Pythagoras followed this advice, returning to Greece with another generation of knowledge pertaining to mathematics, astronomy, and philosophy. For example, though it is evident that the Pythagorean theorem was used in the creation of pyramids thousands of years earlier, it is named after him due to the growing Greek affinity for logic. No one in Egypt would have been named after such a concept, because it was not regarded in that land as important knowledge outside of the occasional launch of a new pyramid. Pythagoras found many followers in Greece, and eventually settled with them outside of Athens to live life truly by numbers. To the Pythagoreans, everything in life was numerical. Mathematics was essentially their religion. Outsiders were generally regarded as blasphemers, unworthy of the beautiful truths of advanced mathematics. Their notion of irrational numbers (numbers that cannot be represented as one whole number divided by another) was one of their greatest secrets, and one member was allegedly drowned for leaking this knowledge.

About two centuries later, Euclid (c. 325–265 B.C.), another Greek mathematician, again enhanced the study of geometry. While little of his work provided any original ideas, Euclid is often regarded as the father of geometry because his systematic methods for representing geometric ideas sculpted the subject into a more manageable form. Using consistent and simple notation, geometry could be studied by all and generalized to fit more and more situations. Euclid's work has provided the basis for communicating geometric ideas ever since.

With the rise of the Roman Empire came a heavy decline in intellectual advancement. During the ensuing 900-year period referred to as the Dark Ages, the works of Euclid and his predecessors were locked up in private libraries, and not until the seventeenth century would geometric thought continue to advance.

In the early seventeenth century, an Italian philosopher and mathematician named René Descartes revolutionized both fields in ways that have yet to be improved upon. In mathematics, he melded the laws of numbers and geometric measurements by conceiving the coordinate plane. By placing geometric figures in a well-labeled grid, algebraic manipulations could be applied directly to geometric figures. In this way, geometric concepts began to be analyzed in new ways, illuminating the concepts of the past and spawning an enormous amount of new theories. The simultaneous consideration of algebraic and geometric properties first introduced by Descartes has come to be known as analytical geometry, and has provided the basis for all scientific endeavors in the four centuries since his life.

In modern times, scientists have realized that the three space dimensions are not truly independent of the time dimension. This concept was popularized by the German-born American physicist and mathematician Albert Einstein (1879–1955) in the early twentieth century. Just as the works of Thales, Pythagoras, Euclid, and Descartes were difficult to grasp when they were first introduced, the idea that time and space interact complicates geometry to a degree that most people are not yet equipped to understand. However, as time advances, so shall the collective human understanding of it as a concrete dimension.

Real-life Applications

POTHOLE COVERS

Millions of potholes, also called manholes, are scattered throughout the world, giving workers access to underground sewer lines. In the United States, these potholes and the large hunks of metal that cover them are almost always round. A pothole cover is a cylinder with a relatively small height. This shape is chosen because of two important properties about circles.

First of all, a circle is one of only two basic two-dimensional geometric figures that cannot fit through an identical figure in three dimensions. That is, if two identical circles (having radii of equal length) are placed in three-dimensional space, one could not be made to pass through the other, no matter how the circles were angled. All but one other basic shape can be lifted and rotated so that it will fit through a copy of itself. A square, for example, can be rotated upward until vertical, and then rotated approximately 45° in any other direction to fit diagonally through a copy of itself because the diagonal of a square is longer than any of its sides (a fact which can be proven using the Pythagorean Theorem). Similar reasoning can be used to show that almost all other basic shapes can be made to fit through an identical shape. The fact that a circle will not slip through an identical circle ensures that heavy round pothole covers do not fall through the openings of potholes and injure the workers below.

An equilateral triangle (a triangle with three sides of equal length and three 60° angles) is the only other basic two-dimensional geometric figure that cannot be made to pass through a copy of itself. If the lengths of the sides of the two identical triangles were not all equal, one of the triangles could be propped up vertically with a side other than the longest on the bottom, and then rotated until the bottom side of the vertical triangle was parallel to a longer side of the horizontal triangle, and fall through.

In some parts of the world, pothole covers are occasionally found in the form of an equilateral triangle and they, of course, never fall through the hole. However, most pothole covers are chosen to be round because of another unique characteristic of circles; they have no corners. A circular pothole cover saves the energy of workers by allowing them to roll the heavy covers out of the way. The three vertices of a triangle also make it more likely to injure someone while it is being moved or sitting on the ground. A circular shape also makes replacing pothole covers easier because they do not need to be rotated in order to line up with the shape of the hole.

ARCHITECTURE

Mathematical reasoning was used to build the Egyptian pyramids over 6,000 thousand years ago, long before geometry was conceived as a fundamental field of mathematics. Since the rise of the early Greek mathematicians, standardized geometric concepts have provided the essential tools for planning and constructing all types of buildings.

All architectural structures—from simple four-walled buildings with flat roofs to elaborate, multipurpose constructions—comprise combinations of geometric figures. The types of curves and shapes used to create an enclosed space are carefully chosen for their effects on function and beauty.

In the design of some buildings, functionality is far more important than artistic expression. For example, when building a silo for the storage of large amounts of grain, a cylinder provides the most efficient use of space. The horizontal cross-section of a cylinder (e.g., the base) is circular, and circles can be shown to have the largest area with respect to perimeter out of all two-dimensional shapes. Because the circular cross-section is extended vertically to form a three-dimensional cylinder, a cylinder requires the least surface are—and therefore the least building materials—to provide a given volume.

Most temples and churches, for instance, are designed to balance the beauty required for paying respect to the religious figures worshiped by the congregation (e.g., painted ceilings, intricate moldings, and stained picture windows) with the function of fitting a large amount of people into a single enclosed space. When congregations first began to grow, the task of building a room of worship large enough to accommodate everyone posed a serious problem. Enclosures had previously been created using only flat walls and ceilings, which greatly limited the space because the flat ceiling would collapse if the supporting walls were too far apart. This obstacle was eventually overcome by the Romans when they began using semicircular arches in their architecture. A semicircular arch is created by stacking half of a circle on top of a rectangle. This structure allows the supporting walls to be further apart because, as gravity pulls on the structure, the semicircle distributes weight so that the arch does not collapse. Soon thereafter, arches were placed one in front of the other to create the walls and roofs of relatively huge halls, giving congregations room to grow. Domes are created by copying and rotating an arch around its highest point and were later added to create even higher ceilings, instilling a sense of awe worthy of a building devoted to gods. The round nature of domes also adds a sense of perfection, an important aspect in many religions. Individual arches were also used as the shapes of windows and doors to increase structural stability and effect beauty.

The arch is just one of the geometric figures integral in the design of various structures. Like an arch, a triangle adds stability by distributing and balancing the downward force of gravity as well as the lateral forces of wind. The triangular configuration of rafters in the roofs of

Intricate geometry contained in glasswork. BETTMAN/CORBIS.

many houses also allows rain and debris to slide down and fall off so that the roof does not have to endure much weight. Sloping roofs also add a familiar aesthetic characteristic to the house's overall appearance.

Rectangles, ellipses, hexagons, and more can be found throughout any human construction, including the landscaping that surrounds many buildings. Most structures made by humans are heavily dependent on geometric figures, from the triangles found in the enormous towers that support elaborate networks of electrical wiring to the visually pleasing dimensions of national monuments. The 630-foot (192-m) tall Gateway Arch in St. Louis provides another example of the abilities of the geometric arch to create a sense of beauty while stretching the limits of architecture. The Washington Monument consists of four trapezoidal sides topped by a pyramid. This gigantic vertical structure is impressive for its size: the square base of the shaft has a width of 55 feet (16.7 m); the shaft is 500 feet (152 m) tall; and the pyramid is 55 feet tall. Its construction required in-depth analysis of lengths, angles, areas, and volumes.

Some structures, such as the pointy onion-shaped roofs common in the regions of the former Soviet Union, do not appear to be associated with basic geometric shapes. Other buildings, including some modern museums, are intentionally designed to appear irregular in shape, seemingly not following any laws of mathematics. However, all of these shapes are defined and thoroughly analyzed using mathematical formulas describing complicated geometric figures in one, two, and three dimensions.

Regardless of the goals of the architect, every aspect of a building is deeply considered. Every consideration—from the orientation of the wood panels on a residential property to the ponderous calculations involved in building a retractable dome over a gigantic sports

The Washington Monument consists of four trapezoidal sides topped by a pyramid. CRAIG AURNESS/CORBIS. REPRODUCED BY CORBIS CORPORATION.

stadium—are modeled and investigated using geometric reasoning.

HONEYCOMBS

Just like geometric figures are used to affect certain functions in a structure, many species of animals seem to utilize a small amount of instinctive geometric knowledge in order to conserve materials and energy. Various species of spiders, for example, spin webs in patterns that maximize the ability to catch bugs while minimizing the amount of silk expended. Birds generally create circular nests because a circle provides the maximum amount of area for a given amount of materials.

Honeybees also display an instinctive understanding of geometry in their use of hexagonal chambers for storing honey in honeycombs. One might think that bees would choose to build chambers with circular holes (cylinders) in their honeycombs because, just as a circle

allows for the most area given a specified perimeter, a cylinder provides the largest volume for a given perimeter. However, placing a group of circles next to each other wastes space because they do not fit together. For example, if a group of cylindrical silos were built side by side and viewed from above, it would be easy to see that space is wasted between the silos where grain cannot be stored. Most two-dimensional geometric figures cannot be used to completely fill a two-dimensional area. In fact, only equilateral triangles, rectangles, and hexagons can fit together with identical figures to completely fill an area.

Of these three space-filling shapes, the hexagon is the most efficient—that is, it uses the least material (wax, in this case) to make a honeycomb having a given volume. It might seem strange to think of an insect using geometric reasoning but over time nature has a knack for uncovering the most logical solution to any problem.

GLOBAL POSITIONING

In the new millennium, a global positioning system (GPS) receiver is all a person needs to determine his or her exact location on Earth. Every day, thousands of campers, hikers, bikers, skiers, hunters, boaters, pilots, and motorists around the globe use these ingenious devices to ensure that they do not get lost.

The global positioning system consists of 27 solar-powered satellites that orbit Earth. Only 24 of the satellites are in operation at all times; the extra three are in orbit in case any of the operational satellites malfunctions. This network of satellites was originally launched by the United States government to aid in military navigation, but was shortly thereafter made available for use by anyone. Each satellite weighs about 4,000 pounds (1,814 kg), orbits at about 12,000 miles (19,312 km) above the surface of Earth, and completes a rotation around Earth approximately twice a day. From any position on the surface of Earth, at least four of these satellites can be detected by a GPS receiver at all times.

In order to understand how a GPS receiver uses satellites to determine its own location, it is helpful to first discuss the process in two dimensions. Imagine, for example, that an explorer is lost somewhere on Earth with a detailed map, but absolutely no idea of her current location. She asks one of the locals if he can help, but the only information she receives is that she is currently 699 miles (1,125 km) from Barcelona. Because she does not know the direction in which Barcelona lies, this only indicates that she is somewhere on the perimeter of a circle with its center in Barcelona and a radius of 699 miles. This is a good start, so she draws this circle on her map. Hoping to

239

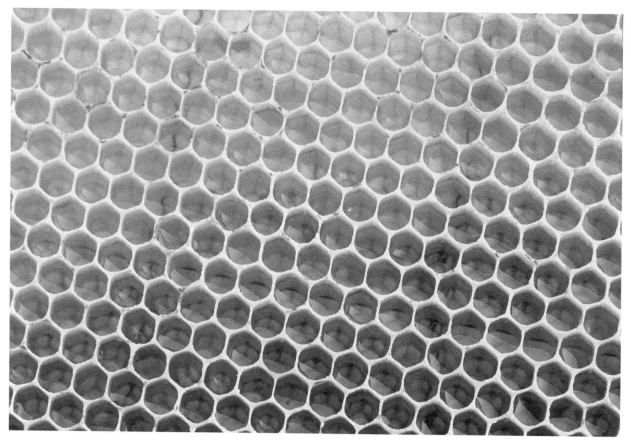

Hexagons in honeycombs. RALPH A. CLEVENGER/CORBIS.

get a better idea of her location, she asks another person for help. This time, she learns that she is currently 680 miles (1,094 km) from Berlin. Again, this means that she could be at any point on a circle around Berlin. After drawing this circle on the map, she notices that there are now two intersecting circles on which she might currently be located. For both of the pieces of information she received to be true, she must be located at one of the two points where the circles intersect. Having two reference points lowers her possible locations from an infinite number of points around a large circle down to only two points. To figure out which of these two points represents her location, she needs one more reference point, which will generate another circle of possible points. The next person she talks to tells her that she is currently 257 miles (413.6 km) from London. She draws a circle around London with a radius of 257 miles and finds that it only intersects one of the two possible points found with the first two circles. This point represents her current location, which turns out to be Paris.

GPS receivers use similar reasoning extended to three dimensions. By measuring the lag in radio waves sent from a satellite, the GPS receiver determines how far the satellite currently is from the receiver; but the direction from which the radio waves approached is not determined. Just like a circle can be staked out on a map, the distance to a satellite allows the GPS receiver to map out a sphere in space. The GPS receiver must be located somewhere on the boundary of this sphere.

Next, the GPS receiver locates another satellite and determines how far away it is, creating another sphere. The intersection of the two spheres is a circle. (Imagine chopping off a thin portion of two oranges to create flat surfaces so that they can be stuck together; the boundary of the area where the two oranges touch is a circle.) Using these two spheres allows the GPS receiver to narrow its location down to a circle, just like using a single reference point on a two-dimensional map creates a circle. Because one more dimension is involved, it takes one more reference point to narrow the location down to a two-dimensional circle.

The GPS receiver then locates a third satellite, mapping out a third sphere. Just as a second circle drawn on a two-dimensional map determines two possible points, the third sphere determines two points in three-dimensional space. At this point, the receiver can approximate its location by using Earth as a fourth sphere. One of the two points created by the first three spheres is always out in space, so the GPS receiver knows that the point that lies on Earth's surface is its approximate location.

Errors can occur in the calculations performed by GPS receivers, due mainly to the effects that Earth's atmosphere has on the speed of the radio waves sent out from the satellites, which throws off the perceived distance to each satellite by a small amount. So, while three satellites are theoretically enough to pinpoint the location of a GPS receiver, these devices always attempt to locate a fourth satellite in order to increase the accuracy of their calculations.

FIREWORKS

Fireworks employ two different types of explosive powder, flash powder and black powder. Flash powder is used to create bright a bluish-white light that can easily be changed to any color by adding certain chemicals. Almost all fireworks that create light include flash powder. Sparklers, for instance, consist of flash powder stuck to a small metal rod. Black powder (also known as gunpowder) is usually used in fireworks to create a loud explosion. Basic firecrackers contain mostly black powder. Some of the larger and more elaborate fireworks available at firework stands near the 4th of July consist of a cylindrical casing that houses combinations of flash powder and black powder. The black powder creates noise and launches projectiles into the air, while the flash powder emits bright light from the main casing and airborne projectiles. Projectile charges that create light during their flight are called flash star pellets, or stars.

Large public gatherings at sports stadiums and other outdoor venues feature enormous displays of fireworks, creating explosions that are audible for miles and visually stunning color patterns that can seem to take up the entire sky. Different patterns of light (e.g., spheres and discs) are achieved by mapping out the desired pattern of the stars inside of the large casing. Most main casings are cylindrical, with the height and size of the base dependent on the intended size and shape of the color pattern. The stars are separated and held in place by black powder, which fills the remaining volume of the casing. To cause a spherical pattern, the stars are equally spaced throughout the three dimensions of their casing. To create a disc that spreads in a plane parallel to the locally flat surface of Earth, the stars are situated in a circular pattern around the center of the casing; if the disc sits at an angle inside of the casing, a similar angle will appear in the explosion.

Spheres and discs are the most logical shapes because they require that the stars are all projected the same distance. Those distances, the radii of the spheres and discs, are determined by the amount of black powder packed between the stars. Therefore, a larger casing usually results in a larger spread of stars because there is more extra room for black powder.

In order to launch the whole thing into the sky, the firework is first placed into a cylindrical tube with black powder in the bottom. The correct volume of black powder is determined based on the mass of the firework and the desired height to be reached. The black powder in the tube is then ignited, shooting the firework upward (and possibly at an angle determined by the angle of the tube) and igniting the firework's main fuse. All of the fuses in the contraption have a pre-calculated length in order to control the timing of the explosions of light and sound.

MANIPULATING SOUND

Using the right materials and a little geometric reasoning, enclosed spaces can be designed to affect sounds in different ways. The muffler attached to the exhaust pipe on a car, for example, is intended to absorb most of the noise produced by the engine, but the thin materials used absorb a relatively small amount of the noise. The majority of the engine noise is canceled out by additional waves of sound produced when the engine's sound waves bounces off the walls and inner structure. Sound from the engine enters the muffler through a pipe, where some of the sound waves escape through circular holes in the side of the pipe, called perforations. The dimensions of the pipe and the size and position of perforations are important to directing the sound against different walls of the muffler. The main part of the muffler is sectioned off into three chambers, and as sound waves bounce in different directions, they cancel each other out. Technically, the frequencies of the sound waves interfere with each other, and the combination of sound waves results in fewer vibrations in the air. Because a muffler cannot consist of, say, padded concrete walls, it must be designed to use sound waves against each other in order to disrupt as much of the noise as possible.

A concert hall, on the other hand, is designed to bounce sound waves around a large volume in order to enhance the sound. The dimensions of the walls, ceiling, and floors surrounding the performers and audience greatly affect the resulting sounds, as do the furniture and anything protruding from the walls or ceiling. The most important consideration when designing a concert hall is the equalization of the different frequencies coming from the stage. The sound waves that produce higher sounding notes do not tend to bounce of surfaces as well as the slower vibrations of lower sounding notes. The overall shape of the hall and the additional surfaces (especially shapes and patterns incorporated into the walls and ceiling) are intended to ensure that the best balance of tones makes it to each listener's ears.

Reverberation is another important aspect of an acoustic space. Reverberation relates to the amount of time that sound waves continue bouncing off of the various surfaces. As a simple interpretation, reverberation defines the amount of echoing effect that will be heard. A mathematical formula, called the Sabine formula, can be used to measure the relative amount of reverberation and to determine how much absorbing surfaces need to be added to or removed from a space to achieve optimal reverberation time.

Acoustic instruments rely on the same types of considerations on a much smaller scale. The sounds produced by any drum that does not rely on electricity depend almost entirely on its shape and size. The dimensions of a piano can greatly affect the quality of sound produced when the keys are pressed. On an acoustic guitar, the strings are attached to the outside of the wooden body; but the dimensions of the body are the main reason that the sounds produced by the strings can be heard clearly. Much of the vibrations coming from the strings enter the body of the guitar through a circular sound hole situated directly beneath the portion of the strings that is most often plucked or strummed. The body is carefully designed so that the entering sound waves are most likely to bounce at angles that enhance the sound in ways similar to sound waves bouncing around a well-designed concert hall.

The geometric dimensions of strings are a major factor in the tones produced by any stringed instrument. The strings are essentially cylindrical, having a small circular cross-section stretched along a third dimension to create a volume. On an acoustic guitar each string is stretched to the same length, so the sounds produced by the different strings are dependent on the area of the circular cross-section (the thickness). A larger cross-sectional area results in a larger volume. Because the strings are made of similar materials, thicker strings also have a greater mass, which causes them to vibrate slower and to produce lower sounds than thinner strings.

Reverting to the single dimension of curves and lines, guitar strings can be thought of as parallel line segments. Strips of metal (called frets) are spaced out along a wooden fret board, or fingerboard. The frets lie perpendicular to the strings, defining important points where they intersect. Pressing a string against one of the frets essentially changes the length of the segment being strummed or plucked, causing the strings to vibrate at different frequencies and altering the resulting tones.

SOLAR SYSTEMS

All things in the physical universe, from molecules to exploding stars, have forms that can be defined geometrically. The laws of the Universe have worked together over the past few billion years to create incredible geometric shapes.

For example, solar systems all over the Universe tend to be relatively planar, like huge spinning discs in space. Basically, as a star begins to be crushed by gravity, pulling in all sorts of nearby materials, it picks up a spinning motion, similar to the spinning motion of tornadoes or water escaping through a drain in a bathtub. (If gravitational shrinkage keeps up for too long, a black hole results.) Things continue to spin, and like a ball of dough spun in the air to create a flat pizza crust, the ensuing solar system expands to a practically flat shape. Because of the relative emptiness of space, things spread out at a somewhat constant speed, creating an elliptical disc expanding in a plane. As the rocks, gas, dust, and other debris spin around the star, they collide and collect together to form planets, moons, comets, and asteroids.

The planets are the largest collections of materials and continue on an elliptical orbit around the star. Because they are so big, planets create their own substantial amount of gravity and attract debris that settles into orbit around them. Sometimes this debris collides, and eventually creates a single satellite around the planet. Planets and moons are suspended in an orbit, so each collision causes the material to spin (similar to the effect of flicking a coin held vertical by one finger) eventually leading to large bodies spinning on a constant axis and taking spherical forms. When the debris collecting around a planet has not collected to form a single satellite, the planets are encircled by planar belts of debris.

As new debris enters the atmosphere, it is attracted to the belt by the spinning forces. It is no coincidence that all of these spinning objects take on elliptical or spherical forms. These heavenly bodies all provide stellar examples of the idea of a radius defining a collection of points equally spaced from a center.

As moons spin around planets that are spinning around a star, all of the orbiting bodies create thousands of constantly changing angles between them. These changes in these angles are rather periodic, and are often studied and accurately predicted from Earth. For example, lunar and solar eclipses, in which Earth and the moon line up to create a straight line with the sun, are marked on many standard calendars.

Until late in the twentieth century, it was thought that Earth's solar system might be the only true solar system in the known Universe. Because most stars are unfathomable distances from Earth, their intense light drowns out any nearby material, no matter how powerful the telescope. However, as planets orbit around the star, they cause it to move around in a relatively small circle. From a viewpoint on Earth, this movement appears as a minute back-and-forth motion. Using mainly the star's wobble as an indicator, astronomers can determine the number of planets, the mass of each planet, and their relative distances from the central star.

Scattered throughout the Universe and confined to planes tilted at various angles, brilliant solar systems and exploding star systems (that may or may not become solar system sometime in the vast future) illustrate nature's affinity for geometric figures and provide marvelous examples of how geometric reasoning continues to light the paths that lead to knowledge previously thought impossible.

Rubik's Cube consists of smaller cubes, where only the outer faces of the outermost cubes can be seen. STEFANO BIANCHETTI/CORBIS.

RUBIK'S CUBE

For any geometric concept, there is an associated puzzle or riddle. While many puzzles have been designed to clearly illustrate a preexisting concept, some sound geometric theories were actually discovered as a solution to such puzzles, or proofs that no solution exists.

Rubik's Cube is possibly the most famous and addictive of all geometric puzzles. It was invented in 1974 by Hungarian sculptor, architectural engineer, and professor, Erno Rubik. Rubik's Cube consists of smaller cubes, where only the outer faces of the outermost cubes can be seen. The original Rubik's Cube has dimensions of $3 \times 3 \times 3$. That is, each edge of the cube is three cube lengths long; each layer of the cube is 3×3 to create a square; and the cube consists of three of these layers; so the measurement in any direction is the length of three smaller cubes. Any layer of the cube contains nine cubes ($3 \times 3 = 3^2 = 9$). There are three layers for a total of 27 smaller cubes ($3 \times 3 \times 3 = 3^3 = 27$).

The Rubik's Cube is a perfect illustration of the reason that numbers like 1, 8, 27, 64, and 125 are referred to as cubic numbers; they can be configured to make perfect cubes. Any eight identical objects can be situated in space to form a cube. This is the geometric interpretation of two raised to the third power. Similarly, if a number of objects can be arranged (in two dimensions) to form a square, that number is called a square number. Square

numbers can be represented algebraically by some other whole number raised to the power of two. This concept is illustrated by the faces of a Rubik's Cube.

Professor Rubik was not actually trying to create a puzzle, but rather to solve a three-dimensional geometric problem that had become a hot topic at the time. The problem was to create a seemingly solid cube consisting of twenty-seven smaller cubes, where any layer could be rotated around its center without disturbing the other layers.

Fascinating mechanics allow any layer to be rotated around its center without causing the rest of the apparatus to fall apart. The 27 smaller cubes can be categorized as a single central cube (which is not actually a cube, but the main component of the complex rotating mechanism); six cubes surrounding the central mechanism; and the 20 cubes with faces that can be seen. The rotation of layers in different directions is enabled by a series of spring-loaded spindles and plastic flanges, in addition to the intricate mechanism in the center of the cube.

A Rubik's Cube provides a concrete example of the geometric concepts of surface area and volume. The area of one face on a small cube is equal to the length of one of its edges squared; and each of the six faces of a $3 \times 3 \times 3$ cube consists of nine smaller faces for a total of 54 visible faces; so the surface area of the entire cube is equal to the area of one small face multiplied by 54. The volume of a small cube is equal to the length of one if its edges cubed; and there are 27 smaller cubes; so the volume of the main cube is equal to the volume of one small cube multiplied by 27. The multitude of mathematical facts that can be illustrated (and even discovered) while playing with a Rubik's Cube is amazing.

Initially and when in solved form, each of the six faces of the cube is its own color: green, blue, red, orange, yellow, or white. As the layers are rotated, the colored faces are shuffled. The goal of the puzzle is to restore each face to a single color after thorough shuffling. Numerous strategies have been developed for solving a Rubik's Cube, all of which involve some degree of geometric reasoning. Some strategies can be simulated by computer programs, and many contests take place to compare strategies based on the average number of moves required to solve randomized configurations. The top strategies can require less than 20 moves.

Possibly the most daunting fact about the $3 \times 3 \times 3$ Rubik's Cube is that 43,252,003,274,489,856,000 different combinations of colors can be created on the faces of the cube. That's more than 43 quintillion combinations, or 43 million multiplied by a million, and then multiplied by a million again. Keep in mind that the original $3 \times 3 \times 3$ cube is among the smallest and least complicated of Rubik's puzzles!

SHOOTING AN ARROW

The aim of archery is to shoot an arrow and hit a target. The three main components involved in shooting an arrow—the bow, the arrow, and the target—are thoroughly analyzed in order to optimize accuracy.

The act of shooting an arrow provides an excellent exploration of vectors (as may be deduced by the fact that vectors are usually represented by arrows in mathematical figures). The intended path of the arrow, the forces that alter this path, and the true path taken by the arrow when released can all be represented as vectors. In fact, the vector that represents the true path taken by the arrow is the sum of the vectors produced by the forward motion of the arrow and the vectors that represent the forces that disrupt the motion of the arrow. Gravity, wind, and rain essentially add vectors to the vector of the intended path, so that the original speed and direction of the arrow is not maintained. When an arrow is aimed directly at a target and then released, it begins to travel in the direction of the target with a specific speed. However, the point at which an arrow is directly aimed is never the exact point hit by the arrow. Gravity immediately adds a downward force to the forward force created by the bow, pulling the arrow down and reducing its speed. Gravity is constant, so the vector used to represent this force always points straight toward the ground with the same magnitude (length). If gravity is the only force acting on an arrow flying toward its target, then the point hit will be directly below the pointed at which the arrow is aimed; how far below depends on the distance the arrow flies. Any amount of wind or rain moving in any direction has a similar affect on the flight of the arrow, further altering the speed and direction of the arrow. To determine the point that the arrow will actually hit involves moving from the intended target in the direction and length of the vectors that represent the additional forces, similar to the way that addition of vectors is represented on a piece of graph paper.

Though the addition of vectors in three-dimensional space is the most prominent application of geometry found in archery, geometric concepts can be unearthed in all aspects of the sport. The bow consists of a flexible strip of material (e.g., wood or light, pliable metal) held at a precise curvature by a taught cord. The intended target and the actual final location of the arrowhead—whether on a piece of wood, a bail of hay, or the ground—can be thought of as theoretical points in space. The most popular target is made of circles with different radial distances from the same center, called concentric circles. If feathers are not attached at precise angles and positions near the rear of the arrow, they will not properly stabilize the arrow and it will wobble unpredictably in flight. In these ways and more, geometric reasoning is essential to every release of an arrow.

STEALTH TECHNOLOGY

Radar involves sending out radio waves and waiting a brief moment to detect the angles from which waves are reflected back. An omnidirectional radar station on the ground detects anything within a certain distance above the surface of Earth, essentially creating a hemisphere of detection range. A radar station in the air (e.g., attached to a spy plane), can send out signals in all directions, detecting any object within the spherical boundary of the radar's range. The direction and speed of an object in motion can be determined by changes in the reflected radio waves. Among other things, radar is used to detect the speed of cars and baseballs, track weather patterns, and detect passing aircraft.

Most airplanes consist almost entirely of round surfaces that help to make them aerodynamic. For example, a cross-section of the main cabin of a passenger plane (parallel to the wingspan or a row of seats) is somewhat circular; so when the plane flies relatively near a radar station on the ground, it provides a perfect reflecting surface for radio waves at all times. To illustrate this, consider someone holding a clean aluminum can parallel to the ground on a sunny day. If he looks at the can, he will be able to see the reflection of the Sun no matter how the can is turned or moved, as long as it remains parallel to the ground. However, if the can were traded for a flat mirror, he would have to turn the mirror to the proper angle or move it to the correct position relative to his eyes in order to reflect the Sun into his face. The difficulty of accurately reflecting the sun using the flat mirror provides the basis for stealth technology.

To avoid being detected by radar while sneaking around enemy territories, the United States military has developed aircraft—including the B-2 Bomber and the F-117 Nighthawk—that are specially designed to reflect radio waves at angles other than directly back to the source. The underside of an aircraft designed for stealth is essentially a large flat surface; and sharp transitions between the various parts of the aircraft create well-defined angles. The danger of being detected by radar comes into play only if the aircraft is directly above a radar station; a mistake easily avoided with the aid of devices that warn pilots and navigators of oncoming radio waves.

Potential Applications

ROBOTIC SURGERY

While the idea of a robot operating on a human body with metallic arms wielding powerful clamps, prodding rods, probing cameras, razor-sharp scalpels, and spinning saws could make even the bravest of patients squeamish, the day that thinking machines perform vital operations on people may not be that far away.

Multiple robotic surgical aids are already in development. One model is already in use in the United States and another, currently in use in Europe, is waiting to be approved by the U.S. Food and Drug Administration (FDA). All existing models require human input and control. Initial instructions are input via a computer workstation using the usual computer equipment, including a screen and keyboard. A control center is also attached to the computer and includes a special three-dimensional viewing device and two elaborate joysticks. Cameras on the ends of some of the robotic arms near or inside the patient's body send information back to the computer system, which maps the visual information into mathematical data. This data is used to recreate the three-dimensional environment being invaded by the robotic arms by converting the information into highly accurate geometric representations. The viewing device has two

goggle-like eyeholes so that the surgeon's eyes and brain perceive the images in three dimensions as well. The images can be precisely magnified, shifting the perception of the surgeon to the ideal viewpoint.

Once engrossed in this three-dimensional representation, the surgeon uses the joysticks to control the various robotic appendages. Pressing a button or causing any slight movement in the joysticks sends signals to the computer, which translates this information into data that causes the precise movement of the surgical instruments. These types of robotic systems have already been used to position cameras inside of patients, as well as perform gallbladder and gastrointestinal surgeries. Immediate goals include operating on a beating heart without creating large openings in the chest.

By programming robotic units with geometric knowledge, humans can accurately navigate just about any environment, from the inside of a beating human heart to the darkest depths of the sea. By combining spacecraft, telescopes, and robotics, scientists can send out robot aids that explore the reaches of the Universe while receiving instructions from Earth. When artificial intelligence becomes a practical reality, scientists in all fields will be able to send out unmonitored helpers to explore any environment, perform tasks, and report back with pertinent information. With the rise of artificial intelligence, robots might soon be programmed to detect any issues inside of a living body, and perform the appropriate operations to restore the body to a healthy state without any human guidance. From the first incision to the final suture, critical decisions will be made by a thinking robotic surgeon.

THE FOURTH DIMENSION

Basic studies in geometry usually examine only three dimensions in order to facilitate the investigation of the properties of physical objects. To say that anything in the Universe exists only in three dimensions, however, is a great oversimplification. As humans perceive things, the Universe has a fourth dimension that can be studied in the same way as the length, width, and height of an object. This fourth dimension is time, and has just as much influence on the state of an object as its physical dimensions. Similar to the way that a cylinder can be seen as a two-dimensional circle extended into a third dimension, a can of soda thrown from one person to another can be seen as a three-dimensional object extending through time, having a different distinct position relative to the things around it at every instant. This is the fundamental concept behind the movement of objects. If there were truly only three dimensions, things could not move or change. But just as a circular cross-section of a cylinder helps to shed light on its three-dimensional properties, studying snapshots of objects in time makes it possible to understand their structure.

As perceived by the people of Earth, time moves at a constant rate in one direction. The opposite direction in time, involving the moments of the past, only exists in the forms of memory, photography, and scientific theory. Altering the perceived rate of time—in the opposite direction or in the same direction at an accelerated speed—has been a popular fantasy in science fiction for hundreds of years. Until the twentieth century, the potential of time travel was considered by even the most brilliant scientists to lie much more in the realm of fiction. In the last hundred years, however, a string of scientists have delved into this fascinating topic to explore methods for manipulating time.

The idea of time as a malleable (changeable) dimension was initiated by the theory of special relativity proposed by Albert Einstein (1879–1955) in the early twentieth century.

An important result of the theory of special relativity is that when things move relative to each other, one will perceive the other as shrinking in the direction of relative motion. For example, if a car were to drive past the woman in the chair, its length would appear to shrink, but not its height or width. Only the dimension measured in the direction of motion is affected. Of course, humans never actually see this happen because we do not see things that move quickly enough to cause a visible shrinking in appearance. Something would have to fly past the woman at about 80% the speed of light for her to notice the shrinking, in which case she would probably miss the car altogether, and would surely have no perception of its dimensions.

Similar to the manner in which the length of an object moving near the speed of light would seem to shrink as perceived by a relatively still human, time would theoretically seem to slow down as well. However, time would not be affected in any way from the point of view of the moving object, just as physical measurements only seem to shrink from the point of view of someone not moving at the same speed along the same path. If two people are flying by each other in space, to both of these people it will seem that the other is the one moving. So while one could theoretically see physical shrinking and a slowing of the watch on the other's arm, the other sees the same affects in the other person. Without a large nearby reference point, it is easy to feel like the center of the universe, with the movement, mass, and rate of time all-dependent upon the local perception.

All of these ideas about skewed perception due to speed of relative motion are rather difficult to grasp because none of it can be witnessed with human eyes, but recall that the notion of Earth as a sphere moving in space was once commonly tossed aside as mystical nonsense. Einstein's theory of relativity explains events in the Universe much more accurately than previous theories. For example, relativity corrects the inaccuracies of English mathematician Isaac Newton's (1642–1727) proposed laws of gravity and motion, which had been the most acceptable method for explaining the forces of Earth's gravity for hundreds of years. Just as humans can now film the Earth from space to visually verify its spherical nature, its path around the sun, and so forth, the future may very well bring technology that can vividly verify the theories that have been evolving over the last century. For now, these theories are supported by a number of experiments. In 1972, for example, two precise atomic clocks were synchronized, one placed on a high-speed airplane, and the other left on the ground. After the airplane flew around and landed, the time indicated by the clock on the airplane was behind that of the clock on the ground. The amount of time was accurately explained and predicted by the theory of relativity. Inconsistencies in experiments involving the speed of light dating back to the early eighteenth century can be accurately accounted for by the theory of relativity as well.

To travel into the past would require moving faster than the speed of light. Imagine sitting on a spacecraft in outer space and looking through a telescope at someone walking on the surface of Earth. New light is continually reflecting off of Earth and the walker, entering the telescope. However, if the spacecraft were to begin moving away from Earth at the speed of light, the walker would appear to freeze because the spacecraft and the light would be moving at the same speed. The same vision would be following the telescope and no new information from Earth would reach it. The light waves that had passed the spacecraft just before it started moving would be traveling at the same speed directly in front of the spacecraft. If the spacecraft could speed up just a little, it would move in front of the light of the past, and the viewer would again see events from the past. The walker would appear to be moving backward as the spacecraft continued to move past the light from further in the past. The faster the spacecraft moved away from Earth, the faster everything would rewind in front of the viewer's eyes. Moving much faster than the speed of light in a large looping path that returned to Earth could land the viewer on a planet full of dinosaurs. Unfortunately, moving faster than the speed of light is considered to be impossible, so traveling backward in time is out of the

Key Terms

Angle: A geometric figure formed by two lines diverging from a common point or two planes diverging from a common line often measured in degrees.

Area: The measurement of a surface bounded by a set of curves as measured in square units.

Cross-section: The two-dimensional figure outlined by slicing a three-dimensional object.

Curve: A curved or straight geometric element generated by a moving point that has extension only along the one-dimensional path of the point.

Geometry: A fundamental branch of mathematics that deals with the measurement, properties, and relationships of points, lines, angles, surfaces, and solids.

Line: A straight geometric element generated by a moving point that has extension only along the one-dimensional path of the point.

Point: A geometric element defined only by an ordered set of coordinates.

Segment: A portion truncated from a geometric figure by one or more points, lines, or planes; the finite part of a line bounded by two points in the line.

Vector: A quantity consisting of magnitude and direction, usually represented by an arrow whose length represents the magnitude and whose orientation in space represents the direction.

Volume: The amount of space occupied by a three-dimensional object as measured in cubic units.

question. The idea of traveling into the future at and accelerated rate, on the other hand, is believed to be theoretically possible; but the best ideas so far involve flying into theoretical objects in space, such as black holes, which would most likely crush anything that entered and might not even exist at all.

The interwoven relationship of space and time is often referred to as the space-time continuum. To those who possess a firm understanding of the sophisticated ideas of special relativity, the four dimensions of the universe begin to reveal themselves more plainly; and to some, the fabric of time is begging to be ripped in order to allow travel to other times. While time travel is not likely to be realized in the near future, every experiment and theory helps the human race explain the events of the past, and predict the events of the future.

Where to Learn More

Books

Hawking, Stephen. *A Brief History of Time: From the Big Bang to Black Holes.* New York: Bantam, 1998.

Pritchard, Chris. *The Changing Shape of Geometry.* Cambridge, UK: Cambridge University Press, 2003.

Stewart, Ian. *Concepts of Modern Mathematics.* Dover Publications, 1995.

Web sites

Utah State University. "National Library of Virtual Manipulatives for Interactive Mathematics." National Science Foundation. April 26, 2005. <http://matti.usu.edu/nlvm/nav/topic_t_3.html> (May 3, 2005).

Graphing

Overview

In its most straightforward definition, graphing is the act of representing mathematical relationships or quantities in a visual form. Real-life applications can range from records of stock prices to calculations used in the design of spacecraft to evaluations of global climate change.

Fundamental Mathematical Concepts and Terms

In basic mathematics, graphs depict how one variable changes with respect to another and are often referred to as charts or plots. The graphs can be either empirical, meaning that they show measured or observed quantities, or they can be functional. Examples of empirical measurements are the speed shown on the speedometer of a car, the weight of a person shown on a bathroom scale, or any other value obtained by measurement. Function plots, in contrast, show pure mathematical relationships known as functions, such as $y = b + m, x,$ or $y = x^2$. In these examples, each value of x corresponds to a specific value of y and y is said to be a function of x.

Mathematicians and computer scientists sometimes refer to graphs in a different sense when they are analyzing possible ways to connect points (also known as vertices or nodes) in space using networks of lines (also known as edges or arcs). The body of knowledge related to this kind of analysis is known as graph theory. Graph theory has applications to the design of many kinds of networks. Examples include the structure of the electronic links that comprise the Internet, determining the most economical route between two points connected by a complicated network of roads (or railroads, air routes, or shipping routes), electrical circuit design, and job scheduling.

In order to accurately represent empirical or functional relationships between variables, graphs must use some method to scale, or size, the information being plotted. The most common way to do this relies upon an idea developed by the French mathematician René Descartes (1596–1650) in the seventeenth century. Descartes created graphs by measuring the value of one variable along an imaginary line and the value of the second variable along another imaginary line perpendicular to the first. Each of the lines is known as an axis, and it has become standard practice to draw and label the axes rather than using only imaginary lines. Other kinds of coordinate systems exist and are useful for special applications in science and engineering, but the

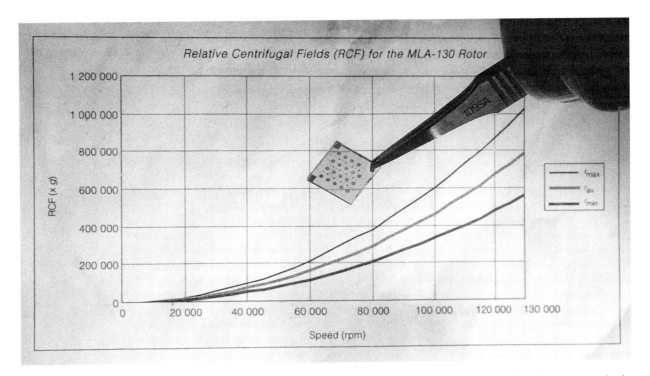

A computer chip (which contains billions of pure light converting proteins) is shown in the foreground. The chip may one day be a power source in electronics such as mobile phones or laptops. In the background is a graph which displays gravity forces that can separate light-electricity converting protein from spinach. Researchers at MIT say they have used spinach to harness a plant's ability to convert sunlight into energy for the first time, creating a device that may one day power laptops, mobile phones and more. AP/WIDE WORLD PHOTOS. REPRODUCED BY PERMISSION.

majority of graphs encountered on a daily basis use a set of two perpendicular axes.

In most graphs, the dependent variable is plotted using the vertical axis and the independent variable is plotted using the horizontal axis. For example, a graph showing measured rainfall on each day of the year would commonly show the rainfall on the vertical axis because it is dependent upon the day of the year and is, therefore, the dependent variable. Time, represented by the day of the year, is the independent variable because its value is not controlled by the amount of rainfall. Likewise, a graph showing the number of cars sold in the United States for each of the past ten years will usually have the years shown along the horizontal axis and the number of cars sold along the vertical axis. There are some exceptions to this general rule. Atmospheric scientists measuring the amount of air pollution at different altitudes or geologists measuring the chemical composition of rocks at different depths beneath Earth's surface often choose to create graphs in which the independent variable (in these cases, altitude or depth) is shown on the vertical axis. In both cases the dependent variable is being measured vertically, so it makes sense to make graphs having the same orientation.

BAR GRAPHS

Bar graphs are used to show values associated with clearly defined categories. For example, the number of cars sold by a dealer each month, the numbers of homes sold in different cities during a certain year, or the amount of rainfall measured each day during a one-year period can all be shown on bar graphs. The categories are shown along one axis and the values are represented by bars drawn perpendicular to the category axis. In some cases bar graphs will contain a value axis, but in other cases the value axis may be omitted and the values indicated by a number just above or next to each bar. The term "bar graph" is sometimes restricted to graphs in which the bars are horizontal. In that case, graphs with vertical bars are called column graphs.

One bar is drawn for each category on a bar graph, and the height or length of the bar is proportional to the value being shown. For example, the following set of numbers could reflect the average price of homes sold in different parts of Santa Barbara County, California, in February 2005: Area 1, $334,000; Area 2, $381,000; Area 3, $308,000; Area 4, $234,000; Area 5, $259,950. If these figures were plotted on a bar graph, the tallest bar would correspond to the price for Area 2. The absolute height of this

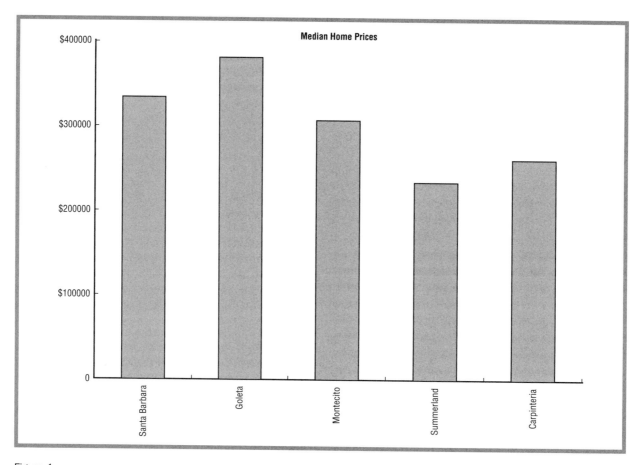

Median Home Prices

$400000

$300000

$200000

$100000

0

Santa Barbara

Goleta

Montecito

Summerland

Carpinteria

Figure 1.

bar does not matter, because the largest value will control the values of all the other bars. The height of the bar for Area 1, which has the second most expensive homes, would be 334,000 / 381,000 = 88% as tall as the bar representing Area 2. Similarly, the bar representing Area 3 would be 308,000 / 381,000 = 81% as tall as the Area 2 bar. See Figure 1, which depicts the bar graph reflecting the average price of homes sold in different parts of Santa Barbara County, California, in February 2005.

Bar graph categories can represent virtually anything for or about which data can be collected. In Figure 1, the categories represent different parts of a county for which real estate sales records are kept. In other cases bar graph categories represent a quantity such as time, such as the rainfall measured in New York City on each day of February 2005, with each bar representing one day.

Scientists and engineers often use specialized forms of bar graphs known as stem graphs, in which the bars are replaced by lines. Using lines instead of bars can help to make the graph more readable when there are many categories; for example, the sizes of the largest floods along the Rio Grande during the past 100 years would require

100 bars or stems. More often than not, the kinds of data collected by scientists and engineers dictate that the categories involve some measure of distance or time (for example, the year in which each flood occurred). As such, they are usually ordered from smallest to largest. Stem graphs can also have small open or filled circles at the end of each stem. Unless the legend for the graph specifies otherwise, the circles are used simply to make the graph more readable and do not have any significance of their own.

Histograms are specialized bar graphs in which each category represents a range of possible values, and the values plotted perpendicular to the category axis represent the number of occurrences of each category. An important characteristic of a histogram is that each category does not represent just one value or attribute, but rather a range of values that are grouped together into a single category or bin. For example, suppose that in a group of 100 people there are 20 who earn annual salaries between $20,000 and $30,000, 40 who earn annual salaries between $30,001 and $40,000, 30 who earn annual salaries between $40,001 and $50,000, and 10 who earn annual

250

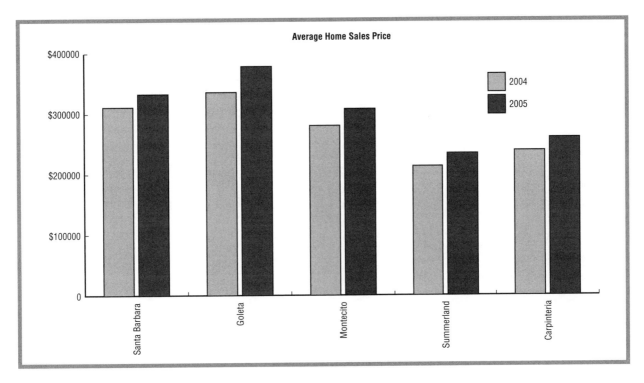

Average Home Sales Price

2004

2005

$400000

$300000

$200000

$100000

0

Santa Barbara

Goleta

Montecito

Summerland

Carpinteria

Figure 2.

salaries between $50,001 and $60,000. The bins in a histogram showing this salary distribution would be $20,000 to $30,000, $30,001 to $40,000, $40,001 to $50,000, and $50,001 to $60,000. The height of each bin would be proportional to the number of people whose salaries fall into that bin. The tallest bar would represent the bin with the most occurrences, in this case the $30,001 to $40,000. The second tallest bar would represent the $40,001 to $50,000 category, and it would be 30/40 = 75% as tall as the tallest bin. The width of each bin is proportional to the range of values that it represents. Therefore, if each class interval is the same size then all of the bars on a histogram will be the same width. A histogram containing bars with different widths will have unequal class intervals.

Some bar graphs use more than one set of bars in order to convey several sets of information. Continuing with the home price example from Figure 1, the bars showing the 2005 prices could be supplemented with bars showing the average home sales prices for the same areas in February 2004. Figure 2 allows readers to quickly compare prices and see how they changed between 2004 and 2005. Each category has two bars, one for 2004 and one for 2005, filled with different colors, patterns, or shades of gray to distinguish them from each other.

A third kind of bar graph is the stacked bar graph, in which different types of data for each category are represented using bars stacked on top of each other. The

bottom bar in each of the stacks will generally have a different height, which makes it difficult to compare values among categories for all but the bottom bars. For this reason, stacked bar graphs can be difficult to read and should generally be avoided.

LINE GRAPHS

Line graphs share some similarities with bar graphs, but use points connected by straight lines rather than bars to represent the values being graphed. As with bar graphs, the categories on a line graph can represent either some kind of measurable quantity or more abstract qualities such as geographic regions.

Line graphs are constructed much like bar graphs. In line graphs, values for each category are known or measured, and the categories are placed along one axis. The values are then scaled along the value axis, and a point, sometimes represented by a symbol such as a circle or a square, is drawn to represent the value for each category. The points are then connected with straight line segments to create the line graph.

One of the weaknesses of line graphs is that they can imply some kind of connection between categories, which may or may not be the intention of the person creating the graph. In a bar chart, each category is represented by a bar that is completely separate from its

Graphs are often used as visuals representing finances. AP/WIDE WORLD PHOTOS. REPRODUCED BY PERMISSION.

neighbors. Therefore, no connection or relationship between adjacent categories is implied by the graph. A line graph implies that the value varies continuously between adjacent categories because the points are connected by lines. If there is no real connection between the values for adjacent categories, for example the home sales prices used in the Figure 1 bar graph example, then it may be better to use a bar graph or stem graph than a line graph.

Like bar graphs, line graphs can be combined to create multiple line graphs. Each line represents a different value associated with each category. For example, a multiple line graph might show different household expenses for each month of the year (rent, heat, water, groceries, etc.) or the income and expenses of a business for each quarter of a particular year. Rather than being placed side-by-side as in a multiple bar graph, however, multiple line graphs are placed on top of each other and the lines are distinguished by different colors or patterns. If only two sets of values are being graphed and their values are significantly different, two value axes may be used. As shown in Figure 3, each value axis corresponds to one of the sets of values and is labeled accordingly.

AREA GRAPHS

Area graphs are line graphs in which the area between the line and the category axis is filled with a color or pattern, and are used when there is a need to show both the values associated with each category and the total of all the values. As Figure 4 shows, the values are represented by the height of the colored area, whereas the total is represented by the amount of area that is colored. If the total area beneath the lines is not important, then a bar graph or line graph may be a better choice. Area graphs can also be stacked if the objective is to show information about more than one set of values. The result is much like a stacked bar graph.

PIE GRAPHS

Pie graphs are circular graphs that represent the relative magnitudes of different categories of data using angular wedges resembling slices of pie. The size of each wedge, which is measured as an angle, is proportional to the relative size of the value it represents.

If the data are given as percentages that add up to 100%, then the angular increment of each wedge is its percentage × 360°, which is the number of degrees in a complete circle. For example, imagine that Store A sells 30% of all computers sold in Boise, Idaho, Store B sells 18%, and all other stores combined sell the remainder. The wedge representing Store A would be 0.30 × 360° = 108° in size. The wedge representing Store B would, by the same logic, be 0.18 × 360° = 65°, and the wedge representing all other stores would (1.00 − 0.30 − 0.18) × 360° = 0.52 × 360° = 187°. Figure 5 depicts a representative pie graph.

The calculations become slightly more complicated if the data are not given in terms of percentages that add up to 100%. Suppose that instead of the percentage of computers sold by the stores in the previous example, only the number of computers sold by each store is known. In that case, the number of computers sold by each store must be divided by the total number sold by all stores to calculate the percentage for that store. If Store A sold 1,500 computers, Store B sold 900 computers, and all other stores combined sold 2,600 computers, then the total number of computers sold would be 5,000. The percentage sold by Store A would be 1,500/5,000 = 0.30, or 30%. Similar calculations produce results of 18% for Store B and 52% for all other stores combined (just as in the previous example).

RADAR GRAPHS

Radar graphs, also known as spider graphs or star graphs, are special types of line graphs in which the values are plotted along axes radiating from a common point. The result is a graph that looks like a radar screen to some people, and a spider or star to others. There is one axis for each category being graphed, so for *n* categories each axis will be separated by an angle of 360°/*n*. A radar graph showing five categories, for example, would have five axes separated by angles of 360°/5 = 72°. The value of each category is measured along its axis, with the distances from the center proportional to the value, and adjacent values are connected to form an irregularly shaped polygon. One of the advantages of radar plots, as shown below in Figure 6 (p. 254), is that they can convey information about the values of many categories using shapes (the polygons created by connecting adjacent values) that can be easily compared for many different data sets.

Multiple radar graphs are constructed much like multiple line graphs, with several values plotted for each category. The lines connecting the values for each category have different colors or patterns in order to distinguish among them.

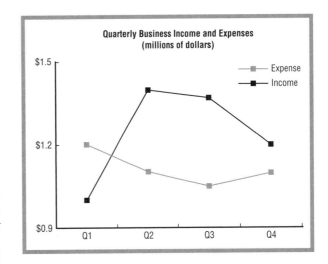

Figure 3.

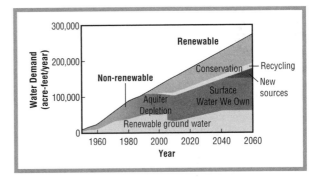

Figure 4: Stacked area graph showing different sources of water (values) by year (categories).

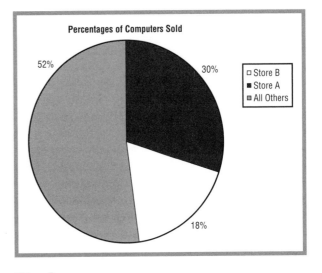

Figure 5.

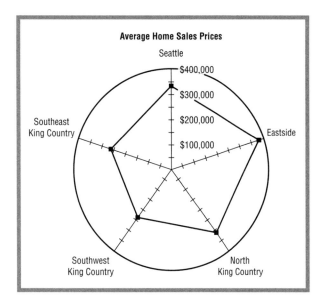

Figure 6.

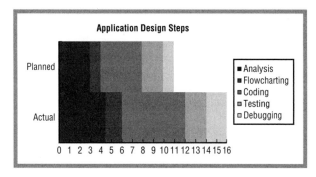

Figure 7.

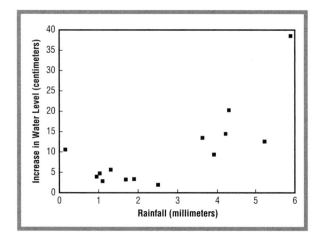

Figure 8.

GANTT GRAPHS

Gantt graphs are used by project managers and others to show job activity over time, which can range from a single workday to a complicated construction project that stretches over several years. The horizontal axis shows time, with units depending on the length of the project. The vertical axis shows resources, which can be anything from the names of people working on the project to different pieces of equipment needed to complete the project. Blocks of time are marked off along the time axis showing how each resource will be used during that time.

PICTURE GRAPHS

Graphs that are intended for general readers rather than scientists or engineers, such as those frequently published in newspapers and magazines, often use artistic symbols to denote the values of different categories. An article about money, for example, might show stacks of currency instead of plain bars in a bar graph. A different article about new car sales might include a graph using a small picture of a car to represent every 100 cars sold by different dealers. These kinds of artistic graphs are usually varieties of bar graphs, although the use of artistic symbols can make it difficult to accurately compare values among different categories. Therefore, they are most useful when used to illustrate general trends or relationships rather than to allow readers to make exact comparisons. For that reason, picture graphs are almost never used by scientists and engineers.

X-Y GRAPHS

X-y graphs are also known as scatterplots. Instead of having a fixed number of categories along one axis, x-y graphs allow an infinite number of points along two perpendicular axes and are used extensively in scientific and engineering applications. Each point is defined by two values: the abscissa, which is measured along the x axis, and the ordinate, which is measured along the y axis. Strictly speaking, the terms abscissa and ordinate refer to the values measured along each axis although in day-to-day conversation many scientists and engineers use the terms in reference to the axes themselves. Each piece of data to be graphed will have both an abscissa and an ordinate, sometimes referred to as x- and y-values.

The most noticeable property of an x-y graph is that it consists of points rather than bars or lines. Lines can be added to x-y plots but they are in addition to the points and not a replacement for them. Line graphs can also have points added as an embellishment and can therefore

Graphing Functions and Inequalities

Continuous mathematical functions and inequalities involving real numbers have an infinite number of possible values, but are graphed in much the same way as x-y graphs containing a finite number of points.

Consider the function $y = x^2$. The first step is to determine the range of the x axis because, unlike a finite set of points that have a minimum and maximum x value, functions can generally range over all possible values of x from $-\infty$ to $+\infty$. For this example, allow x to range from 0 to 3 ($0 \geq x \geq 3$). Next, select enough points over that range to produce a smooth curve. This must be done by trial and error, and becomes easier once a few graphs are made. Seven points will suffice for this example: 0, 0.5, 1, 1.5, 2, 2.5, and 3. These values will be the abscissae. Substitute each abscissa into the function (in this case $y = x^2$) and calculate the value of the function for that value, which will produce the ordinates 0, 0.25, 1, 2.25, 4, 6.25, and 9. Finally, plot a point for each corresponding abscissa and ordinate, or (0,0), (0.5,0.25), (1,1), (1.5,2.25), (2,4), (2.5,6.25), and (3,9).

Because a continuous function has values for all possible values of x, not just those for which values were just calculated, the points can be joined using a smooth curve. Before computers with graphics capabilities were widely available, this was done using drafting templates known as French curves, or thin flexible strips known as splines. The French curve, or spline, was positioned so that it passed through the graphed points and used as a guide to draw a smooth curve. A smooth curve can also be approximated by calculating values for a large number of points and then connecting them with straight lines, as in a line graph. If enough points are used, the straight line

segments will be short enough to give the appearance of a smooth curve. Computer graphics programs follow a digital version of this procedure, calculating enough sets of abscissae and ordinates to generate the appearance of a continuous line. In many cases the programs use sophisticated algorithms that minimize the number of points by evaluating the function to see where values change the most, plotting more points in those areas and fewer in parts of the graph where the function is smoother.

To plot an inequality, temporarily consider the inequality sign ($<$, $>$, $\geq$, $\leq$) to be an equal sign. Decide upon a range for the abscissae, divide it into segments, and calculate pairs of abscissae and ordinates in the same manner as for a function. If the inequality is $>$ or $<$, then connect the points with a dashed line and indicate which side of the line represents the inequality. For example, if the inequality is $y > x^2$, then the area above the dashed line should be shaded or otherwise identified as the region satisfying the inequality. If the inequality had been $y < x$, then the area beneath the dashed line would satisfy the inequality. In cases of $\geq$ or $\leq$ inequalities, the two regions can be separated by a solid line to indicate that points exactly along the line, not just those above or below it, satisfy the relationship.

Graphs of functions can also be used to solve equations. The equation $4.3 = x^2$, for example, is a version of the equation $y = x^2$ described in this sidebar. Therefore, it can be solved by graphing the function $y = x^2$ over a range of values that includes x = 4.3 (for example, $4 \geq x \geq 5$) and reading the abscissa that corresponds to an ordinate of 4.3. In this case, the answer is x = 2.07.

be confused with x-y graphs under some circumstances. Line graphs and x-y graphs, however, have some important differences. First, the categories on a line graph do not have to be numbers. As described above, line graphs can represent things such as cities, geographic areas, or companies. Each value on a line graph must correspond to one of a finite number of categories. The abscissa of a point plotted on an x-y graph, in contrast, must always be a number and can take on any value. Second, the lines on a line graph must always connect the values for each category. If lines are added to an x-y graph, they do not have to connect all of the points. Although they can connect all

of the points, especially in cases where there are only a few points on the graph, lines connecting the data points are not required on x-y graphs. Lines can, for example, be used to show averages or trends in the data on an x-y graph. Figure 8 represents an x-y graph. Adding lines to connect all of the points in an x-y graph can be very confusing if there are a large number of points, and should be done only if it improves the legibility of the graph.

To create an x-y graph, first move along the x-axis to the abscissa and draw an imaginary line perpendicular to the x-axis and passing through the abscissa. Next, move

Graphing Fallacies

Some people believe that graphs don't lie because they are based on numbers. But, the way that a graph is drawn and the numbers that are chosen can deliberately or accidentally create false impressions of the relationships shown on the graph. Scientists, engineers, and mathematicians are usually very careful not to mislead their readers with fallacious graphs, but artists working for newspapers and magazines sometimes take liberties that accidentally misrepresent data. Dishonest people may also deliberately create graphs that misrepresent data if it helps them to prove a point.

One way to misrepresent data is to create a graph that shows only a selected portion of the data. This is known as taking data out of context. For example, if the number of computers sold at an electronics store increases by 100 computers per year for four years and then decreases by 25 computers per year during the fifth year, it is possible to make a graph showing only the last year's information and title the graph, "Decreasing Computer Sales." Actually, though, sales have increased by $4 \times 100 - 25 = 375$ computers over the five years, so the fifth year represents only a small change in a longer term trend. It is true to state that computer sales fell during the fifth year but, depending on how the graph is used, it may be misleading to do so because it presents data out of context.

Another way to misrepresent data is by choosing the limits of the vertical axis of the graph. Imagine that a survey shows that men working in executive jobs earned an average salary of $100,000 per year and that women working in executive jobs earned an average salary of $85,000 per year. If these two pieces of information were plotted on a graph with an axis ranging from zero to $100,000, it would be clear that the women earned an average of 15% less than the men. But, if the axis were changed so that it ranged only from $80,000 to $100,000 it might appear to the casual reader than women earned only about 25% as much as men. Because the information conveyed by a graph is largely visual, many readers will not notice the values on the axis and base their interpretation only on the relationships among the lines, bars, or points on the graph. Some irresponsible graph-makers even eliminate the ordinate axis altogether and use bars or other symbols that are not proportional to the values that they represent.

Sometimes it is the data themselves that are the problem. A graph showing how salaries have increased during the past 50 years may show a tremendous increase. If the salaries are adjusted for inflation, however, the increase may appear to be much smaller.

---

along the y-axis to the ordinate, then draw an imaginary line perpendicular to the y-axis. Draw a small symbol at the location where the two imaginary lines intersect. Repeat this procedure for each of the points to be graphed. The symbols used should be the same for all of the points in each data set, and can be circles, squares, rectangles, or any other simple shape. If more than one data set is to be shown on the same graph, choose a different symbol or color for the points in each set.

The abscissa and ordinate values of points on x-y graphs created for scientific or engineering projects are sometimes transformed. This can be done in order to show a wide range of values on a single set of axes or, in some cases, so that points following a curved trend are graphed as a straight line. The most common way to transform data is to calculate the logarithm of the abscissa or ordinate, or both. If the logarithm of one is plotted against the original arithmetic value of the other, the graph is known as a semi-log graph. If the logarithms of both the abscissa and ordinate are plotted, the result is

a log-log graph. The logarithms used can be of any base, although base 10 is the most common, and the base should always be indicated. At one time, base 10 logarithms were referred to as common logarithms and denoted by the abbreviation log. Base e logarithms ($e = 2.7183...$) were referred to as natural logarithms and denoted by the abbreviation ln. This practice fell out of favor among some scientists and engineers during the late 1900s. Since then, it has been common to use log to denote the natural logarithm, and $\log_{10}$ to denote the base 10, or common, logarithm.

A map with points plotted to indicate different cities or landmarks can be considered to be a special kind of x-y graph. In this case, the abscissa and ordinate of each point consist of its geographic location given in terms of latitude and longitude, universal transverse Mercator (UTM) coordinates, or other cartographic coordinate systems. Likewise, the outline of a country or continent can be thought of as a series of many points connected by short line segments.

The underlying principles of x-y plots can be extended into the third dimension to produce x-y-z plots. Points are plotted along the z axis following the same procedure that is used for the x and y axes. One difficulty associated with x-y-z plots is that two-dimensional surfaces such as pieces of paper have only two dimensions. Complicated geometric constructions known as projections must be used to create the illusion of a third dimension on a flat surface. Therefore, x-y-z plots of large numbers of points are practical only if done on a computer, which allows the plots to be virtually rotated in space so that the data can be examined from any perspective.

BUBBLE GRAPHS

Bubble graphs allow three-dimensional data to be presented in two-dimensional graphs, and are in many cases useful alternatives to x-y-z graphs. For each data point, two of the three variables are plotted as in a normal x-y graph. The third variable for each point is represented by changing the size of the point to create circles or bubbles of different sizes. One important consideration is the way in which the bubble size is calculated. One way is to make the diameter of the circle proportional to the value of the third variable. Because the area of a circle is proportional to the square of its radius, doubling the radius or diameter will increase the area of the circle by a factor of 4. Therefore, doubling the diameter may mislead a reader into believing that one bubble represents a value four times as large as another when the person creating the graph intended it to represent a value only twice as large. In order to create a circle with twice the area, the radius or diameter must be increased by a factor of 1.414 (which is the square root of 2). Figure 9 is representative of a bubble graph.

A Brief History of Discovery and Development

The graphing of functions was invented by the French mathematician and philosopher René Descartes (1596–1650) in 1637, and the Cartesian coordinate system of x-y (and sometimes z) axes used to plot most graphs today bears his name. Ironically, however, Descartes did not use axes as known today or negative numbers when he created the first graphs.

Commercially manufactured graph paper first appeared in about 1900 and was adopted for use in schools as part of a broader reform of mathematics education. Leading educators of the day extolled the virtues

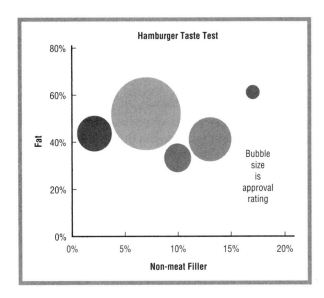

Figure 9.

of using so-called squared paper or paper with squared lines to graph mathematical functions. As the twentieth century progressed, students and professionals came to have a wide range of specialized graph paper available for use. The selection included graph paper with preprinted semi-log and log-log axes, as well as paper designed for special kinds of statistical graphs.

Digital computers were invented in the middle of the twentieth century, but computers capable of displaying even simple graphs were rare until personal computers became common in the 1980s. So-called spreadsheet programs, in particular, represented a great advance because they allowed virtually anyone to enter rows and columns of numbers and then examine relationships among them by creating different kinds of graphs. Handheld graphing calculators appeared in the 1990s and were quickly incorporated into high school and college mathematics courses. At about the same time, sophisticated scientific graphing and visualization programs for advanced students and professionals began to appear. These programs could plot thousands of points in two or three dimensions.

Real-life Applications

GLOBAL WARMING

Most scientists studying the problem have concluded that burning fossil fuels such as coal and oil (including gasoline) during the twentieth century has caused the amount of carbon dioxide, carbon monoxide, and other gasses in Earth's atmosphere to increase, which has in turn led to a warming of the atmosphere and oceans. Among the tools that scientists use to draw their conclusions are graphs showing how carbon dioxide and temperature change from day to day, week to week, and year to year. Although actual measurements of atmospheric gasses date back only 50 years or so, paleoclimatologists use other information such as the composition of air bubbles trapped for thousands of years in glacial ice, the kinds of fossils found buried in lake sediments, and the widths of tree rings to infer climate back into the recent geologic past. Data collected over time are often described as time series. Time series can be displayed using line graphs, stem graphs, or scatter plots to illustrate both short-term fluctuations that occur from month to month and long-term fluctuations that occur over tens to thousands of years, and have provided compelling evidence that increases in greenhouse gasses and temperatures measured over the past few decades represent a significant change.

FINDING OIL

Few oil wells resemble the gushers seen in old movies. In fact, modern oil well-drilling operations are designed specifically to avoid gushers because they are dangerous to both people and the environment. Geologists carefully examine small fragments of rock obtained during drilling and, after drilling is completed, lower instruments down the borehole to record different rock properties. These can include electrical resistivity, natural radioactivity, density, and the velocity with which sound waves move through the rock. All of this information helps to determine if there is oil thousands of feet beneath the surface, and is plotted on special graphs known as geophysical logs. In most cases, the properties are measured once every 6 inches (15.2 cm) down the borehole, so depth is the category (or abscissa) and each rock property is a value (or ordinate). Unlike most line graphs or x-y graphs, though, the category axis or abscissa is oriented vertically with the positive end pointing downward because the borehole is vertical and depth is measured from the ground surface downward. Geophysical logs are plotted together on one long sheet of paper or a computer screen so that geologists can compare the graphs, analyze how the rock properties change with depth, and then estimate how much oil or gas there is likely to be in the area where the well was drilled. If there is enough to make a profit, pipes and pumps are installed to bring the oil to the ground surface. If not, the well is called a dry hole and filled with cement.

GPS SURVEYING

Surveyors, engineers, and scientists use sensitive global positioning system (GPS) receivers that can determine the locations of points on Earth's surface to an accuracy of a fraction of an inch. In some cases, the information is used to determine property boundaries or to lay out construction sites. In other cases, it is used to monitor movements of Earth's tectonic plates, the growth of volcanoes, or the movement of large landslides. GPS users, however, must be certain that their receivers can obtain signals from a sufficient number of the 24 global positioning system satellites orbiting Earth in order to make such accurate and precise measurements. This can be difficult because the number of satellites from which signals can be received in a given location varies from place to place throughout the course of the day. Professional GPS users rely on mission-planning software to schedule their work so that it coincides with acceptable satellite availability. Two of the most important pieces of information provided by mission-planning software are bar graphs showing the number of satellites from which signals can be received and the overall quality or strength of the signals, which is known as positional dilution of precision (PDOP). A surveyor or scientist planning to collect high-accuracy GPS measurements will enter the latitude and longitude of the project area, information about obstructions such as tall buildings or cliffs, and the date the work is to take place. The mission-planning software will then create a graph showing the satellite coverage and PDOP during the course of that day, so that fieldwork can be scheduled for the most favorable times.

BIOMEDICAL RESEARCH

Genetic and biomedical research generate large amounts of data, particularly related to genetic sequences or genomes. Researchers in these fields use specialized graphing programs to visualize genetic sequences of different organism, including computer programs that can simultaneously display information about two different organisms and graphically illustrate which genes are present in both. Phylogenetic tree graphs, which have a branching structure, are used to illustrate the relationships between groups of many different organisms. Other biomedical scientists have developed new ways to construct multidimensional graphs to represent similarities between proteins. The field of biomechanics combines physics with biology and medicine to analyze how physical stresses and forces affect living organisms. Sophisticated scientific visualization software is used to analyze computer models simulating the stresses developed in the bones of athletes or in the blood vessels of people suffering heart attacks.

Technical Stock Analysis

Some investors rely on hunches or tips from friends to decide when they should buy or sell stock. Others rely on technical analysis to spot trends in stock prices and sales that they hope will allow them to earn more money by buying or selling stock at just the right time. Technical stock analysts use different kinds of specialized graphs to depict information that is important to them. Candlestick plots use one symbol for each day to show the price of the stock when the market opened, the price when it closed, and the high and low values for the stock during the course of the day. This is done by using a rectangle to indicate opening and closing prices, with vertical lines extending upward and downward from the box to indicate the daily high and low prices. The result is a symbol that looks like a candle with a wick at each end. The color of the box, usually red or green, indicates whether the closing price of the stock was higher or lower than the opening price.

Day-to-day fluctuations in stock price can be smoothed out using moving average or trend plots that remove most, if not all, of the small changes and let investors concentrate on trends that persist for many days, weeks, or even months. Moving averages calculate the price of a stock on any given day by averaging the prices over a period of days. For example, a five-day moving average would calculate the average price of the stock over a five-day period. The "moving" part of moving average means that different sets of data are used to calculate the average each day. The five-day moving average calculated for June 5, 2004, will use a different set of five prices (for June 1 through June 5) than the five-day moving average calculated for June 6, 2004 (June 2 through June 6).

The volume, or number, of shares sold on a given day, is also important to stock analysts and can be shown using bar charts or line graphs.

PHYSICAL FITNESS

Many health clubs and gyms have a variety of computerized machines such as stationary bicycles, rowing machines, and elliptical trainers that rely on graphs to provide information to the person using the equipment. At the beginning of a workout, the user can scroll through a menu of different simulated routes, some hilly and some flat, that offer different levels of physical challenge. As the workout progresses, a bar graph moves across a small screen to show how the resistance of the machine changes to simulate the effect of running or bicycling over hilly terrain. In other modes, the machine might monitor the user's pulse and adjust the resistance to maintain a specified heartbeat, with the level of resistance shown using a different bar graph.

AERODYNAMICS AND HYDRODYNAMICS

The key to building fast and efficient vehicles—whether they are automobiles, aircraft, or watercraft—lies in the reduction of drag. Using a combination of experimental data from wind tunnels or water tanks and the results of computational fluid dynamics computer simulations, designers can create graphs showing how factors such as the shape or smoothness of a vehicle affect the drag exerted by air or water flowing around the vehicle.

Experiments are conducted or computer simulations run for different vehicle shapes, and the results are summarized on graphs that allow designers to choose the most efficient design for a particular purpose. In some cases, these are simply x-y graphs or line graphs comparing several data sets. In other cases, the graphs are animated scientific visualizations that allow designers to examine the results of their experiments or models in great detail.

COMPUTER NETWORK DESIGN

Computer networks from the Internet to the computers in a small office can be analyzed using graphs showing the connectivity of different nodes. A large network will have many nodes and sub-nodes that are connected in a complicated manner, partly to provide a degree of redundancy that will allow the network to continue operating even if part of it is damaged. The United States government funded research during the 1960s on the design of networks that would survive attacks or catastrophes grew into the Internet and World Wide Web. A network in which each computer is connected to others by only one pathway, be it a fiber optic cable or a wireless signal, can be inexpensive but prone to disruption. At the other end of the spectrum, a network in which each computer is connected to every other computer is almost

Scientific Visualization

Scientific visualization is a form of graphing that has become increasingly important since the 1980s and 1990s. Advances in computer technology during those years allowed scientists and engineers to develop sophisticated mathematical simulations of processes as diverse as global weather, groundwater flow and contaminant transport beneath Earth's surface, and the response of large buildings to earthquakes or strong winds. Likewise, computers enabled scientists and engineers to collect very large data sets using techniques like laser scanning and computerized tomography. Instead of tens or hundreds of points to plot in a graph, scientists working in 2005 can easily have thousands or even millions of data points to plot and analyze.

Scientific visualizations, which can be thought of as complicated graphs, usually contain several different data sets. A visualization showing the results from a computer simulation of an oil reservoir, for example, might include information about the shape and extent of the rock layers in which the oil is found, information about the amount of oil at different locations in the reservoir, and information about the amount of oil pumped from different wells. A visualization of a spacecraft reentering Earth's atmosphere might include the shape of the spacecraft, colors to indicate the temperature of the outside of the spacecraft, and vectors or streamlines showing the flow of air around the spacecraft. Animation can also be an important aspect of scientific visualization,

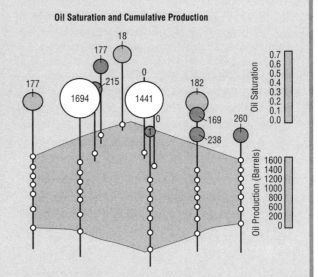

Oil Saturation and Cumulative Production

Figure A: Scientific visualization, especially for problems in which the values of variables change over time such as representations of data related to oil drilling depicted above, are an increasingly important ways to understand and depict data.

especially for problems in which the values of variables change over time. Visualization software available in 2005 typically allows scientists to interactively rotate and zoom in and out of plots showing several different kinds of data in three dimensions.

always prohibitively expensive even though it may be the most reliable. Therefore, the design of effective networks balances the costs and benefits of different alternatives (including the consequences of failure) in order to arrive an optimal design. Because of their built-in redundancy and complexity, large computer networks are impossible to comprehend without graphs illustrating the degrees of interconnection between different nodes. Applied mathematicians also use graph theory to help design the most efficient networks possible under a given set of constraints.

Potential Applications

The basic methods of graphing have not changed over the years, but continually increasing computer capabilities give scientists, engineers, and businesspeople powerful and flexible graphing tools to visualize and analyze large amounts of data. Likewise, scientific visualization tools provide a way to comprehend the voluminous output of supercomputer models of weather, ocean circulation, earthquake activity, climate change, and other complicated natural processes. Ongoing technology development is concentrated on the use of larger and faster computers to better visualize these kinds of data sets, for example using transparent surfaces and advanced rendering techniques to visualize three-dimensional data. Computer-generated movies or animations will also allow visualization of changes in three-dimensional data sets over time (so-called four-dimensional analysis). The design and implementation of user-friendly interfaces will also continue, bring powerful visualization technology within the grasp of more people.

Where to Learn More

Books

Few, Stephen. *Show Me the Numbers: Designing Tables and Graphs to Enlighten.* Oakland, CA: Analytics Press, 2004.

Huff, Darrell. *How to Lie with Statistics.* New York: W.W. Norton, 1954.

Tufte, E.R. *The Visual Display of Quantitative Information.* Cheshire, CT: Graphics Press, 1992.

Web sites

Friendly, Michael. "The Best and Worst of Statistical Graphics." Gallery of Data Visualization. 2000. <http://www.math.yorku.ca/SCS/Gallery/> (March 9, 2005).

Goodman, Jeff. "Math and Media: Deconstructing Graphs and Numbers." How Numbers Tell a Story. 2004. <http://www.ced.appstate.edu/~goodmanj/workshops/ABS04/graphs/graphs.html> (March 9, 2005).

National Oceanic and Atmospheric Administration. "Figures." Climate Modeling and Diagnostics Laboratory. <http://www.cmdl.noaa.gov/gallery/cmdl_figures> (March 9, 2005).

Weisstein, E.W. "Function Graph." Mathworld. <http://mathworld.wolfram.com/FunctionGraph.html> (March 9, 2005).

Imaging

We each process hundreds or thousands of manufactured images every day, including those displayed by books, magazines, computers, digital cameras, signage, TVs, and movies. Images are an important form of communication in entertainment, war, science, art, and other fields because a human being can grasp more information more quickly by looking at an image than in any other way.

Fundamental Mathematical Concepts and Terms

Most of the images we see have been either altered or created from scratch using computers. Computers process images in "digital" form, that is, as collections of digits (numbers). A typical black-and-white digital image consists of thousands or millions of numbers laid out in a rectangular array like the squares on a checkered tablecloth. (The numbers are not stored this way physically in the computer, but they are organized as if they were.) To turn this array of numbers into a visible image, as when making a printout or displaying the image on a screen, a tiny, visible dot is created from each number. Each dot is called a picture element or "pixel." A color image of the same size consists of three times as many numbers as a black-and-white image because there are three numbers per pixel, one number for the brightness of each color channel. The three colors used may be the three primary colors (red, yellow, blue), the three secondary colors (cyan, magenta, yellow), or the colors of the popular RGB scheme (red, green, blue). By adding different amounts from each color channel, using the three numbers for each pixel as a recipe, a pixel of any color can be made.

A rectangular array of numbers is also called a "matrix." An entire field of mathematics—"matrix algebra"—is devoted to working with matrices. Matrix algebra may be used to change the appearance of a digital image, extract information from it, compare it to another image, merge it with another image, and to affect it in many other ways. The techniques of Fourier transforms, probability and statistics, correlation, wavelets, artificial intelligence, and many other fields of mathematics are applied to digital images in art, engineering, science, entertainment, industry, police work, sports, and warfare, with new methods being devised every year.

In general, we are interested in either creating, altering, or analyzing images.

A Brief History of Discovery and Development

The relationship between images and mathematics began with the invention of classical geometry by Greek thinkers such as Euclid (c. 300 B.C.) and by mathematicians of other ancient civilizations. Classical geometry describes the properties of regular shapes that can be drawn using curved and straight lines, namely, geometric figures such as circles, squares, and triangles and solids such as spheres, cubes, and tetrahedra. The extension of mathematics to many types of images, not just geometric figures, began with the invention of perspective in the early 1400s. Perspective is the art of drawing or painting things so as to create an illusion of depth. In a perspective drawing, things that are farther from the artist are smaller and closer together according to strict geometric rules. Perspective became possible when people realized that they could apply geometry to the space in a picture, rather than just to shapes such as circles and triangles. Today, the mathematics of perspective—specifically, the group of geometric methods known as trigonometry—are basic to the creation of three-dimensional animations such as those in popular movies like *Jurassic Park* (1993), *Shrek* (2001), and *Star Wars Episode II: Attack of the Clones* (2003).

Real-life Applications

CREATING IMAGES

Because a digital image is really a rectangular array matrix full of numbers, we can create one by inserting numbers into a matrix. This is done, most often in the movie industry, by cooking up numbers using mathematical tools such as Euclidean geometry, optics, and fractals. A digital image can also be created by scanning or digitally photographing an existing object or scene.

ALTERING IMAGES

The most common way of altering a digital image is to take the numbers that make it up and apply some mathematical rule to them to create a new image. Methods of this kind including enhancement (making an image look better), filtering (removing or enhancing certain features of the image, like sharp edges), restoration (undoing damage like dust, rips, stains, and lost pixels), geometric transformation (changing the shape or orientation of an image), and compression (recording an image using fewer numbers). Most home computers today contain software for doing all these things to digital images.

Sports Video Analysis

Video analysis is the use of mathematical techniques from probability, graph theory, geometry, and other areas to analyze sports and other kinds of videos. Sports video analysis is a particularly large market, with millions of avid watchers keen for instant replays and new and better ways of seeing the game.

Traditionally, the only way to find specific moments in a video (or any other kind) of video was to fast-forward through the whole thing, which is time-consuming and annoying. Today, however, mathematics applied to game footage by computers can automatically locate specific plays, shots, or other moments in a game. It can track the ball and specific players, automatically extract highlights and statistics, and provide computer-assisted refereeing. Soon, three-dimensional computer models of the game space constructed from multiple cameras will allow the viewer to choose their own viewpoint from which to view the game as if from the front row, floating above the field, following a certain player, following the ball, or wherever. Some software based on these techniques, such as the Hawk-Eye program used to track the ball in broadcast cricket matches, is already in commercial use.

Video analysis in sports is also used by coaches and athletes to improve performance. Mathematical video analysis can show exactly how a shot-putter has thrown a shot, or how well the members of a crew team are pulling. By combining global positioning system (GPS) information about team players' exact movements with computerized video analysis and radio-transmitted information about breathing and heart rates, coaches (well-funded, high-tech, and "math savvy" coaches, that is) can now get an exact picture of overall team effort.

ANALYZING IMAGES

Analyzing an image usually means identifying the objects in it. Is that blob a face, a potato, or a bomb in the luggage? If it's a face, whose face is it? Is that dark patch in the satellite photograph a city, a lake, or a plowed field? Such questions are answered using a wide array of mathematical techniques that reduce images to representation of pixels by numbers that are then subject to mathematical analysis and operations.

OPTICS

Mathematics and imaging formed another fruitful connection with the growth of modern mathematical optics starting in the 1200s. Mathematical optics is the study of images are formed by light reflecting from curved mirrors or passing through one or more lenses and falling on any flat or light-sensitive surface such the retina of the eye, a piece of photographic film, or a light-sensitive circuit such as is used in today's digital cameras. Mathematical optics makes possible the design of contacts, eyeglasses, telescopes, microscopes, and cameras of all kinds. Advanced mathematics are needed to predict the course of light rays passing through many pieces of glass in high-quality camera lenses, and to design lens shapes and coatings that will deliver a nearly perfect image.

MEDICAL IMAGING

For the better part of a century, starting in the 1890s, the only way to see anything inside of a human body without cutting it open was to shine x rays through it. Shadows of bones and other objects in the body would cast by the x rays on a piece of photographic film placed on the other side of the body. This had the disadvantages that it could not take pictures of soft tissues deep in the body (because they cast such faint shadows), and that the shadows of objects in the path of the x-ray beam were confusingly overlaid on the x-ray film. Further, excessive x-ray doses can cause cancer. However, the spread of inexpensive computer power since the 1960s has led to an explosion of medical imaging methods.

Due in part to faster computers, it is now possible to produce images from x-rays and other forms of energy, including radio waves and electrical currents, that pass through the body from many different directions. By applying advanced mathematics to these signals, it is possible to piece together extremely clear images of the inside of the body—including the soft tissues. Magnetic resonance imaging (MRI), which places the body in a strong magnetic field and bombards it with radio waves, is now widely available. A technique called "functional MRI" allows neurologists to watch chemical changes in the living brain in real time, showing what parts of the brain are involved in thinking what kinds of thoughts. This has greatly advanced our knowledge of such brain diseases as Alzheimer disease, epilepsy, dyslexia, and schizophrenia.

COMPRESSION

Imagine a square digital image 1,000 pixels wide by 1,000 pixels tall—all one solid color, blue. That's 1,000 × 1,000 or 1 million blue pixels. If each pixel requires 3 bytes (one byte equals eight bits, that is, eight 1s and 0s), this extremely dull picture will take up 3 million bytes (megabytes, MB) of computer memory. But we don't need to waste 3 MB of memory on a blue square, or wait while they transmit over the Web. We could just say "blue square, 1,000 pixels wide" and have done with it: everything there is to know about that picture is summed up by that phrase. This is an example of "image compression." Image compression takes advantage of the redundancy in images—the fact that nearby pixels are often similar—to reduce the amount of data storage and transmission time taken up by images. Many mathematical techniques of image compression have been developed, for use in everything from space probes to home computers, but the most of the images that are received and sent over the World Wide Web are compressed by a standard method called JPEG, short for Joint Photographic Experts Group, first advanced in 1994.

JPEG is a "block encoding" method. This means that it divides the image up into blocks 8 by 8 pixels in size, then records as much of the image redundancy in that block as it can in a series of numbers called "coefficients." The coefficients that don't record as much redundancy are thrown away. This allows a smaller group of numbers (the coefficients that are left) to record most of the information that was in the original image. An image can then be reconstructed from the remaining coefficients. It is not quite as sharp as the original, but the difference may be too slight for the eye to notice.

RECOGNIZING FACES: A CONTROVERSIAL APPLICATION

Human beings are expert at recognizing faces. We effortlessly correct for different conditions of light and shadow, angles of view, glasses, and even aging. It is difficult, however, to teach a computer how to do this. Some progress has been made and a number of face-recognition systems are on the market.

The mathematics of face recognition are complex because faces do not always look the same. We can grow beards or long hair, don sunglasses, gain or lose weight, put on hats or heavy makeup, be photographed from different angles and in different lights, and age. To recognize a face it is therefore not enough to just look for matching patterns of image dots. A mathematical model of whatever it is that people recognize in a face—what it is about a face that doesn't change—must be constructed, if possible. Face-recognition software has a low success rate in real-life settings such as streets and airports, often wrongly matching people in the crowd with faces in the records or failing to identify people in the records who are in the crowd.

Face on Mars

In 1976, two spidery robots, *Viking 1* and *Viking 2*, became the first spacecraft to successfully touch down on the rocky soil of Mars. Each lander had a partner, an "orbiter" circling the planet and taking pictures. Images and other data from all four machines were radioed back to Earth.

One picture drew public attention from the first. It had been taken from space by a Viking orbiter, and it looked exactly like a giant, blurry face built into the surface of Mars (See Figure 1.)

Notice the dots sprinkled over the image. These are not black spots on Mars, but places where the radio signal transferring the image from the Viking orbiter as a series of numbers was destroyed by noise. However, one dot lands on the "nose" of the Face, right where a nostril would be; one lands on the chin, looking like the shadow of a lower lip; and several land in a curve more or less where a hairline would be. These accidents made the image look even more like a face.

Some people erroneously decided that an ancient civilization had been discovered on Mars. Scientists insisted that the "face" was a mountain, but a better picture was needed to resolve any doubt. In 2001 an orbiter with an better camera than Viking's did arrive at Mars, and it took the higher resolution picture of the "face" shown in Figure 2.

In this picture, the "Face" is clearly a natural feature with no particular resemblance to a human face. Thanks to mathematical processing of multiple images, we can now even view it in 3-D.

In later releases of Viking orbiter images in the 1970s the missing-data dots were "interpolated," that is, filled in with brightness values guessed by averaging surrounding pixels. Without its dots, and seen in more realistic detail, the "Face" does not look so face-like after all.

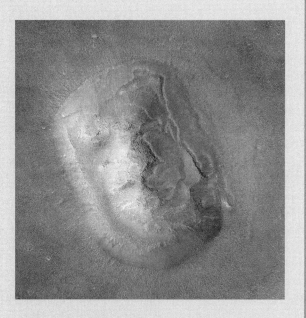

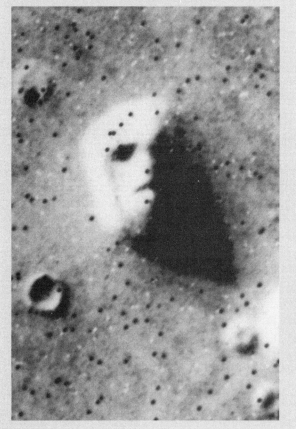

Figure 1 (top). NASA/JPL/MSSS.
Figure 2 (bottom). 1989 ROGER RESSMEYER/NASA/CORBIS.

Using face-recognition systems to scan public spaces is politically controversial. At the Super Bowl game in Tampa, Florida, in 2001, for example, officials set up cameras to scan the fans as they went through the turnstiles. The videos were analyzed using face-recognition software. A couple of ticket scalpers were caught, but no serious criminals. Face-recognition technology has not been used again at a mass sporting event, but is in use at several major airports, including those in Boston, San Francisco, and Providence, Rhode Island.

Critics argue that officials might eventually be able to track any person's movements automatically, using the thousands of surveillance cameras that are being installed to watch public spaces across the country. Such a technology could be used not only to catch terrorists (if we knew what they looked like) but, conceivably, to track people for other reasons.

Face-recognition systems may prove more useful and less controversial in less public settings. Your own computer—which always sees you from about the same angle, and in similar lighting—may soon be able to check your identity before allowing you to spend money or access secure files. Some gambling casinos already use face-recognition software to verify the identities of people withdrawing winnings from automatic banking machines.

FORENSIC DIGITAL IMAGING: SHOEPRINTS AND FINGERPRINTS

Forensic digital imaging is the analysis of digital images for crime-solving. It includes using computers to decide whether documents are real or fake, or even whether the print of a shoe at a crime scene belongs to a particular shoe. Shoeprints, which have been used in crime detection even longer than fingerprints, are routinely photographed at crime scenes. These images are stored in large databases because police would like to know whether a given shoe has appeared at more than one crime scene. Matching shoe prints has traditionally been done by eye, but this is tedious, time-consuming, and prone to mistakes. Systems are now being developed that apply mathematical techniques such as fractal decomposition to the matching of fresh shoeprints with database images—faster and more accurately than a human expert. Fingerprints, too, are now being translated into digital images and subjected to mathematical analysis. Evidence that will stand up in court can sometimes now be extracted from fingerprints that human experts pronounced useless years ago.

DANCE

Dance and other motions of the human body can be described mathematically. This knowledge can then be used to produce computer animations or to record the choreography of a certain dance. In Japan, for example, the number of people who know how to dance in traditional style has been slowly decreasing. Some movies and videos, however, have been taken of the older dances. Researchers have applied mathematical techniques to these videos—some of which have deteriorated from age and are not easy to view—in order to extract the most complete possible description of the various dances. It would be better if the dances could be passed down from person to person, as they have in the past, but at least in this way they will not be completely forgotten. Japanese researchers, who are particularly interested in developing human-shaped robots, also hope to use mathematical descriptions of human motion to teach robots how to sit, stand, walk—and dance.

MEAT AND POTATOES

The current United States beef-grading system assigns a grade or rank to different pieces of beef based on how much fat they contain (marbling). Until recently, an animal had to be butchered and its meat looked at by a human inspector in order to decide how marbled it was. However, computer analysis of ultrasound images has made it possible to grade meat on the hoof—while the animal is still alive. Ultrasound is any sound too high for the ear to hear. It can be beamed painlessly into the body of a cow (or person). When this is done, some of the sound is reflected back by the muscles and other tissues in the body. These echoes can be recorded and turned into images. In medicine, ultrasound images can reveal the health of a human fetus; in agriculture, mathematical techniques like gray-scale statistical analysis, gray-scale spatial texture analysis, and frequency spectrum texture analysis can be applied to them in order to decide the degree of marbling.

Different mathematics are applied to the sorting of another food item that often appears at mealtime with meat: potatoes. Potatoes that are the right size and shape for baking can be sold for higher price, and so it is desirable to sort these out. This can either be done hand or by passing them down a conveyer belt under a camera connected to a computer. The computer is programmed to decide which blobs in the image are potatoes, how big each potato is, and whether the potatoes that are big enough for baking are also the right shape. All these steps involve imaging mathematics.

STEGANOGRAPHY AND DIGITAL WATERMARKS

For thousands of years, people have been interested in the art of secret messages (also called "cryptography," from

Key Terms

Matrix: A rectangular array of variables or numbers, often shown with square brackets enclosing the array. Here "rectangular" means composed of columns of equal length, not two-dimensional. A matrix equation can represent a system of linear equations.

Pixel: Short for "picture unit," a pixel is the smallest unit of a computer graphic or image. It is also represented as a binary number.

the Greek words for "secret writing"), and computers have now made cryptography a part of everyday life; for example, every time someone uses a credit card to buy something over the Internet, their computer uses a secret code to keep their card number from being stolen. The writing and reading of cryptographic or secret messages by computer is a mathematical process.

But for every code there is a would-be code-breaker, somebody who wants to read the secret message. (If there wasn't, why would the message be secret?) And a message that looks like it is in a secret code—a random-looking string of letters or numbers—is bound to attract the attention of a code-breaker. Your message would be even more secure if you could keep its very existence a secret. This is done by steganography (from the Greek for "covered writing"), the hiding of secret messages inside other messages, "carrier" messages, that do not appear secret at all. Secret messages can be hidden physically (a tiny negative under a postal stamp, or disguised as a punctuation mark in a printed letter) or mathematically, as part of a message coded in letters, numbers, or DNA. Digital images are particularly popular carriers. We send many images to each other, and an image always has an obvious message of its own; by drawing attention to itself, an image diverts suspicion from itself. But a digital image may be much more than it appears. The matrix of numbers that makes it up can be altered slightly by mathematical algorithms to convey a message while changing the visible appearance of the image very little, or not at all. And since images contain so much more binary information than texts such as letters, it is easier to hide longer secret messages in them.

You do not have to be a spy to want to hide a message in an image. People who copyright digital photographs want to prevent other people from copying them and using them for free, without permission; one way to do so is to code a hidden owner's mark, a "digital watermark," into the image. Software exists that scans the Web looking for images containing these digital watermarks and checking to see whether they are being used without permission.

ART

Digital imaging and the application of mathematics to digital images have proved important to the caretaking of a kind of images that are emphatically not digital, not a mass of numbers floating in cyberspace, not reproducible by mere copying of 1s and 0s: paintings of the sort that hang in museums and collections. Unlike digital images, these are physical objects with a definite and unique history. They cannot be truly copied and may often be worth many millions of dollars apiece. The role of digital imaging is not to replace such paintings, but to aid in their preservation.

The first step is to take a super-high-grade digital photograph of the painting. This is done using special cameras that record color in seven color bands (rather than the usually three) and take extremely detailed scans. For example, a fine-art scanner may create a digital image 20,000 × 20,000 pixels (color dots) large, which is 400 million pixels total. But each pixel has seven color bands, so there are actually seven times this many numbers in the image record, about 2.8 billion numbers per painting. This is about 100 times larger than the image created by a high-quality handheld digital camera.

Once this high-grade image exists, it has many uses. Even in the cleanest museum, paintings slowly dim, age, and get dirty, and so must eventually be cleaned up or "restored." A digital image shows exactly what a painting looks like on the day it was scanned; by re-scanning the painting years later and comparing the old and new images using mathematical algorithms, any subtle changes can be caught. By applying mathematical transformations to the image of a painting whose colors have faded, experts can, in effect, look back in time to what the painting used to look like (probably), or predict what it will look like after cleaning. Also, famous paintings are often transported around the world to show in different museum. By re-imaging a painting before and after transport and comparing the images, any damage during transport can be detected.

Where to Learn More

Web sites

"Computer Technology in Soccer." Soccer Performance.org. <http://www.soccerperformance.org/specialtopics/companalsystems.htm> (October 16, 2004).

Johnson, N.F. "Steganography and Digital Watermarking." 2002. <http://www.jjtc.com/Steganography/> (October 16, 2004).

"Privacy and Technology: Q&A on Face-Recognition." American Civil Liberties Union. Sep. 2, 2003. <http://www.aclu.org/Privacy/Privacy.cfm?ID=13434&c=130> (October 16, 2004).

Kimmel, R., and G. Sapiro. "The Mathematics of Face Recognition." Society for Industrial and Applied Mathematics (SIAM). SIAM News, Volume 36, Number 3, April 2003. <http://www.siam.org/siamnews/04-03/face.htm> (October 16, 2004).

Wang, J.R., and N. Parameswaran. "Survey of Sports Video Analysis: Research Issues and Applications." Australian Computer Society. Conferences in Research and Practice in Information Technology, Vol. 36. M. Piccardi, T. Hintz, X. He, M.L. Huang, D.D. Feng, J. Jin, Eds. 2004. <http://crpit.com/confpapers/CRPITV36Wang.pdf> (October 16, 2004).

Overview

It is often said that we live in the Information Age. Computer enthusiasts sometimes speak as if we were now being fed and housed by the "information economy," or as if we were all racing down the "information highway" toward a perfect society. But what, exactly, is "information"? We all know that disks and chips store it, and that computers process it, and that is supposed to be a good thing to have lots of—but what is it?

The answer is given by information theory, a branch of mathematics founded in 1948 by American telephone engineer Claude Shannon (1916-2001). Shannon discovered how to measure the amount of information in any given message. He also showed how to measure the ability of any information-carrying channel to transmit information in the presence of noise (which disrupts and changes messages). Information theory soon expanded to include error-correction coding, the science of transmitting messages with the fewest possible mistakes.

Shannon's ideas about information have proved useful for many things besides telephones. Information theory enables designers to make many kinds of message-handling devices more efficient, including compact disc (CD) players, deep-space probes, computer memories, and other gadgets. Information theory has also proved useful in biology, where the DNA molecules that help to shape us from birth to death turn out to be written in code, and in economics, where information processing is key to making money in a complicated, competitive world. Error-correction coding also enables billions of files to be transferred over the Internet every day with few errors.

Fundamental Mathematical Concepts and Terms

The central idea of information theory is information itself. In everyday speech, "information" is used to mean "useful knowledge"; if you have information about something, you know something useful or significant about that thing. In mathematics, however, the word has a much narrower meaning.

Shannon began with the simple idea that whatever information is, messages carry it. From this he derived a precise mathematical expression for the information in any given message. Every system that transmits a message has, Shannon said, three parts: a sender, a channel, and a receiver. If the sender is a talker on one end of a phone line, the phone line is the channel and the listener at the far end is the receiver.

Information Theory

We imagine that the sender chooses messages from a collection of possible messages and sends them one by one through the channel. Say that N stands for the number of messages that the sender has to choose from each time. If the message is a word from the English language, N is about 600,000. Often the message is a string of ones and zeroes. Ones and zeroes are often used to represent messages because they are easy to handle. Each one or zero is called a "binary digit" (or "bit," for short). If a binary message is M bits long, then the number of possible messages, N, equals 2^M. This is because there can be only 2^M different strings of ones and zeroes M digits long. For example, if the message could be any string 3 bits long ($N = 3$), then $M = 8$, because there are $2^3 = 8$ different 3-bit strings, as shown in Table 1.

000	101
100	110
010	111
001	011

Table 1.

The sender chooses a message at random. Here, random means that all N messages are equally likely, just as, when you flip a fair coin, heads and tails are equally likely. If all N messages are equally likely, the chance or probability of each message being sent is $1/N$. For example, if we flip a coin to choose whether to send 1 or 0 (1 for heads, 0 for tails), then $N = 2$ (there are two possible messages) and the probability of each message is 1/2 (because $1/N = 1/2$).

From the sender's point of view, the situation is simple: choose a message and send it. From the receiver's point of view, things are less simple. The receiver knows that a message is coming, but they do not know which one. They are therefore said to have uncertainty about what message will be sent. Exactly how much "uncertainty" they have depends on N. That is, the more possible messages there are (the larger N is), the harder it is for the receiver to guess what message will be sent.

The receiver's uncertainty is important because it tells us exactly much they learn by receiving a message. If there is only one possible message—say, if the sender can only send the digit "0", over and over— then the receiver can always "guess" it ahead of time, so they learn nothing by receiving it. If there are two possible messages ($N = 2$), then the receiver has only a 50–50 chance of guessing which will be sent, and definitely learns something when

a message is received. If there are more than two possible messages ($N > 2$), then the receiver's chance of guessing which message will be sent is less than 50–50.

The harder it is to guess a message before getting it, the more one learns by getting it. Therefore, the receiver's uncertainty tells us how much they learn—how much information they gain—from each message. Messages chosen at random from large message-sets are harder to guess ahead of time, so the receiver learns more by receiving them; they convey more "information."

Now assume that a message has been chosen from the list of N possibilities, sent, and correctly received. The receiver's uncertainty about this particular message has now been reduced to 0. This reduction in uncertainty corresponds, as we have seen, to a gain in information. This, then, is information theory's definition of information: Information is what reduces uncertainty. We will label information H, as is customary.

The information that the receiver derives from a single message, H, depends on the number of possible messages, N. Bigger N means more uncertainty: more uncertainty means more information gained when the message arrives. To signify the dependence of information on N, we write H as a "function" of N, like so: $H(N)$. (This is pronounced "H of N".) A function is a rule that relates one set of numbers to another set. For example, if we write $f(x)$, we mean that for every number x there is another number, f, related to it by some rule; if the rule is, for example, that f is always twice x, we write $f(x) = 2x$. Likewise when we write $H(N)$, we say that for every N there is another number, H, related to it by some rule. Below, we'll look at exactly what this rule is.

H, which stands for the amount of information in a single message, has units of "bits." Similarly, numbers that record distances have units of feet (or meters, or miles) and numbers that record time intervals have units of seconds (or hours, or days). The bit is defined as follows: If a message consisting of a single binary digit is received, and that message was equally likely to be a 1 or a 0, then 1 bit of information has been received.

To find out what the function or rule is that relates the numbers H and N, we first introduce an imaginary wrinkle. Let us say that the sender is picking messages from two groups of possibilities, like two buckets of marbles. One group of possible messages has N_1 choices (Bucket Number 1, with N_1 marbles in it) and the other has N_2 choices (Bucket Number 2, with N_2 marbles in it). (The small "1" and small "2" attached to N_1 and N_2 are just labels that help us tell the two numbers apart.) Now imagine that the sender picks a message from the first group and sends it, then picks a message from the second group and sends it, like grabbing one marble from Bucket Number 1 and a second marble from Bucket Number 2.

This sender is really sending messages (or picking marbles) in *pairs*. How many such pairs could there be? If we call the number of possible pairs N, then $N = N_1 N_2$. This is easy to see with simple groups of messages. If the first message is a 0 or 1 (a single binary digit), then $N_1 = 2$, and if the second set is a *pair* of binary digits, the four possible messages are 00, 10, 01, and 11, so $N_2 = 4$. Choosing one message from each set allows eight (that is, $N_1 \times N_2$) possible pairs, as shown in Figure 1.

It is easy to prove to yourself that these really are the only possible message pairs—just try to write one down that isn't already on the list.

How much information does one of these message-pairs contain? To give a specific number we would have to know the correct rule for relating H and N, that is, the function $H(N)$, which is what we're still looking for. But we can say one thing right off the bat: $H(N)$ should agree with common sense that the information given by the two messages together is the sum of the information given by the two messages separately. It turns out that this common-sense idea is the key to finding $H(N)$. Saying that the information in the two messages adds can be written as follows: $H(N) = H(N_1) + H(N_2)$.

But we also know, as shown above, that $N = N_1 \times N_2$. We can therefore rewrite the previous equation a little differently: $H(N_1 N_2) = H(N_1) + H(N_2)$. It may not seem like we've proven much by writing this equation, but it is actually the key to our whole problem. Because it has the form it has, there is only one possible way to compute the information content of a message, that is, one possible rule or function. Mathematicians have shown, using techniques too advanced to go over here, that there is only one function that satisfies $H(N_1 N_2) = H(N_1) + H(N_2)$, namely $H(N) = \log_2 N$.

The expression "$\log_2 N$" means "the base 2 logarithm of N," namely, that power of 2 which gives N. For example, $\log_2 8 = 3$ because $2^3 = 8$, and $\log_2 = 4$ because $2^4 = 16$. (See the entry in this book on Logarithms.) The graph of $H(N) = \log_2 N$ is shown in Figure 2.

As we saw earlier, the number of different messages that can be sent using binary digits (ones and zeroes) is $N = 2^M$. So, for example, if we send messages consisting of 7 binary digits apiece, the number of different messages is $N = 2^7 = 128$. Using 2^M for N in $H(N) = \log_2 N$, we get a new expression for $H(N)$: $H(M) = \log_2 2^M$.

But $\log_2 2^M$ is just M, by the definition of the base-2 logarithm given above, so $H(M) = \log_2 2^M$ simplifies to $H(M) = M$. This is just a straight line, the simplest of all

Message 1	Message 2	Eight possible message pairs
0	00	0 00
0	10	0 10
0	01	0 01
0	11	0 11
1	00	1 00
1	10	1 10
1	01	1 01
1	11	1 11

Figure 1.

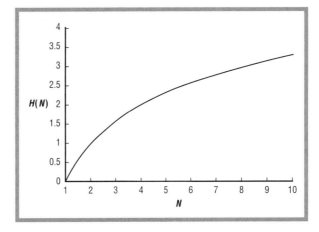

Figure 2. The information content of a single message selected from N equally likely messages: $H(N) = \log_2 N$. Units of $H(N)$ are bits.

functions, as shown in Figure 3. The equation $H(M) = M$ not only looks simple, it has a simple meaning: a message written using M equally likely binary digits conveys M bits of information. This is why we use "bit," an abbreviation of "binary digit," as the unit of information.

It is important to remember that while the "bit" is the unit of all information, not all information is in the form of "binary digits" (e.g., ones and zeroes). For example, the letters in this sentence are not binary digits, but they contain information.

UNEQUALLY LIKELY MESSAGES

A bit is the amount of information conveyed by the answer to the simplest possible question, that is, a question with two equally likely answers. When a lawyer in a courtroom drama shrieks "Answer yes or no!" at a witness, they are asking for one bit of information. Paul Revere's famous scheme for anticipating a British raid

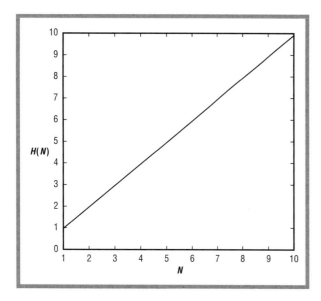

Figure 3. Bits of information, H, in a message, shown as a function of the number of binary digits in that message, N.

from Boston, as described by Henry Longfellow (1807–1882) in the famous poem beginning "Listen my children, and you shall hear / Of the midnight ride of Paul Revere," sought to convey one bit of information:

> [Revere] said to his friend, "If the British march
> By land or sea from the town tonight,
> Hang a lantern aloft in the belfry arch
> Of the North Church tower as a signal light,—
> One, if by land, and two, if by sea;
> And I on the opposite shore will be,
> Ready to ride and spread the alarm . . ."

Strictly speaking, this was a one-bit message only if the British were equally likely to come by land or by sea.

But what if they were not? What if the British were, say, five times as likely to come by land as by sea? So far we've talked about messages selected from equally likely choices, but what if the choices aren't equally likely?

In that case, our rule for the information content of a message must become more complicated. It also becomes more useful, because it is usually the case that some messages are more likely than others. In transmitting written English, for example, not all letters of the alphabet are equally likely; we send the letter "e" about 1.36 times more often than the next most common letter, "i."

Let's say that the sender has three messages to choose from, only now each message has a different chance or probability of being sent. The probability of an event is written as a number between 0 and 1: smaller probability

numbers mean less-likely events, larger numbers mean more-likely events. Say that the probability of the first message on the sender's list is p_1, that of the second message is p_2, and that of the third message is f^3. The amount of information per message is, in this case, given by the following equation: $H(N) = - p_1\log_2 p_1 - p_2\log_2 p_2 - p_3\log_2 p_3$ bits.

If there were more than 3 possible messages, there would be more terms to subtract, such as $p_4\log_2 p_4$, $p_5\log_2 p_5$, and so on up to as many terms as there were possible messages.

These equations are the heart and soul of information theory. Using it, we can calculate exactly how much information, H, any message is worth, if we know the probabilities of all the possible messages. This is best explained by working out a simple example.

Paul Revere had two possible messages to deliver, "land" or "sea," so in his case N = 2. We will call p_1 the probability that the message would be "land," and p_2 the probability that it would be "sea". In this case then, $H(N) = - p_1\log_2 p_1 - p_2\log_2 p_2$ bits. If both messages are equally likely, then p_1 and p_2 both equal 1/2 and so we have

$$H(2) = -\left(\frac{1}{2}\ \log_2\ \frac{1}{2}\right) - \left(\frac{1}{2}\ \log_2\ \frac{1}{2}\right) \text{bits}$$

which works out to $H(2) = 1$ bit. This, we already knew: Where there are two equally likely messages, sending either one communicates 1 bit of information.

But if the probabilities of the two messages are not equal, less than 1 bit is communicated. For example, if $p_1 = .7$ and $p_2 = .3$ then $H(2) = - (.7 \log_2 .7) - (.3 \log_2 .3) = .88129$ bits.

This agrees with common sense, which tells us that if the American revolutionaries had known beforehand that the British were more than twice as likely to come by land than by sea (probability .7 for land, only .3 for sea), they would have had a pretty good shot at guessing what was going to happen even without getting the message from the church tower (and so that message wouldn't have told them as much as in the equal-probability case). If the revolutionaries had known that the British were sure to come by land, then p_1 would have equaled 1 (the probability of a certain event), p_2 would have equaled 0 (the probability of an impossible event), and the message would have communicated no information, zero bits: $H(2) = - (1 \times \log_2 1) - (0 \times \log_2 0) = 0$ bits.

And that makes sense too. A message that conveys 0 bits is one that you don't need to receive at all.

INFORMATION AND MEANING

The assignment of Revere's colleague in the church tower was to send a single binary digit: "one, if by land, and two, if by sea." If the person in the tower had written "Land" or "Sea" on paper, instead of putting up lights, the message would have contained more bits of information—about 18.8 bits for "Land" and 14.1 bits for "Sea," taking each letter as worth $\log_2 26 = 4.7$ bits (because there are 26 letters in the alphabet)—yet the message would have *meant*—the same thing. This seems like a contradiction: More information does not necessarily provide greater knowledge. Why not?

The answer is that the everyday sense of the word "information" is different from the mathematical sense. The everyday sense is based on meaning or importance. If a message is meaningful, that is, tells us something important, we tend to think of it as having more information in it. Mathematically, however, this isn't true. How much information a message contains has nothing to do with how meaningful that message is. The answers to 1-bit, "Yes-No" questions like "Shall we surrender?" or "Will you marry me?", which are very important, contain only 1 bit of information. On the other hand, a many-bit message, say a hundred 1s and 0s picked by flipping a coin, may have no meaning at all. Meaning and information are not the same thing.

The Paul Revere statue in Paul Revere Plaza in the North End neighborhood of Boston. The spire of the famous Old North Church is seen in the background. According to information theory, Paul Revere's famous scheme for anticipating a British raid from Boston, as described by Henry Longfellow in the famous poem beginning "Listen my children, and you shall hear / Of the midnight ride of Paul Revere," sought to convey approximately 1 bit of information. AP/WIDE WORLD PHOTOS. REPRODUCED BY PERMISSION.

A Brief History of Discovery and Development

Information theory dates from the publication of Claude Shannon's 1948 paper, "A Mathematical Theory of Communication." A few scientists had suggested using a logarithmic measure of information before this, but Shannon—who was famous for riding a unicycle up and down the hallways of Bell Laboratories—was the first to hit on the necessary mathematical expressions. He defined "information," distinguished it from meaning, and proved several important theorems about transmitting it in the presence of noise (random signals that cause erroneous messages to be received).

palm pilots, global positioning system units, and laptops all rely information theory to operate efficiently.). Information theory is also applied to electronic communications, computing, biology, linguistics, business, cryptography, psychology, and physics. It an essential branch of the mathematical theory of probability.

Real-life Applications

A great deal of work has been done on information theory since Shannon's 1948 paper, applying and extending his ideas in thousands of ways. Few of us go through a single day without availing ourselves of some application of information theory. Cell phones, MP3 players,

COMMUNICATIONS

Paul Revere was not only a revolutionary conspirator, but part of a communications channel. Once he had seen whether one light or two was burning in the church tower, it was his job to deliver that one precious bit of information to its final destination, the revolutionary militia at Concord, Massachusetts.

However, he was captured by a British patrol before getting there. He was thus part of what engineers call a "noisy channel." All real-world channels are noisy, that is, there is some chance, large or small, that every message will suffer damage or loss before it gets to its intended receiver. Before Shannon, engineers mostly thought that the only way to guarantee transmission through a noisy channel was to send messages more slowly. However, Shannon proved that this was wrong. Every channel has a certain capacity, that is, a rate at which it can send information with as few errors as you please *if* you are allowed to send a certain number of extra, redundant bits—information that repeats other parts of your message—along with your actual message.

Shannon showed how to calculate channel capacity exactly. With this tool in hand, engineers have known for half a century how to make every message channel as good as it needs to be, squeezing the most work possible out of every communications device in our increasingly gadget-dependent world: optical disc drives, cell phones, optical fibers carrying hundreds of thousands of telephone calls through underground pipes, radio links with deep-space probes, file transfers over the Internet, and so on.

Around A.D. 1200, the Chinese were able to invent primitive rockets without knowing calculus or Newton's Laws of Motion, but without mathematics they could never have built truly huge rockets such as the Long March 2F booster that lifted the first Chinese astronaut into space in May 2004. Likewise, communications devices and digital computers were invented before information theory, but without information theory engineers could not build such machines as well (and as cheaply) as we do today. In rocketry, communications, powered flight, and many other fields, the early steps depend mostly on creative spunk but later improvements depend on mathematics.

Today applications of information theory are literally everywhere. Every cubic inch of your body is at this moment interpenetrated by scores or hundreds of radio signals designed using information theory.

Physics and Information

From the very beginning there has been a connection between physics and information theory. Shannon's rule for calculating information was nearly identical to the expression in statistical physics for the entropy of a system (a measure of its disorder or randomness), as was pointed out to Shannon before he published his famous 1948 paper. One physicist even advised Shannon to call his new measure "entropy," not "information," because "most people don't know what 'entropy' really is. If you use 'entropy' in an argument you will win every time!"

Nor is the connection between physics and information merely a matter of look-alike equations. In 1951, the physicist L. Brillouin proved the amazing claim that there is an absolute lower limit on how much energy it takes to observe a single bit of information. Namely, to observe one bit takes at least $kBT\log_e 2$ ergs of energy, where T is temperature in degrees Kelvin, k and B are constants (fixed numbers) from physics, and $\log_e 2$ equals approximately .693. The precise value of this very small number is not important: what is important is what it tells us. One of the things it tells us that it is impossible to have or process an infinite amount of information. That would, by Brillouin's theorem, take an infinite amount of energy; but there is only a limited amount of energy in the whole Universe.

INFORMATION THEORY IN BIOLOGY AND GENETICS

Most of the cells in your body contain molecules of DNA (deoxyribonucleic acid). Each DNA molecule is shaped like a long, narrow ribbon or zipper that has been twisted lengthwise like a licorice stick. Each side of the zipper has a row of teeth, each tooth being a cluster of atoms. There are four kinds of zipper teeth in DNA, the chemicals guanine, thymine, adenine, and cytosine (always called G, T, A, and C for short).

These teeth of the DNA zipper are lined up in groups of three: AGC, GGT, TCA, and so on for many thousands of groups. Each group of three teeth is a code word bearing a definite message. Also, each type of zipper tooth is shaped so that it can link up with only one other kind of zipper tooth: A and T always zip together, and G and C always zip together. Therefore, both sides of the zipper bear the same series of messages, only coded with different chemicals: thus, AGCGGT zips together with TCGCCA. If you know what one side of the zipper looks like, you can say what the other side must look like.

DNA is usually zipped up so that the two opposite sets of teeth are locked together. Sometimes, however, DNA gets partly unzipped. This happens whenever the cell needs to read off some of the messages in the DNA, such as when the cell needs to make a copy of itself or to refresh its stores of some useful chemical. The unzipping is done by special molecules that move down the DNA, separating the two sides like the slide on an actual zipper. When a section of DNA has been unzipped, other molecules move along it and copy (or "transcribe") its three-letter code words. These code words order the cell to string certain molecules ("amino acids") together like beads on a necklace. These strung-together amino acids are the very complex molecules called proteins, which do most of the microscopic, chemical work that keeps us alive. Proteins are produced from step-by-step instructions in DNA much as a cook bakes a cake from step-by-step instructions (a recipe) in a cookbook. The exact same three-letter DNA code is used in the cells of every living thing on Earth, from people to pine trees.

Biologists have found it helpful to view each three-letter DNA code word as a message. Since there are four choices of letter (A, C, G, and T) and three letters per word, there are $4^3 = 64$ possible words that DNA might send. According to information theory, each DNA code word could contain up to $\log_2 64 = 6$ bits of information. Actually, some words are used by DNA to mean the same thing as other words, so the DNA code only codes for 20 different amino acids, not 64. Each DNA code word therefore contains about $\log_2 20 = 4.32$ bits. There are about three billion pairs of molecular zipper teeth (base pairs) in a complete set of human DNA molecules. These three billion pairs could encode, at most, one billion three-letter words, each conveying 4.32 bits. Therefore, the most information that the human DNA could contain is about 4.32 billion bits. A standard 700 MB CD-ROM also contains about this much information.

Thus, an entire CD-ROM's worth of information is packed by Nature into a chemical speck too small to be seen without a powerful microscope—a set of human DNA molecules. Most of the cells in the human body contain these molecules.

Accordingly, the "recipe" for a human being requires about as much information storage space as it would take to record 80 minutes of dance hits.

Seeing the DNA-to-protein system in terms of information theory has helped biologists understand evolution, aging, growth, and viruses such as AIDS. Biologists have also applied information theory to molecules other than DNA and to the brain.

ERROR CORRECTION

Every message has some chance of not getting through or of getting through with damage, like a letter that is delivered with a corner torn off or with a letter "O" smeared into a letter "Q." Here is another problem that begs for a clever solution.

Once again, Paul Revere is ahead of us. Revere's task was to deliver his one-bit message to the town of Concord, Massachusetts. On the way there, he stopped at Lexington and shared the message with two other men,

William Dawes and Samuel Prescott. All three set off for Concord; all three were captured by the British. Revere was released without his horse. Dawes and Prescott made a break for it, but Dawes fell off his horse. Only Prescott got through. If Revere had headed straight for Concord by himself, the message would never have been delivered. Sending the message three separate times, by three separate riders, an example of "triple redundancy," got this one-bit message through this very noisy channel.

Today we send messages using electrons and photons rather than horses, but triple redundancy (sending a message three times) is still an option. For instance, instead of 101, we can send 111000111. If this message is damaged by electronic noise (static), then the receiver will receive a different message, for example, 011000111. In this case, noise has changed the first bit from a 1 to a 0. By looking at the first three bits the receiver knows, first of all, that an error must have happened, because all three bits are not the same. Triple redundancy thus has the power of *error detection*. Second, the receiver can decide whether the first three bits were a 1 or a 0 in the original message since there are two 1's and only one 0; thus 011 decodes to 1, which is correct. Triple redundancy also, therefore, has the power of *error correction*. In particular, if no more than one bit out of every three is changed by noise, the entire message can still get through correctly. If we were to send triple-redundant messages forever, we could send an infinite number of bits despite an infinite number of errors, as long as the errors didn't happen too fast!

In practice this scheme isn't used because it would be wasteful. It forces us to send three times as many bits as there are in the original message, but there are only a few simple errors that it can find and fix. If two bits that are close to each other get flipped by noise, we can find the error but our fix may be wrong: for instance, if 111 gets changed to 001 or 010, we will know that an error has happened (because the three bits are not all the same, as they should be), but by majority vote we will decode the received word incorrectly to 0, rather than 1. Errors that are near each other, as in this example, are called "burst" errors.

There are several ways to handle burst errors. The simplest that is used in many real-world codes is termed "interleaving." Interleaving takes one chunk of a message and slips its bits between the bits of another chunk, like two halves of a deck of cards being shuffled together. For example, we may want to transmit the message 01. We first create the two triply redundant words 000 and 111, then interleave them to get 010101. If two bits right next to each other get changed anywhere in this six-bit string, our simple code can both detect and correct them. If the second and third bits, for instance, are both changed during transmission, 010101 is turned into 001101. De-shuffling this (taking the first, third, and fifth bits first, then the second, fourth, and sixth bits) gives us 010011. By majority vote, the first three bits decode to 0 and the second three bits decode to 1—which is our original message, 01. This shows the power of combining repetition with interleaving.

If *three* bits in a row are flipped, this code will fail. Every code has its limits. Nevertheless, Shannon's channel capacity theorem guarantees that we can always drive the corrected error rate down to any specific level we want, for a channel with a given amount of noise, by adding redundant bits. In the interleaved code with repetition that we've been considering, for example, we could correct longer burst errors simply by repeating each bit more than three times (e.g., six, or 10, or a 1,000 times). More redundancy, and more interleaving, would offer more protection. This is a general property of all error-correcting codes: You can never get something for nothing—but you can get something for something, namely, reliability for redundancy.

Information theorists call a code "perfect" if its level of error correction is bought for the least possible redundancy. Error correction coding is that branch of information theory that concerns itself with getting the most bang for the bit, that is, with inventing practical, real-world codes that are as close as possible to perfect. Many error-correcting codes have been developed. They are used in virtually all consumer electronics devices that transmit, code, or decode digital information: digital telephone links, DVDs, audio CDs, computer hard drives, CD-ROMs, and more.

Sometimes the error rate that needs to be dealt with is low to begin with. For example, the industry standard for computer hard drives before error correction is one error for every billion bits read to or from the spinning magnetic discs inside the drive. This is a very quiet (low-noise) channel, but still too noisy for a computer. Modern computers read and write many billions of bits to and from their hard drives, so one error per billion bits might result in scrambled documents, crashed programs, messed-up money transfers, and e-mail sent to wrong addresses.

To prevent this, all computer hard drives use error-correcting schemes belonging to a family of codes called the Reed-Solomon codes. Reed-Solomon codes (named after their two inventors, who published the idea in 1960) are nearly "perfect" in the sense that they give almost the maximum amount of error correction possible for the number of redundant bits they add. When using an error-correcting code you cannot fit as much data onto a hard drive because of the redundant bits inserted by the code, but reading and writing from the drive suffers very *very* few errors.

Various versions of the Reed-Solomon code are used not only for computer hard drives but for audio CDs, DVDs, digital videotape, digital cable TV, digital cameras, and virtually every other commonplace digital data storage-and-retrieval device.

The most heroic deed of coding-theory history so far was performed by the Voyager spacecraft. Launched by the United States in 1977, *Voyager 1* traveled to Jupiter and Saturn and *Voyager 2* sailed past Jupiter, Saturn, Uranus, and Neptune. The two probes took sharp color pictures of scores of mysterious moons that are nothing but fuzzy points of light as seen from the Earth, even through powerful telescopes.

The Voyagers sent their pictures and other data back to the Earth as strings of ones and zeros. But they didn't have much power to do it with, so by the time a Voyager's signal arrived on Earth it was extremely faint. The largest radio antennas on Earth could only gather about 10^{-16} watts of power from a *Voyager 2* signal originating near Neptune, 4.6 hours away at the speed of light. This much power, if collected for three billion years, would light a 40-watt light bulb for less than one second. A digital watch uses billions of times as much power. Yet these ghostly signals from Voyager signals were detected, and color images of far-distant worlds were reconstructed from them.

This feat would have been impossible without error correction. For the Jupiter-Saturn leg of the journey, the Voyagers used a sophisticated error-correcting code called a Golay (24,12,8) self-dual code; *Voyager 2*, for its 1986 flyby of Uranus, was re-programmed to use an even more complex code, actually one code wrapped up inside another. The "outer" code was a Reed-Solomon code related to those used in CD players and other consumer electronics. Each of *Voyager 2*'s code words contains 32 bits, of which 26 bits are redundant bits. That is, over 80% of each Voyager code word consists of error-correction bits. These redundant bits enable computers on Earth to fix single and burst errors and even to replace "erasures" or completely lost bits.

The Voyagers are still transmitting data from deep space, leaving the solar system farther and farther behind with each passing second. And their messages are still surviving the trip thanks to error-correction codes designed on Earth using the principles of information theory.

Potential Applications

QUANTUM COMPUTING

Ordinary information is encoded in objects or signals that behave in a more or less reliable way. When a laser beam burns a microscopic pit on a music CD, recording a single bit of information, that pit stays put. At least, it stays put until something physical, say a steel fork wielded by a younger sibling, comes along and changes it by force. Furthermore, each pit only means one thing, zero or a one, never both at once or some third, undefined thing.

That's ordinary information. There is also a thing called "quantum" information. Even computer scientists call it "weird" and are, according to the journal *Science*, "increasingly confused about how it works" (Sep. 14, 2001, Vol. 293, p. 2026). But the basic idea is not hard to understand.

Quantum information is still information, but it is called "quantum" because it is stored not by microscopic pits in a CD, or by voltages, magnetic fields, ink on paper, or any other physical object that has a definite state (there or not there, on or off, etc.). Instead, it is stored by objects like individual atoms, electrons, or photons. Such small objects, instead of obeying the physical laws of our everyday, human-scale world, obey the laws of the kind of physics called "quantum mechanics," which deals with the properties of atoms or smaller objects. The properties of such small objects are, by the standards of our everyday experience, simply crazy. For example, a basketball cannot spin to the left and the right at the same time, but an electron, according to quantum mechanics, can. If you can't see this in your mind, don't worry, physicists can't either. But they describe it mathematically, and the math always works, so we take it as a fact.

If, then, we use spin to store bits (say, leftward spin to store a 1 and rightward spin to store a 0), we can store a 1 and a 0 at the *same time* in the *same electron* (or atom). Physicists call a bit stored in this way a "qubit" (pronounced KYOO-bit), short for "quantum bit."

One reason computer scientists get excited when they think of qubits is that you can, because of superposition of states (the ability to spin left and right at the same time), run a single computation on quantum information and get twice as many answers as if it had been run on a conventional computer. And that's only the beginning. Groups of cubits can be linked to each other at a distance by the quantum effect that physicists call "entanglement."

With a boost from entanglement, N qubits can do the computational work not just of $2N$ but of 2^N conventional bits. That makes computation with qubits not just twice as fast as conventional computation, but *exponentially* faster.

But there is a good reason why we don't all have qubit-based PCs and Macs on our desks, and that is that it is not easy to teach an individual atom to sit up, roll over, and bark ones and zeroes on command. Researchers are approaching the problem by using electromagnetic fields to bottle up small numbers of ultracold atoms, then zapping them gently with laser beams, a procedure that does not involve not the kind of hardware you easily can scoot out and buy at the local electronics store. Further, qubits tend to "leak" or evaporate, losing their information. So ticklish is quantum computation that although it has been studied for over 30 years, the first demonstration of a true quantum computation was not made until 2002, and even that only involved a few bits. Progress remains slow but the potential payoffs are great, so research continues. Computation using quantum information teases us with the possibility that computers may someday billions or even trillions of times faster than today's.

Where to Learn More

Periodicals

Bennet, Charles H., and David P. DiVincenzo. "Quantum Information and Computation," *Nature*, 16 March 2000, Vol. 404, pp. 247–255.

Seife, Charles. "The Quandary of Quantum Information," *Science*, 14 September 2001, Vol. 293, pp. 2026-2027.

Shannon, C.E. "A Mathematical Theory of Communication," *The Bell System Technical Journal*, Vol. 27, pp. 379-423, 623-656, July, October, 1948. Available online at <http://cm.bell-labs.com/cm/ms/what/shannonday/shannon1948.pdf.>

Websites

Touretzky, David S. "Basics of Information Theory." Computer Science Department, Carnegie Mellon University <http://www-2.cs.cmu.edu/~dst/Tutorials/Info-Theory/> (September 28, 2004).

Inverse

Overview

If something is transformed by a sequence of events, the inverse is another sequence of events that will bring it back to the start. The nature of the inverse steps, compared to the steps that are applied, allows identification of how the system is related to its starting conditions under the transformation. The definition of the inverse is of fundamental importance and is probably one of the first mathematical entities that would be sought in many problems. The nature of the inverse allows one to understand how the transformation will act when trying to manipulate it. This concept is important in understanding the algebra that can be applied to the transformation in a mathematical analysis. The nature of the inverse has implications and applications that span all areas of mathematics, science, and engineering, and will often be defined in many different forms.

Fundamental Mathematical Concepts and Terms

DEFINITION OF AN INVERSE

If a sequence of actions is applied, then by definition, the inverse is the sequence of events that undo these effects, returning to the start. For example, imagine a length of rope, the sequence of events that will be applied include tying a knot in the rope. The inverse of this operation will then be the sequence of events that untie the knot and leave the rope as it was originally. This is where the term inverse comes from, as we invert what has been done. If R is the series of actions that were applied to something, F, then the inverse sequence is written by adding a superscript: $R^{-1}(RF) = F$.

THE MULTIPLICATIVE INVERSE

It is now possible to investigate some interesting forms of the inverse. Consider if the reverse of the above is true. In this case, we swap the order of our actions by applying the inverse first, followed by the original sequence of events: $R(R^{-1}F)$. If the result is also identical when nothing is done, the events are their own inverse. They are said to be a multiplicative inverse, or that they commute. For example, consider rotating a photograph to the right through some angle and then its inverse, to the left, so that it is back to where it started. Now, swap the order of the events by applying the inverse, rotating the photograph to the left this time, followed by the original rotation to the right. This will again leave the photograph unchanged. This is written as $R^{-1}(RF) = R(R^{-1}F) = F$.

OPERATIONS WITH MORE THAN ONE INVERSE

Now consider tossing a coin. If the coin always starts as heads before it is flipped, there are two possible outcomes and hence, two inverses. Approximately fifty percent of the time, the inverse action would be to do nothing, as the coin lands on heads. For the other fifty percent of the time, the inverse action would be to turn the coin over after it lands on tails. However, it is not possible to know which inverse to choose before the coin is flipped, as at this point, both outcomes are just as likely. In this case, the multiplicative inverse does not hold true. The multiplicative inverse only holds for actions that have one possible inverse.

OPERATIONS WHERE THE INVERSE DOES NOT EXIST

Sometimes the steps needed to find the inverse may be so complicated that we can assume that it does not exist. An example involves hitting a pack of balls on a snooker table forcefully. After hitting the pack, the balls will be spread all over the table. The inverse of this would be to give all the balls a shove in the opposite direction so that they all roll back into a pack, just as if a film of the shot were run in reverse. However, if we tried to do this in reality, small errors in the velocity and imperfections on the surface of the balls and table would be magnified as they moved in reverse, with the result that no matter how hard you tried, the balls would never form the original pack shape again. In this case, the inverse will be impossible to perform in reality.

INVERSE FUNCTIONS

Consider a set of points along the x axis, 3, 5 and 9. The function $y = f(x) = 3x + 2$ will transform these points to the following coordinates (x,y): (3,11), (5,17), and (9,29).

An inverse function would have the effect of returning each y axis value to the original x axis value. Looking at the formula, it is seen that this is achieved if it is rearranged by adding 2 to both sides and dividing by 3, as in $x = f^{-1}(y) = (y - 2) / 3$.

If the function and its inverse function are substituted for each other, they are seen to leave the coordinates unchanged as expected, $f^{-1} f(x) = (3(x + 2) - 2) / 3 = x$.

If a graph of the function is made, the effect of the inverse function will be to reflect all the points, (x,y) along the line $y = x$ (see Figure 1).

Now consider the function $f(x) = x^2 + c$.

By definition, a function can only have one result and functions such as $y = x^2$ are said to not have an

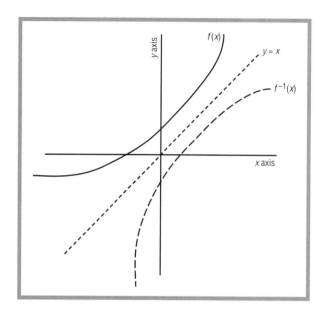

Figure 1: Reflection along $y = x$ of the inverse function.

inverse. It can also be seen that $y = x^2$ will not satisfy the multiplicative inverse relation. This is not quite as bad as it sounds because in real life applications, boundary conditions can be put on the range of values a function can take. In this case, we may find that in our analysis x can only take positive values; maybe it is a radius or mass where negative values would make no sense. It is then safe to ignore the negative square roots and define an inverse within these boundaries.

A Brief History of Discovery and Development

An inverse operation is such a fundamental idea that examples are found all through the history of mathematics, geometry, and science. One of the more historically interesting uses is in code breaking. All throughout history, codes have been invented and deciphered in attempts to gain some form of political or military advantage. In the time of the French king Henry the IV (1553–1610), Spain had ciphers that they changed regularly and was thought impossible to break. Henry IV gave the problem to a mathematician called Vieta, who figured out how to decode it and it was used by the French for two years to read Spanish documents. It was so successful that King Philip II of Spain (1527–1598) complained to the pope that the French were using sorcery to read their communications.

Another interesting fact is the development of negative numbers, used as the inverse of addition. It would

Inverse Square Law

The inverse square law is used to describe a field generated by a point charge, where the field is free to extend in every direction into free space. The field generated by a uniform sphere will be identical to the point charge field and at large enough distances; non-uniform shapes will often become good approximations to a point source as well. This means that the inverse square law accurately describes or gives very good approximations to many fields generated in nature. Examples are the gravitational fields generated by stars and planets, sources of radiation such as light and heat, and the electromagnetic forces such as those between atoms.

At a given distance, r, from a source the inverse square law is given as the intensity of the source A, divided by the area of a sphere,

$$I = \frac{A}{4\pi r^2}$$

For example, if one moved two feet away from a spherical light source, the intensity of the light would drop by a factor of four. To invert this, and keep the light intensity the same as before, the intensity of the light source would need to be increased by a factor of four. This explains the reason why a lighthouse focuses the light into a beam of light. This is not a point source, as the light has been kept in a narrow beam, and the intensity will only fall off as the light beam spreads due to focusing imperfections or atmospheric effects and will be visible to much greater distances.

seem that such a common concept would have been used since ancient times, but in fact it was not until 1545 that they came into common use. It was Gerolano Cardano (1501–1576), an Italian mathematician, who first showed that negative numbers could be used as an extension of our number system and that this was useful in the calculation of debts for example.

Real-life Applications

CRYPTOGRAPHY

Cryptography is the science of encoding information so that it can be transmitted secretly without an eavesdropper decoding it. Consider a simple cipher when encoding a letter to a friend by swapping all the letters and spaces by numbers. The friend has a similar sheet of numbers to letters that they use to decode your message. This sheet is called a key. In this case, the inverse steps are equal to the original number of steps you used to encode the message. However, an eavesdropper can decode your message by hand or writing a computer program that randomly assigns letters to numbers and scans for words in a dictionary. This is repeated until parts of a word is found that you can guess, and as more words are found, it will become possible for the eavesdropper to regenerate the key and decode your message. It is possible to use more complex functions to encode your message, but it can be shown that an eavesdropper will always be able to find the inverse function in a length of time that gets shorter the more the transmitted text becomes larger than the key. By sophisticated use of mathematics and computers, it is obvious this is not the most secure method to transmit data, and that it becomes weaker the more it is used.

A method called public key encryption is very common nowadays, and can get around this problem wherever information needs to be sent securely over unsecured lines. For example, Internet banking and automatic teller bank machines use public key encryption. Public key encryption works as follows. Imagine Bob wants to receive information from Alice. Bob has a special function that is split into two parts, the public function and private function, and generates keys using these functions. Bob sends the public key to Alice over an ordinary unsecured line and keeps the private key. Alice encodes the information she wants to send to Bob with the key using her public function and sends the resulting sequence of code back to Bob over the unsecured line. This public function has no inverse, so even if an eavesdropper were to gain access to the function, the public key alone provided no information.

When Bob receives the encoded data from Alice, he combines the private and public keys, to generate the inverse function. By running the information through this new function, the message from Alice is recovered. If Alice wants to receive information from Bob, then the reverse scenario is used and a completely secure two-way conversation has been set up over public lines. The reason this system works is that there is no analytical way to generate the inverse from the public key, but it can be generated with the combination of public and private keys. The only way it is thought possible to break this encryption would be to use the theoretical quantum computer. Until such a device exists, this form of encryption will remain theoretically impossible to break.

Multiplicative Inverse

As described previously, the multiplicative inverse is satisfied by actions that only have one possible inverse as $R^{-1}(RF) = R(R^{-1}F) = F$.

An obvious example of such a function is multiplication, which has the inverse of division. Consider multiplication by the number 3, which has the inverse $\frac{1}{3}$ as in $(\frac{1}{3}.3).F = (3.\frac{1}{3}).F = F$. Multiplication by $\frac{1}{3}$ is really division by 3—the inverse of multiplication by 3.

Another example is addition. The inverse of this is subtraction. Consider the action of adding 3. Here $R = +3$ and the inverse R^{-1} is subtracting 3: $(F + 3) - 3 = (F - 3) + 3 = F$.

Even though this system was known about for many years, it was only the invention of such functions by the English mathematician Clifford Cocks in the 1970s that made it possible. It was promptly made a state secret until it was independently invented by several American mathematicians in 1976. The American government made the code military property and legal battles ensued over about the encryption method's future. However, control of the encryption method was finally defeated in 1991 after versions of the code for Pretty Good Privacy (PGP) were published. In some countries, this method of encoding information is still illegal.

NEGATIVES USED IN PHOTOGRAPHY

When a black and white photograph is taken with a camera, light falling on the film will turn the film black, and areas where no light falls will remain clear. After the photograph has been taken, the film is chemically fixed so that it no longer responds to light and the image has been recorded. However, on examining the film, the image will be the inverse of what is desired; dark areas will be light and light areas dark. In this case, the photographic film is called a negative. In order to produce the final image, the areas of light and dark must be reversed. This is done with a special camera that will pass light through the negative and onto photographic sensitive paper. By retaking the image again in this way, the areas of light and dark are inverted and the original image is returned. The second step of re-exposing the negative of the image is equivalent to taking the inverse. Although technically more complicated, the process is the same for producing a color image.

THE BRAIN AND THE INVERTED IMAGE ON THE EYE

The eye works by passing light through a single lens and focusing it on the retina at the back of the eye. Due to the nature of optics, this image is upside down. The reason humans do not see the world upside down is that the brain inverts the image, therefore, the world is seen right-side up. It was thought that this process was hard-wired into the human brain for many years until a series of experiments were conducted that suggested otherwise. In these experiments, subjects wore a special mask over their eyes for 24 hours a day for several weeks. This mask placed lenses in front of the subjects' eyes that caused them to see the world upside down. For a period of time, the subjects were naturally disorientated and confused, but after a while, started to see the world the correct way again, suggesting that the human brain had the ability to re-configure itself to cancel the effect of the mask.

FLUID MECHANICS AND NONLINEAR DESIGN

In the snooker ball example discussed previously, it was seen that after hitting a pack of snooker balls, there existed no inverse operation where we could give the balls a shove and they would roll back into the formation of a pack as any tiny error would be magnified and ensure that the balls were always distributed randomly over the table. This has an important consequences for industrial design.

Consider using a computer to optimize the flow of fuel from a nozzle for an engine you are trying to design. At low velocities the flow from the nozzle will be smooth, called laminar flow and the computer simulation will accurately reproduce the flow. However, the laws of fluid dynamics are non-linear, i.e, they do not have a simple inverse and the flow at future moments is not related to the flow to past moments in a simple way. As a consequence any errors in the measurements will start to multiply rapidly and at a certain flow rate this is seen as a turbulent flow where any predictions from the computer will rapidly deviate from the real situation. In this case, it is necessary to find a specific approximating model to the nozzle that you are trying to design.

In industry, there are many examples of this situation, such as modeling the flow and timing of inks from printer heads, gas flow in exhaust systems to the effects of rain and dirt on glass. Even though the physical laws of each situation may be written down in a few lines, the nonlinearity means that special and often highly technical models have to be constructed to investigate each situation and this can involve much time and investment making the details of such models closely guarded industrial secrets.

ANTI-SOUND

Sound is a series of pressure waves in air. Sound that is not loud enough to cause damage to the hearing can still cause discomfort and irritation after long periods of exposure. In modern design keeping sound down to minimal levels is a key concern in commercial passenger transport, such as ships, trains, planes and cars. By reducing the noise most passengers will experience a more present and relaxing journey. Another area where sound levels need to be kept under control is in military equipment where the high performance needed will often result in very noisy equipment but large amounts of sound absorbent material is undesirable due to weight and space concerns.

One solution to reduce sound in these environments is to generate anti-sound. This involves wearing earphones or fitting the passenger compartment with loudspeakers that generate a special form of sound. Speakers will record the sound waves in the compartment and a computer will calculate the exact pressure waves that will cancel these pressure waves, called inverse or anti-sound. When the anti-sound plays through the speakers it will reduce the sound that reaches the ear of the passenger. Although conceptually simple, in practice these systems are very difficult to make, for example real time anti-sound generation requires rapid computing and playback times, and can be very expensive and sensitive to changes in its environment. Some systems get around this by using a digital recording to playback the anti-sound. In practice, it will be impossible to completely cancel all the background sound, but the systems have been shown to be very effective at reducing background noise levels.

STEALTH SUBMARINE COMMUNICATIONS

Communicating with submarines is not as easy as communicating with space craft. Unlike space and air, water is a much denser medium. In pointing radio signals at submarines, they will be rapidly absorbed by the water. Another problem with submarine communications is that one of their main strengths comes from being hidden. For example, submarines make up a key part in the nuclear defense systems of both the East and West, and can remain hidden off a coastline or under an ice flow for many months at a time, if needed, before engaging their target. An exposed submarine is vulnerable to missiles from other submarines, surface ships, and aircraft. Submarines are also vulnerable to attack, as explosions under water can be far more devastating even if there is not a direct hit due to the large pressures on the hull. For this reason, high-power focused radio transmissions are a problem, as the direction of the beam can give the position of the submarine away to the enemy.

One system for communicating with submarines relies on a series of radio antennas that transmit at extremely low frequencies (ELF). It is a physical property of radio waves that their absorption by water increases with frequency. By using ELF transmissions at around 76 Hertz, the radio signal can penetrate the water to depths of hundred of feet as opposed to a similar VHF transmissions would only penetrate to a depths of around 10 feet (3 m).

When setting up a system like this, there are multiple transmitters to give a wide coverage. Having multiple transmitters of such low frequency can cause blank spots in radio transmissions in areas where a signal would be expected. This is because the waves from each transmitter interfere with each other, adding to the overall power at one point or diminishing the overall power at another, forming an interference pattern. This can be thought of as throwing two stones into a small pond. After the initial splash, the ripples will form a stationary pattern on the water's surface, with some areas having higher ripples and some areas that look calmer. The interference pattern generated with such low-frequency transmissions will leave blank spots in a signal that can be larger then the submarine, and a careful design of the antennas is needed to give constant communications with the submarine.

Blank spots in the reception can be shifted if the timing of the transmitted signals, called the phase of the transmission, is changed in one or more of the transmitters. Now imagine if the submarine is in a certain position in the sea and it transmitted a signal. This signal would spread out and reach all of the transmitters slightly different times, i.e., with different phases. If a signal is transmitted back to the submarine as the inverse of the way a signal is received, then the signals from the transmitters can be made to interfere with each other and build up as they approach the submarines location. This effect that can be thought of as the inverse of ripples formed from throwing a stone into a pond, in this case, the radio ripples move toward the center of the pond, building up into the original splash, at the location of the submarine.

As the submarine moves, the phase of the transmitters can be changed to refocus the transmission to the new location. At other points on, the radio signals will be more difficult to detect, making the system more secure against eavesdropping.

STEREO

Having two ears, the human brain can figure out the location of a sound by the time delay in the signal as it reaches either ear. For example, if a sound reaches the left ear first and then the right ear a fraction of a second later, the brain interprets the source closer to the left side and this is the feeling we would have from this source.

To record a sound in stereo, two microphones are needed and these are placed side-by-side at about the distance of the human head in front of the artist. Having the two microphones means that not only is the sound recorded, but the delay phase of the sound is also recorded. In a modern studio, equipment exists that can add this delay electronically. In this manner, artists can record themselves with high-quality microphones and pickups onto separate channels, and the studio can electronically add phases to each channel and make the artists sound as if they had been playing at different positions in a group.

To play back sound in stereo, two speakers are placed in front of the listener to invert the effect of the recording. During playback, the timing or phase of the sound waves from either speaker will interfere with each other, some waves adding to each other (called constructive interference) and some waves canceling each other (called destructive interference). The interference of the waves as they reach the ear will result in time delays in the sound between each ear, which the brain will reconstruct as positions of sound sources. It is possible to hear the musicians playing from different positions in the group as if they were in the room with the listener. For this system to work properly, the locations of the speakers and listeners, and reducing sound reflections from the walls of the room will all be crucial factors in how well the stereo effect is reconstructed. It can take much time and investment to construct the ideal listening environment, especially in environments such as cinemas, where there are many listeners to consider.

Where to Learn More

Web sites

Fischer, Charlie. "Public Key Encryption." <http://www.krellinst.org/UCES/archive/modules/charlie/pke/> (April 9, 2005).

United States Navy. "Extremely Low Frequency Transmitter Site, Clam Lake, Wisconsin." *Navy Fact File* <http://enhttp://www.elfrad.com/clam.htm> (April 9, 2005).

Wicks, J. "Details about the functional inverse." North Park University. <http://campus.northpark.edu/math/PreCalculus/Functions/Functions/Algebra/> (April 5, 2005).

Iteration

Iteration is a process of calculation that is repeated again and again, each time improving the accuracy of the result using the output of each step for the input of the next iteration. By repeating the process and analyzing successes and errors, humans and even some machines can improve at performing the action.

Fundamental Mathematical Concepts and Terms

An example of problem solving by the process of iteration could be found in the following set of computer programming instructions: "While a number is less than 15, continue printing that number but increase its value by one. Upon arriving at the number 15, activate the programmed screensaver." These instructions could well relate to the set up of a screensaver to come on after a computer screen was been inactive for 15 minutes.

Let's look at the nuts and bolts of the above set of instructions in a bit more detail. If the starting number was two, for example, it would be less than 15, and so the statement would be judged as being true (i.e. less than 15). That would be a signal to go through the process again, and again, until the number was 14. At that point the number plus one would no longer be less than 15. That would be the signal for the computer to switch to another set of instructions that activates the screen saver.

Real-life Applications

ITERATION AND SPORTS

You do not have to be a big basketball fan, or even have much of a knowledge of the game, to recognize the name Michael Jordan. Michael Jordan was an awe-inspiring basketball player. One big reason was talent, but another was iteration. A particularly relevant example of this process was Jordan's routine at the end of each practice. Jordan would shoot the basketball at the hoop from various locations, near to the basket and further away. His practice was not over until he sank 100 shots in a row. If he made 92 baskets and missed one, he would start over again. This routine, which must have been frustrating on a less-than-accurate day, is analogous to mathematical iteration. By performing the action over and over again, using the results of each shot to refine the performance of the next shot, Jordan fine-tuned a technique that predictably put the ball through the hoop.

Similarly, Tiger Woods has used thousands of hours of golf practice to perfect a golf swing that is consistent from one day to the next. This consistency propelled him to become the number one ranked golfer in the world in 2003.

But even the best golfer in the world likes to tinker with his swing, to try out slightly different changes in hopes of producing a swing that is even better than before. That is the essence of iteration. By repeating an action again and again, changes can be evaluated and, if they are successful, can be incorporated into the action.

ITERATION AND BUSINESS

Not surprisingly, iteration is a favorite buzzword of computer programmers. Computer programmers often make available trial software programs on the Internet, called beta versions. Beta programs are a form of practice versions of a new program. Usually this iteration of a program has more features built in that presumably will make people want to buy it and use it instead of the current version of the program. The purpose of a beta version program is to encourage people to try the software, figure out its good points and, most importantly, discover what needs changing or what does not work. The software programmers can then change the beta version to produce the final improved program that is widely sold.

In another business application, iteration is an important feature of accomplishing a project that involves a large team of people. Again, in the realm of computing, an ideal example of iteration involves extreme programming, or XP. Like an extreme sport, XP is a difficult-to-accomplish form of programming that often involves dozens of programmers. These programs are updated frequently and made available much more often than, for example, a program for a video game that might be updated once every three or four years.

In the XP iteration process, the total project is usually broken into chunks. Each chunk can have a back-and-forth process where the component of the program is written, tested, and returned for tinkering. A tight schedule allows the iteration process for each chunk of the project to be accomplished by a deadline, so that all the chunks can be put together to produce the final product. As well, the back-and-forth contact between people that is part of this type of iteration allows for better tracking of minute details in the frenzy of the project.

ITERATION AND CREATIVITY

Creativity involves the ability to look at something in a different way, to find a new idea. A necessary part of creative thinking is gathering information, and then trying to put that information together in a new way. This is where

DJ Kool Herc, the Jamaican-born DJ considered the father of hip-hop. It was Herc, at parties in the early 1970s, who began playing the instrumental segments of songs over and over again, a form of musical iteration, while speaking in rhyme over them. AP/WIDE WORLD PHOTOS. REPRODUCED BY PERMISSION.

iteration comes in. New product ideas come to the forefront after cycles of inventing a design, testing the design, and, as usually happens, discovering and fixing problems.

Iteration in creative product design can be illustrated by considering a new CD from a popular music group. The tracks on the CD do not usually happen in one recording session in the studio. When the band first starts to record a song, the musicians, songwriters, and producer may have different ideas of what the final version will sound like. Different versions (iterations) of the song are tried out, discussed, and changed until the artist is pleased with the final version. The final track that is heard on the CD is often very different from what a band member thought it sound like months before.

People who are known for their creative approach to their work often say that the process they use to come up with all those great ideas is very structured. They take the same approach to each problem, knowing that doing the steps in much the same order (there has to some flexibility in how things are done) helps their mind get ready to think. In other words, their whole approach to being creative involves iteration. Similar actions are repeated.

ITERATION AND COMPUTERS

In a computer program, iteration is the recycling of a set of instructions, known as looping. A single iteration is one pass of the instructions. Once the set of instructions has been written, a computer will quickly pass through the loop over and over again without making mistakes (unlike humans).

Another real-world example of iteration in computing is a macro. A macro is the putting together of a series of commands that responds to one signal (like the pressing of a designated key on the keyboard). Macros are not necessary to do work on a computer, but they make time go more quickly. Instead of typing in the same commands over and over, this iteration is taken care of by the one action of pressing that designated key.

Iteration and Nature

In nature, repeating patterns are common. From the spirals of a seashell to the icy beauty of a snowflake, to the many hexagons of the honeycomb of a beehive, a basic unit is repeated again and again to produce the final structure. This repeat of the basic unit is iteration.

A cutting-edge example of iteration in the laboratory is molecular cloning. Molecular cloning involves creating a genetic twin from the genetic material obtained from a living creature. Experiments in plants and animals have shown that scientists have not yet perfected the cloning process. When they do, then cloning will be a living example of iteration. Whether this form of iteration is desirable or not is being debated at the present time.

Where to Learn More

Books

Woods, T. *How I Play Golf*. New York: Warner Books, 2001.

Web sites

Peterson, Ivars. "Ivars Peterson's Math Trek: Candy for Everyone" *Mathematical Association of America* <http://www.maa.org/mathland/mathtrek_1_10_00.html> (September 5, 2004).

Wells, D. "Iteration Planning." *Extreme programming* <http://www.extremeprogramming.org/rules/iterationplanning.html> (September 3, 2004).

Overview

Linear mathematics deals with linear equations. An equation is "linear" if it consists of a sum of variables or unknowns, each of which is multiplied by some number or constant (examples will be given below). Many real-world problems in physics, engineering, business, chemistry, biology, and other fields are described by linear equations. Computers are use to solve linear equations in groups or "systems," making possible many kinds of medical and scientific imaging, realistic video games, cheaper design of cars and other products, and the more efficient management of money.

Fundamental Mathematical Concepts and Terms

Linear equations are called "linear" (line-like) because the simplest kind of linear equation—one having two variables—describes a straight line. For example, the equation $2x_0 + 3x_1 = 4$ describes the straight line depicted in Figure 1.

Here x_0 and x_1 are "variables," meaning that they stand for any numbers we like; the small 0 and 1 are labels to tell them apart by. For each x_0 we choose, there is one and only one x_1 that makes $2x_0 + 3x_1 = 4$ true. For example, if we set x_0 equal to 0, then x_1 must be 4/3 because:

$$x_0 \text{ value}$$
$$\downarrow$$
$$2 \times 0 + 3 \times \frac{4}{3} = 4$$
$$\uparrow$$
$$x_1 \text{ value}$$

We can also use letters to stand for the fixed numbers that multiply x_0 and x_1. If we replace 2 and 3 in $2x_0 + 3x_1 = 4$ with the symbols a_1 and a_2, and replace 4 with b—where these new letters can stand for any fixed numbers we like—we get a general-purpose linear equation in two variables: $a_1x_1 + a_2x_2 = b$.

We can extend this to as many multipliers (also known as "coefficients") and variables as we like. The equation is still called a "linear" equation no matter how many variables we add. Here is the form of linear equation involving 3 variables and 3 coefficients:

$$a_1x_1 + a_2x_2 + a_3x_3 = b.$$

We have already seen how a two-variable linear equation describes a line. A three-variable linear equation

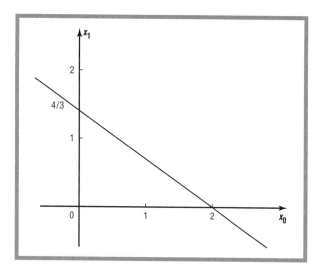

Figure 1: Graph of the equation $2x_0 + 3x_1 = 4$.

Figure 2: A "two-by-two" matrix.

describes a plane, a set of points resembling a stiff sheet of paper tilted in space.

In general, a linear equation containing n variables and n coefficients looks like this:

$$a_1x_1 + a_2x_2 + a_3x_3 + a_4x_4 + \ldots + a_nx_n = b.$$

The three dots in the middle of the equation stand for all the terms between the fourth term and the nth term that we don't want to bother to write down. In real-world applications, linear equations containing dozens or even millions of terms are common.

Linear equations can be combined into groups or systems. A system of linear equations is a group of two or more equations that involve the same variables. The following is a system of two linear equations involving the two variables, x_0 and x_1:

$$2x_0 + 3x_1 = 4$$
$$x_0 + 9x_1 = 0$$

The "solution" of a system of linear equations is that set of numbers which, if plugged in for the variables, makes every equation in the system true at the same time. In this example, the solution of the system is $x_0 = 12/5$, $x_1 = -4/15$. This solution is unique; that is, each equation considered by itself is true for many values of x_0 and x_1, but only at $x_0 = 12/5$, $x_1 = -4/15$ are both equations true.

If you graph the two equations in this system as lines on paper, the solution of the system will be the point where the two lines intersect. Every system of equations has a single, unique solution (like this system), or no solutions, or an infinite number of solutions. Among

systems that consist of two lines, like the ones that we've just been looking at, those that have a single, unique solution are lines that intersect (one point in common); those that have no solutions consist of parallel lines (no points in common); and those with an infinite number of solutions consist of two equations for the exact same line (all points in common).

Systems of equations can also be written as matrix equations. A matrix is a rectangular array of numbers or variables with square brackets around it. It is named according to how high and how wide it is. For example, the matrix shown in Figure 2 is a 2×2 ("two by two") matrix because it is 2 entries tall and 2 entries wide. The matrix depicted on Figure 3 is a 2×3 ("two by three") matrix because it is 3 entries tall and 2 entries wide. A matrix can be added to, subtracted from, or multiplied by other matrices. It can also be multiplied by numbers, variables, and vectors, which are special matrices only 1 entry wide. Vectors containing three entries, as depicted in Figure 4, are particularly useful in science, engineering, and computer animation because each three-entry vector can specify a point, force, velocity, or acceleration in three-dimensional space.

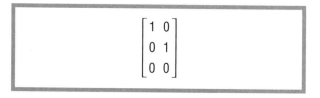

Figure 3: A "two-by-three" matrix.

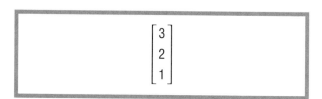

Figure 4: Vector containing three entries.

The system of two equations shown earlier can be written as a 2×2 matrix multiplied by a vector and set equal to a second vector. That is,

$$\begin{matrix} 2x_0 + 3x_1 = 4 \\ x_0 + 9x_1 = 0 \end{matrix} \text{ means the same as } \begin{bmatrix} 2 & 3 \\ 1 & 9 \end{bmatrix} \begin{bmatrix} x_0 \\ x_1 \end{bmatrix} = \begin{bmatrix} 4 \\ 0 \end{bmatrix}.$$

When there are only two or three variables in a system of equations, as in this example, there is no advantage in using the matrix form. But when systems involve many variables, as in most real-life applications, the matrix form is more efficient and revealing. Computers are well-suited to calculating with matrices, and are often used to solve systems whose matrices contain millions of entries. In creating medical images of the inside of the body, searching for oil reserves, predicting global climate change, designing new drug molecules, maximizing profits, and many other applications, the solution of large matrix equations by computer is key.

Because the solution of systems of linear equations is so important in our high-technology society, most of the examples of linear math given below involve the solution of such systems.

Real-life Applications

EARTHQUAKE PREDICTION

Science foresees no way of preventing earthquakes, which occur when whole sections of the Earth's crust, many miles across and weighing billions of tons, slip past east other. These forces are too great to control. However, knowing when and where earthquakes are likely to happen, and how strong they are going to be, would make better preparedness possible and reduce the loss in lives and money caused by major quakes.

It is not yet possible to predict most earthquakes, but with the help of large systems of linear equations solved by computers, scientists are making rapid progress. The basic method, called "finite element modeling," is a common one in manufacturing, science, medical imaging, and other fields today. In finite element modeling, a mathematical model or image of an object or volume of space is built up using either triangles (for flat models) or tetrahedra (four-pointed pyramids, for three-dimensional models). The triangles or tetrahedra are called "elements" and fit together into a web or network called a "mesh." One or more separate variables (like x_0, x_1, and so forth used above) are assigned to each element, and linear equations involving these variables are written so that they approximate the laws of physics that apply in that

Linear equations

Linear equations that involve two variables, such as $2x + 3y = 4$, describe straight lines. That is, if you graph any of the x, y pairs that satisfy the equation, you will find that they all lie on the same line on the paper—and, likewise, that every point on that line satisfies the equation. A linear equation that involves not two but three variables, such as $2x + 3y + 7z = 4$, graphs a plane in three-dimensional space.

Linear equations appear everywhere in science, technology, and business. If you are selling sneakers at x dollars of profit a pair, you know that if you sell 20 pairs of sneakers you will make double the money than if you sell 10 pairs, namely, $20x$ dollars rather than $10x$ dollars. Here the relationship of pairs sold to total profit is described by a linear equation: number of pairs sold (call it a) times profit per pair (x) equals total profit (p), $ax = p$. Anyone running a lemonade stand knows this much linear math by instinct.

But not everything in real life is linear. For example, you might make more profit per pair of sneakers if you sell a million pairs than if you sell only a hundred. In this case, the equation describing your total profit in terms of sales will not be a linear equation.

Nor are all linear equations as bare-bones as $ax = p$. If you are selling two types of sneaker, one of which makes x_1 dollars of profit per pair while the other makes x_2 dollars, a different linear equation arises. Say you sell a_1 pairs of the first type of sneaker and a_2 of the second type. Then your total profit p is given by the sum of the profits from each type: $a_1x_1 + a_2x_2 = p$. This is also a linear equation, but it involves two variables. In real business and industry, equations of this type involving variables are common.

area of space. For earthquake prediction, meshes containing 100 million tetrahedra or more are created that represent parts of Earth's crust containing earthquake faults. Equations are constructed using these meshes that describe how shock waves move through the rock and soil. Supercomputers are then used to solve the resulting systems of millions of linear equations; the solution shows what an earthquake will look like.

Linear Inequalities

Equations express equalities, such as $1 + 2 = 3$. We can also write *in*equalities, expressions that say that one thing is less (or greater) than some other. Four signs are used to express inequality: $<$ (less than), $\leq$ (less than or equal to), $>$ (greater than), and $\geq$ (greater than or equal to). For example, the expression $a > b$ reads "a is greater than b." The expression $c \leq a$ reads "c is less than or equal to a."

A linear inequality is a linear equation with its equals sign replaced by an inequality sign. The linear equation, $x + y = 2$, for example, can become the linear inequality $x + y \leq 2$.

The linear equation $x + y = 2$ describes a straight line—and that's why it's "linear" (See Figure A.)

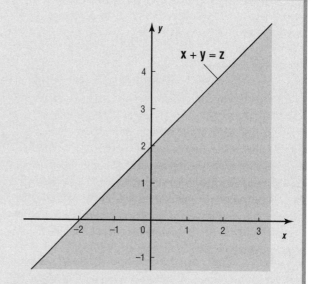

Figure B: Graph of x + y = 2.

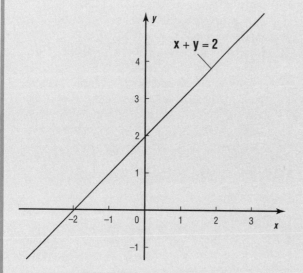

Figure A: Graph of x + y = 2.

The linear inequality $x + y \leq 2$ also describes a set of points. The line $x + y = 2$ is part of that set because the "less than or equal to" sign includes an equals sign.

The inequality is also true for all points below the line, namely the gray area in Figure B.

Linear inequalities arise often in real life. Consider a factory that can make two kinds of computer chip, Chip One and Chip Two, but cannot make both at the same time. The time needed to make a batch of Chip One is 5 minutes, so the time to make c_1 batches of Chip One is $5c_1$ minutes. The time needed to make a batch of Chip Two is 10 minutes, so the time to make c_2 batches of Chip Two is $10c_2$ minutes. But there are only 1,440 minutes in a day. Adding the time spent in one day making Chip One to the time spent making Chip Two, we have the linear inequality: $5c_1 + 10c_2 \leq 1,440$. This example is simple but not far-fetched. Linear inequalities that express limits or constraints on time, material, or other valuables appear constantly in the solution of real-world business and finance problems. Such problems are often solved using the technique called "linear programming."

RECOVERING HUMAN MOTION FROM VIDEO

There is much interest today in teaching computers how to track human motion from video cameras. To track human motion successfully, a computer must be able to pick human forms out of all the other information in a moving image, like the video of a dance or a football game. It should then be able to describe what the human being has done in words, or be able to move a mathematical model or virtual puppet to re-create the motion it has observed. The ability to track and then describe or reproduce human motion mathematically is used in video games, virtual reality, formation analysis in sports, and in other ways.

One method used to track motion is the identification of "feature points" on the subject—idealized dots or spots on the surface of the subject's body, say on their helmet or elbow or knee. These feature points are then tracked in the video images recorded by several video cameras. Since each video image is two-dimensional (flat), the location of each feature point in the image at any one time can be described by two numbers, an x coordinate that says how far from the left-hand edge of the image the point is and a y coordinate that says how high up it us from the bottom edge of the image. If there are, say, 40 feature points on the subject, there are then 80 numbers that describe where the feature points are at a particular moment of time in the video from a single camera, and $3 \times 80 = 240$ numbers to describe where the feature points in the video from 3 cameras. These numbers are put into a matrix. Using linear mathematics methods of matrix algebra, this matrix is separated into two matrices, an M matrix that describes how the cameras are pointing (which we don't really care about) and an S matrix that describes the true arrangement in space of the feature points. The S matrix records how the subject is positioned in space at that moment. A whole series of S matrices, one for each frame of video, describes how the subject's body moves through space over time. Motion capture and analysis using linear algebra is used in computer-animated movies such as *Polar Express* (2004), where live actors' motions were recorded by computers and then used to animate digital figures.

VIRTUAL TENNIS

The ongoing explosion in computer power makes possible the crafting of "virtual" worlds in which a game-player, scientist, or other user can experience the illusion of movement and exploration. In most virtual-world or virtual-reality systems, a headset replaces the scene that the user would otherwise see with computer-generated scenes.

But the sense of touch is not so easy to fool. One approach, in virtual tennis, is to have a computer read position information from a racket grip held in the player's hand. A rod is also attached to the "racket." When the player sees a ball coming in the virtual world which the headset shows to them, they swing at the ball. The computer senses the forces that the player's hand exerts on the racket grip, as well as the position of the racket in three-dimensional space, and calculates whether they player is going to succeed in hitting the virtual ball. If they do, the computer sends a shock along a rod connected to the player's racket so that they can feel the impact of the ball—which does not physically exist—hitting the racket. Such systems are already becoming commercially available.

A check sorting machine separates checks at high speed at the Unisys Corp. check-processing facility. Banks use linear programming to process checks more efficiently. In particular, they want to minimize "float." AP/WIDE WORLD PHOTOS. REPRODUCED BY PERMISSION.

All the computations performed by the computer in such a game involve vectors and linear mathematics. The position of the racket in space is characterized by a set of three-dimensional vectors; the force of the player's grip, the velocity of the ball, and other variables are also represented by vectors. Furthermore, the computer must calculate what racket positions are "feasible" for the system, that is, what positions the rod and wiresattached to the racket can allow. This is done using matrix algebra.

LINEAR PROGRAMMING

Linear inequalities (see sidebar) are important to the problem-solving method known as "linear programming." In a linear programming problem, linear equalities and linear inequalities are combined into a system (that is, they all involve the same variables or unknowns).

Maximizing Profits

The technique known as "linear programming" combines linear equations with linear inequalities (see sidebar) to find the best way of using limited resources. It is used mostly by large organizations, such as corporations or the military, to minimize operating costs.

Banks use linear programming to process checks more efficiently. In particular, they wanted to minimize "float." Float is the amount of money represented by uncancelled checks—checks that have been received by the bank but for which the money has not yet been collected. Float is detrimental to profit because it represents money in limbo; the bank cannot make money *on* that money (invest it) until the check has cleared.

What should a bank do to minimize float without spending so much doing it that the cure is worse than the disease? When checks are received they are "encoded," that is, marked with magnetic ink by a machine. This is the first step in clearing the check. Banks realized they needed to encode checks as quickly as possible without hiring too many machines and clerks, so mathematicians and computer specialists set up a linear programming problem to model the situation. That is, they organized float, encoding machines costs, wages and hours for clerks, and other relevant variables as a set of linear equations and inequalities, and solved this system using linear algebra. The solution showed banks how many full- and part-time clerks to assign to how many shifts on how many machines in order to minimize float. Although there are increasingly high-tech ways to digitize information and handle checks, many financial institutions still use linear programming to save money and increase profits.

dollars in the bank available for investment. Each linear inequality is then converted into a linear equality. For example, the inequality $50x_1 + 12x_2 \leq 100$ really says that $50x_1 + 12x_2$ is less than 100 by some unknown amount (maybe 0). This is the same as saying that an $50x_1 + 12x_2$ plus an unknown quantity equals 100. If we name this third unknown M, we can turn the inequality into an equation: $50x_1 + 12x_2 + M = 100$. When all linear inequalities have been turned into linear equations, we then use matrix algebra methods (which are described in many textbooks) to solve the system and find out the best way to run our business.

Linear programming is used by real-life organizations, especially businesses and the military. An example is the use of processing stations in semiconductor manufacturing plants. These plants make the circuit-covered "chips" that run all complex electronic devices, including computers. Many thin layers of material have to be built up on each chip, and each layer requires many stages of optical and chemical processing. In fact, more resources are consumed in making the tiny chips in a desktop computer than in making all the rest of the computer put together. Manufacturers are therefore keen to use their chip-making factories efficiently.

A processing station in a chip factory is a large, complex device that performs one step at a time in the chip-making process. Instead of having hundreds of stations, one for every step, it is cheaper to re-fit each station (change some of its parts) occasionally so that it can do a different step. But refitting a station takes time; it would be unprofitable to refit a station every single time it performed a step. How many batches of chips should a station process before being refitted for another step? Linear programming is used to answer this question, telling the manufacturer how to schedule steps and stations for maximum profit.

LINEAR REPRODUCTION OF MUSIC

If a musician plays two notes in a recording studio, one twice as loud as the other, you want two notes come out of your stereo's speakers so that the one is twice as loud as the other. If graphed on paper, this relationship between live performance and ideal playback is a straight line—a linear function. A great deal of mathematical design work goes into making sound-reproduction systems as linear as possible.

But nonlinearity—electronic behavior that is *not* linear—has its uses, too. The rough sound of a rock guitar is produced by feeding an electrical signal derived from the guitar's strings into a circuit that does not respond linearly. That is, the original signal looks like a complicated wave or series of up-and-down wiggles; when two

This system is then solved, using the methods of linear algebra, to find the "optimum" (i.e., best possible) way of mixing ingredients, manufacturing items, transporting supplies, or allotting other resources.

The first step in a linear programming problem is to define a linear equation that describes something which we want to minimize (expenses, say) and as many linear inequalities as we need to describe the bounds on our resources: for instance, that there are only so many minutes in a day, or pounds of Ingredient Z available, or

Key Terms

Linear algebra: Includes the topics of vector algebra, matrix algebra, and the theory of vector spaces. Linear algebra originated as the study of linear equations, including the solution of simultaneous linear equations. An equation is linear if no variable in it is multiplied by itself or any other variable. Thus, the equation $3x + 2y + z = 0$ is a linear equation in three variables.

Linear equation: An equation whose left-hand side is made up of a sum of terms, each of which consists of a constant multiplying a variable, and whose right-hand side consists of a constant.

Linear programming: A method of optimizing an outcome (e.g., profit) defined by a linear equation but constrained by a number of linear inequalities. The inequalities are recast as linear equation and the resulting system is solved using matrix algebra.

Matrix: A rectangular array of variables or numbers, often shown with square brackets enclosing the array. Here "rectangular" means composed of columns of equal length, not two-dimensional. A matrix equation can represent a system of linear equations.

System of equations: A group of equations that all involve the same variables.

Vector: A quantity consisting of magnitude and direction, usually represented by an arrow whose length represents the magnitude and whose orientation in space represents the direction.

wiggles, one twice as big as the other, are fed into a non-linear circuit, the larger wiggle does not come out twice as big but gets flattened or chopped off at the top and bottom. This happens because the circuit cannot produce a signal above or below a certain limit. The resulting sound is, technically speaking, "distorted"—but sometimes, that's exactly what we want.

Where to Learn More

Books

Budnick, Frank S. *Finite Mathematics with Applications.* New York: McGraw-Hill, 1985.

Lay, David C. *Linear Algebra and its Applications,* 2nd ed. New York: Addison-Wesley, 1999.

Logarithms

Overview

A logarithm is the power to which a number (usually termed the base number) must be raised to equal a target number.

Fundamental Mathematical Concepts and Terms

In base 10 systems, 2 is the logarithm of 100 because $10^2 = 10 \times 10 = 100$. The number 2 in this example is the exponent of the base number 2 that yields 100. Accordingly, in base 10 the log of 100 is 2.

Because logarithms are so common in mathematics, there are different ways to develop an understanding of them and this will often cause some confusion. However, at their most basic level they can be thought of as a set of rules that allow one quantity to be converted into another to simplify a problem. This idea is the same as multiplication, which is a simplification of the operation of repeated addition; it is easier to say 50×5 rather than write $50 + 50 + 50 + 50 + 50$. Logarithms are effectively the next step, the simplification of repeated multiplication or division. Logarithms have their own form of mathematical notation that can only be manipulated in strict accordance with a set of rules. Once these rules are understood the work of manipulating logarithms is carried by the notation itself, operations no more complex than multiplication, addition, subtraction and division are used to manipulate terms in an equation and generate the desired result.

Before trying to use the mathematical notation of logarithms, it is helpful to understand some aspects of mathematical notation itself. Consider the simple operation of multiplication. The common notation for multiplication is the $\times$ symbol.

Using a symbol to stand for repeated addition, called multiplication, is a form of shorthand. For example, $2 \times 3 = 2 + 2 + 2$. Obviously, the notation reduced the amount of work needed to express repeated addition; it would be hard work if you had to write out $3{,}200 \times 563$ as $3{,}200 + 3{,}200 + 3{,}200 \ldots$ some five hundred sixty three times. However, this notation is more than just a shorthand; it allows us to manipulate quantities that were not possible before and extend our range of mathematical tools. For example, consider multiplying two fractions together. Even though it is not possible to write this out directly, it is still possible to find the answer, $0.5 \times 0.7 = .35$. It is even possible to throw away the numbers and replace them with letters that represent any number that you can think of. Here x means any number we can think of multiplied by three. y now represents the answer, $3 \times x = y$. In equations like this the multiplication symbol, $\times$, is often dropped and letters and numbers that are next to each other are understood to be multiplied. If we set $x = 2$, remembering this is just a number used for example, our equation y is found to be equal to $3 \times 2 = 6$. Again, if $x = 5$, then y is equal to 15.

One of the great powers of this notation is that it allows the terms, such as x and y, to change places. This is done by noticing that any operation performed to on one side of the equal sign must be repeated on the other side of the equal sign. This is because the values are equal and what we do to one side should balance the effect on the other side.

Let's consider an example of this for the formula $3 \times x = y$. This formula may give us some property of a material, and we conduct an experiment where we measured the values of y of that material. Can we find the value of x? Yes, simply find the value of x by dividing both sides by three: $(3 \times x)/3 = y/3$ so this gives $x = y/3$. By dividing the equation by 3 on both sides we have eliminated the term in front of the x and given the equation in terms of y. This process is called rearranging an equation.

THE POWER OF MATHEMATICAL NOTATION

As multiplication was the extension of repeated addition, so raising to powers, or simply powers, is the extension of repeated multiplication. Consider the number 5 multiplied by itself four times; this can be written in shorthand by putting the number of times the multiplication is to be repeated as a smaller number 4 to the top right of the 5. Here are some examples, $5 \times 5 \times 5 \times 5 = 5^4$ is the same as $5 \times 5 \times 5 \times 5 = 5^4$. To read this notation out loud we say base five raised to the power of four. Another example is $2 \times 2 \times 2 \times 2 \times 2 \times 2 = 2^6$ read as base two raised to the power of six. There are a couple of points to remember about this notation. The first is that the power is also known as the exponent and raising to a power is also known as exponentiation. Exponentiation is not to be confused with the exponential function, e^x, discussed later. Another point to note is that numbers raised to the power of two or three are often read as squared or cubed. For example, $5 \times 5 = 5^2$ is read as five squared and $8 \times 8 \times 8 = 8^3$ is read as eight cubed.

As division is the opposite of multiplication, logarithms can be thought of as the opposite of exponents. They come in two common forms. The first form is written as $\log_{10}$, read as log base 10. The base here is related to the base of the powers, as we shall soon see. $\log_{10}$ is so common that in texts and the buttons on most calculators the base 10 is dropped and it simply reads as log or lg. The other form is read as the natural logarithm, and is written as ln, this is identical to "log e." Logs to any other base, such a base 2, are written as $\log_2$, etc.

POWERS AND LOGS OF BASE 10

Now students should try to get a feel for some values in base 10. Using a scientific calculator, they can try the following, $\log (1,000) = 3$. This tells us that 1,000 can be repeatedly divided by 10 three times: $1,000 /10 /10 /10 = 1$, which is true. Another way to look at this is the logarithm has told us that the number 1000 has three zeros after the one. Now raise base 10 to the power of 3 and we are back with the number we started from, $10 \times 10 \times 10 = 10^3 = 1,000$. Here we see the relation between the base of the logarithm and the base of the power. This reflects the relationship of logarithms as repeated division and powers as repeated multiplication. Raising the logarithm to the power like this is called an anti-logarithm, and it gives us back the number we started with. Here is another example, $\log (10,000) = 4$. Again, this shows us that 10,000 can be repeatedly divided by 10 exactly four times, or to view it another way, there are four zeros after the 1. Raising this logarithm to the power of base 10 gives us back our number, $10^4 = 10,000$.

For any number made from one followed by a number of zeros the log will always equal the number of zeros if we use logarithms with base 10.

As with the previous multiplication example, our definitions of this notation allow us to extend this idea of repeated multiplication and division to more than just shorthand, because we can now use fractional values. Students should try the following, $\log (5,246) = 3.7198283$. Even though this cannot be written out as an exact repeated division by ten as we did before, it still tells us how 5,246 would divide into 10 in an abstract sense, about four times. If we raise this to base ten do we get the answer back as expected? $10^{3.7198283} = 5,246.0002$. Almost, but what about the small fraction after the number? (Depending on places and the calculator, the exact fraction may differ.) The digits after the decimal place are not important and are there because the calculator cannot store numbers to infinite precision. However, they can safely be ignored as the error is not in the digits in which we are interested. This will always be found to be true and we obtain the correct answer of 5,246. The notation has allowed the extension of the mathematical idea of repeated multiplication to be taken beyond the simple idea of a shorthand.

LOGARITHMS TO OTHER BASES THAN 10

What about logarithms with a base other than 10, such as, $\log_2 (256) = 8$? You do not find a $\log_2$ button on your calculator because logarithms to bases other than 10 can always be expressed as $\log_{10}$ using the following formula:

$$\log N\ (y) = \log_{10}(y)/\log_{10}(N)$$

Here y is the value of the log and N is the value of the base. So, to solve the previous equation, $\log_2 (256) = \log_{10}(256)/\log_{10}(2) = 2.40824/0.30103 = 8$. As a check, $2^8 = 256$, as expected.

Logarithms to the base 2 are common in computing where a computer will represent numbers by a series of 1s or 0s internally. Arithmetic performed in base two is called binary.

POWERS AND THEIR RELATION TO LOGARITHMS

Let us consider replacing the numbers with letters as was done with multiplication. Again the letter x can take any value, $y = 10^x$. If we apply log to the terms on both sides of the equals sign we can now find x. Check the method used for rearranging the formula, $3x = y$, if you do not understand this step, $\log(y) = \log(10^x) = x\ x = \log(y)$. This shows the effect of the log was to cancel the base of the power, 10, in $y = 10^x$. This is the same in any base and generally can be written $\log_N (N^x) = x$. Notice that the base must be the same in both parts. For instance, the following formula is wrong, $\log_2 (3^x)$ is not equal to x but this is correct, $\log_3 (3^x) = x$. This rule allows us to cancel the base of a power by multiplication with a logarithm. This is useful for extracting the power x.

THE ALGEBRA OF POWERS AND LOGARITHMS

When two powers of the same base are multiplied, the repeated multiplication is effectively extended. This is identical to the base raised to the sum of the powers. If they are divided, then the powers are just subtracted. $N^x \times N^y = N^{(x+y)}\ (N^x)/(N^y) = N^{(x-y)}$.

While thinking about these relations, students can see that we have combined multiplication and addition and vise-versa for division and subtraction. Using the logarithm to extract the powers, shown previously, allows the addition and subtraction parts to be extracted, $\log_N (N^x \times N^y) = x + y = \log_N (N^x) + \log_N (N^y)$ and if we set $N^x = A$ and $N^y = B$ then $\log_N (A \times B) = \log_N (A) + \log_N (B)$. Likewise $\log_N (A/B) = \log_N (A) - \log_N (B)$. The rules shown here are the reason that logarithms allow us to reduce the complexity of large lists of multiplications (or divisions) down to simple addition (or subtraction).

These basic properties of logarithms were critical in the development of science and industry over the past three centuries.

LOG TABLES

Suppose you want to multiply numbers so large that it will take a while to complete the computation by hand. It would be faster if there were some sort of table to look up the answer. If we wanted the answer to be accurate to four digits, a simple solution would be to make a table, called a matrix, with the rows and columns corresponding to the numbers between 1 and 9,999. If we picked a row, say 50 and column say 26, were they crossed we would find the answer for the multiplication, in this case 1300. Picking a row and column would show us the multiplied answer quickly.

Now any multiplication can simply be looked up in our table. For numbers outside the range 1 to 9,999 we can still find the answer by moving the decimal point until they are in this range, reading from the table and finally moving the decimal place back by an opposite number of steps.

The problem with this basic system, that makes it unworkable, is the number of entries needed will be huge. For four digits accuracy each edge of our square table would have 10,000 numbers. This gives us least 100,000,000 entries ($10,000^2$). If the print is very small that is still enough to fill 20 thick books.

Another problem is seen as we increase the number of digits accuracy needed. The square shape of our table, the matrix, will rapidly start to get bigger with each digit added. Most scientific and engineering calculations work at seven digits accuracy. This works out to be more than a million thick books to store our table and we have not even considered division.

History of Logarithms

Logarithms were invented in the seventeenth century by John Napier, a Scottish Barron. During that time in Scottish history the country was undergoing major religious and political upheaval. In this climate academic study was not held in high regard. Later in life Napier considered his greatest publications were his theological works, with his mathematical works as a secondary interest. The development of logarithms at this time came from a need to simplify the computations of repeated multiplications and divisions. These computations were

It is no understatement that the invention of logarithms changed the world. Their usefulness in industry and science was soon realized, and the system rapidly spread around the world. The invention of the electronic calculator in the 1950s allowed complex calculations to be performed by the simple push of a button.

Real-life Applications

COMPUTER INTENSIVE APPLICATIONS

Although the system of using log tables is not in common use since the invention of electronic calculators it has found new life in computer intensive applications. The desktop calculator will have to run a program to multiply numbers together. This process is so fast we cannot see it and to us it looks instant. However, it still takes a certain amount of time, and the time needed increases with the complexity and amount of calculations. In a computer-intensive application, in which millions of numbers have to be multiplied every second and speed is critical, this can become a problem. Some examples are interactive 3D computer games, and software used in spacecraft and aircraft. Here, the small time the computer takes to calculate the numbers will rapidly increase and can become substantial.

This can be reduced if a set of log and anti-log tables are wired into the computers memory, and the computer only has to look up values instead of running a program to find the answer. This technique to improve performance under heavy arithmetic load is called a log lookup table.

USING A LOGARITHMIC SCALE TO MEASURE SOUND INTENSITY

Decibels (dB) are used as a measure of sound level. They are common markings for stereos, televisions, and other audio equipment and are based on a log10 scale. The faintest sound we can hear is called the threshold of hearing. Its value is tiny, about a 0.3 billionths change in air pressure. The scale is given as dB = log (Number of times greater than threshold of hearing) $\times$ 10. A normal conversation is 60dB or, remembering to divide by the 10 from right of the formula, $10^6 = 1,000,000$. This is one million times louder than the faintest sound you can hear. We can safely hear sounds to around 90 dB, the level of an orchestra, before damage to the ear starts to occur, but we can still hear sounds louder than that. The levels of the front row of a rock concert can reach 110 dB. After this, there is pain and instant deafness. This gives the human ear an amazing range of about 100 billion times the faintest sound it can detect.

---

common in the calculations of astronomical charts used by the navy and the shipping industry and religious charts used by the church, three of the most powerful institutions in Britain at the time.

The power of the system of logarithms comes from its simplification of computational steps involved. By converting terms that were to be multiplied or divided to logarithms from a table, they could then simply be added or subtracted, and the result read from another table. The system was compact and flexible enough that the two tables needed to perform the steps could be listed in a few pages at the back of a book. Another device, called a slide rule, reduced computational time further by allowing the user to read the answer directly by simply moving the two rulers on the device. For 300 years this device was commonly used by scientists and engineers, just as hand-held calculators are today.

Tourists watch as tsunami waves hit the shore near in Penang, northwestern Malaysia on December 26, 2004. Although the earthquake causing the tsunami initially measured at 8.0 on the Richter scale, scientists later found that the data indicated an undersea earthquake measuring 9.0. The difference of one point is not insignificant. On the logarithmic Richter scale, each whole number increase means that the magnitude of the quake is ten times greater than the previous whole number. Thus, an earthquake with a magnitude of 9.0 has ten times the force of one with a magnitude of 8.0; an earthquake of 9.0 has 100 times the intensity of the 7.0 earthquake, etc. AP/WIDE WORLD PHOTOS. REPRODUCED BY PERMISSION.

ESTIMATING THE AGE OF ORGANIC MATTER USING CARBON DATING

The atmosphere is continuously being bombarded by radiation from space. In the upper atmosphere, the radiation from space has enough energy to change atoms of nitrogen into carbon. Carbon created this way is called carbon 14 and is different to the majority of the carbon we see around us called carbon 12. Carbon 14, unlike carbon 12, is unstable and will slowly decay back to nitrogen over a period of many thousands of years. The rate of production of carbon 14 in the atmosphere can be shown to be stable for a very long period of history, and this allows us to measure the age of dead organic matter.

All life on Earth is made from carbon, and during the course of an organism's life it will absorb small amounts of carbon 14. When the organism dies, it will stop absorbing carbon 14. So by measuring the ratio of carbon 14 to carbon 12 that is present and using the law of exponential decay of a radioactive source and their logarithms, scientists can calculate the age of the material.

DEVELOPING OPTICAL EQUIPMENT

No matter how pure a material is made, as light passes through it a small majority will always be scattered or absorbed. This is an exponential effect and logarithms are therefore used extensively in the design of optical equipment. Just a few examples are cameras, optical fibers, and the design of television screens.

USE IN MEDICAL EQUIPMENT

Certain cancers can be treated by passing radioactive beams though the body. A machine with a radioactive source is rotated around the patient. Only at the center of this rotation will the radiation be constant. Moving away from the center, the radiation will only pass through the patient periodically as the machine makes each rotation.

The location of the cancer is carefully mapped, and the absorption of the radiation through the body and the absorption by various organs is then calculated. This requires the use of logarithms due to the exponential nature of this absorption. The aim of the surgeons is to locate the cancer and manipulate the intensity of the beam over each rotation so that minimum damage is caused to the surrounding organs and maximum damage is caused to the tumor.

DESIGNING RADIOACTIVE SHIELDING FOR EQUIPMENT IN SPACE

Outside the protection of the Earth's atmosphere we enter a highly radioactive environment. Spaceships, satellites, and spacesuits must all able to absorb and disperse this energy to protect the delicate equipment and astronauts from damage. The absorption must be balanced against weight not too massive to launch.

Absorption of different types of radiation and in different materials involves calculations using logarithms due to the exponential nature of absorption.

SUPERSONIC AND HYPERSONIC FLIGHT

During supersonic and hypersonic flight, the air flow over the craft behaves very differently than at slower flight speeds. Logarithms are used in the design and fuel requirements.

Potential Applications

CRYPTOGRAPHY AND GROUP THEORY

Cryptography is the science of encoding information in such a way that an eavesdropper cannot intercept and decode a message. Modern methods rely on a mathematical phenomenon that some formulas are practically impossible to invert. This means that information encoded by such a formula cannot simply be decoded by rearranging the terms of the formula to reverse the processes.

Two parties can generate and swap unique keys which will unlock the message encrypted by a formula like this. However, if an eavesdropper were to try to decode the encryption by setting a machine up between the two parties without the keys, he would have to invert the formula used to encrypt the information.

One set of such functions that show these properties come from an abstract area of mathematical research that studies the relations between objects called group theory. Certain groups can be given properties that act like exponentials and logarithms. The calculation of the exponential part of these groups is very simple, and the calculation of the logarithmic part is very hard. This property can be exploited in cryptography. Studies of this branch of mathematics are important in the future development of faster and more secure algorithms.

Where to Learn More

Books

Durbin, John R. *College Algebra.* New York: John Wiley & Sons, 1985.

Morrison, Philip and Phylis Morrison. *Powers of Ten: A Book About the Relative Size of Things in the Universe and the Effect of Adding Another Zero.* San Francisco: Scientific American Library, 1982.

Periodicals

Curtis, Lorenzo. "Concept of the exponential law prior to 1900." *American Journal of Physics* 46(9), Sep. 1978, pp. 896–906 (also available at <http://www.physics.utoledo.edu/~ljc/explaw.pdf>.

Web sites

SOS Math! "Introduction to logarithms." <http://www.sos-math.com/algebra/logs/log1/log1.html> (February 1, 2005).

Logic is a set of rules by which decisions and conclusions are either derived or inferred from a set of statements. Logic can be mathematical or predicate (dealing with statements and sentences). Logic is also a set of rules by which computers handle data, and circuit logic dictates how many devices operate.

However, a logical decision or a belief may or may not be correct. Logic is more of a set of rules to follow in reaching a decision. An example of a logical, but incorrect, bit of reasoning is the following: If I believe that sheep have a wool coat and that all sheep are mammals, then it could make logical sense for me to believe that all mammals have wool coat. The conclusion is incorrect, but it is logically drawn.

Fundamental Mathematical Concepts and Terms

Over twenty-four centuries ago, the idea of logic was explored and developed about the same time in China, India, and Greece. The Greek philosopher Aristotle (384 B.C.–322 B.C.) was important in the creation of logical systems.

REASONING

Logic does not necessarily lead to the truth. What logic does do is to allow us to look at an argument and to decide if the reasoning is valid or not valid. Logic also points out how we can come to believe something that is not true (even though that sounds illogical).

PROPOSITION AND CONCLUSION

The starting point of a logical line of thought is called the proposition (or the statement). A proposition is the real meaning of the sentence (or the equation, as it can be written in mathematical language also). The meaning can be expressed in different ways and still mean the same thing. For example "Today is Friday" and "Yesterday was Thursday" are the same proposition, while "My name is Brian" is a different proposition.

A proposition is always true or false, although it is sometimes unknown which proposition is true and which is false. "There is life on Mars" is an example of a proposition that may or may not be true; we have yet to find out.

Logic proceeds from the starting point of the proposition to the conclusion in a series of steps that are related

Proposition	Steps	Conclusion
True	Support the conclusion	Always true
True	Do not support the conclusion	Can be true or false
False	Support the conclusion	Can be true or false
False	Do not support the conclusion	Can be true or false

Table 1.

to each other. That is, one step is followed by a step that supports it.

Here is an example of a logical series of steps:

- Today is Friday.
- My library books were due Thursday.
- My library books are overdue.

Here is an example of a series of steps that is not logical:

- The moon is full.
- There are clouds in the sky.
- My cat has a hairball.

From the proposition, the steps that proceed to the conclusion can be set up so that the steps guarantee that the conclusion is true. This is a good style to use when debating. There is no middle ground with this type of approach. Either all the steps lead to a single conclusion or they do not.

A number of different outcomes can still result, depending on whether the steps from the proposition to the conclusion support this conclusion. Table 1 summarizes these various possibilities.

A less rigid style is when the steps from the proposition to the conclusion support the likelihood of the conclusion. In this style a conclusion does not have to be true, it is just likely to be true. Points can be presented that support the conclusion, but the conclusion could still be debatable. This style of logic is used in many courtrooms by lawyers trying to defend their clients from charges brought against them.

Real-life Applications

BOOLEAN LOGIC

Many persons do most of their banking while sitting at their desk. This is possible since they can hook up to the local bank's Web site, research bank accounts, and then use the computer directions built into the site to shift money from one account to another, pay bills, and look at the action in each account over whatever time period is desired.

These activities are pretty human-like. How can computers do them? The answer is something called Boolean logic.

Boolean logic is named after the Irish mathematician George Boole (1815–1864). From an early age, Boole showed a talent for languages and teaching. When he was 20, Boole began to teach himself mathematics. He proved to be talented at this as well, publishing papers in the leading math journals of the day. When he was 34 years old, he was appointed chair of mathematics at Queens College in Cork, Ireland. He taught there for the rest of his life.

In 1854, when he was only 39 years old, Boole published a paper called "An Investigation into the Laws of Thought, on Which are founded the Mathematical Theories of Logic and Probabilites." The ideas in this paper became the basis of Boolean logic.

One niche that Boolean logic has filled beautifully is the task of sifting through vast amounts of information to find those bits of information that are desired.

FUZZY LOGIC

Fuzzy logic is a way of making computers behave in a way that is similar to the way humans think. Often, we are able to use information that is not really clear or precise to make decisions that are definite. We can relate the imprecise (fuzzy) information with what we already know to make a decision.

Here is an example. You are driving your car on a crowded, four-lane freeway. The speed limit is 65 mph (105 km/h). As is usually the case, traffic is moving faster, at an average speed 70 mph (113 km/h). You know that it

Boolean Logic and Computer Searches

Boolean logic links the common parts of different pieces of information. This feature makes Boolean logic widely used in Internet search engines. For example, if there was no Boolean logic and information from the Internet on the trigonometry and homework problems was desired, the Internet search for every word would show all the documents that separately mention "trigonometry" or "homework." This would probably result in a huge number of sites to search, making the search nearly meaningless. Because of Boolean logic, however, a search can be done to look for those documents that contain "trigonometry" AND "homework." This number of sites will be much less, and the sites will be more likely to have something to do with homework related to trigonometry rather than homework related to all subjects.

Boolean logic even allows a search to focus on one word and not another. To use the above example, the following search could be done: "trigonometry" AND "homework" NOT "advanced." This would allow the search engine to zero in on those site that were about teaching methods of trigonometry at a basic level as opposed to sites that discussed advanced trigonometry.

of fuzzy logic, the computer inside a video camera is able to keep focusing even when the camera is jostled. As another example, fuzzy logic makes it possible to program a microwave oven to cook differently sized and types of foods perfectly with the touch of one button.

The logic of fuzzy logic can be summed up as IF X AND Y THEN Z. It is the 'if' and 'and' that makes things less precise.

The following example may help to make this fuzziness clearer. A conventional oven operates on the basis of exact temperature. A thermometer in the oven can cut off the power to the oven's heater when the oven reaches whatever temperature has been selected, and will kick the heater back into action when the temperature falls below another set value. This occurs no matter what is in the oven.

A microwave with a fuzzy logic temperature control does not rely on exact temperatures. Instead, the process is like this: "IF (the process is too cool) AND (the process is getting colder) THEN (add more heat)", or "IF (the process is too hot) AND (the process is getting colder) THEN (heat it up now)."

Companies have leapt on fuzzy logic as a way of making products that will perform better for people. Self-focusing cameras and video recorders, washing machines that can adjust the strength of cleaning power to how much dirt is in the clothes being washed, the controls to car engines, anti-lock braking systems in vehicles, banking programs, programs that allow people to do stock market trades—all these would not exist if not for fuzzy logic.

is safest for you and those around you to drive "with the traffic." But what exactly does driving "with the traffic" mean?

Watching other drivers, you realize that driving "with traffic" is done different ways. Some drivers will drive more slowly and stay in the right hand lane. Other drivers will speed and zig-zag their way between cars and lanes. Usually, the different styles mesh together to make a smooth flow of traffic. When they do not, there a traffic accident can occur.

Fuzzy logic was conceived by Lotfi Zadeh, a professor of electrical engineering at the University of California at Berkley, and was first proposed in a 1965 paper. From its humble beginnings, fuzzy logic has expanded to assume an important role in our daily lives. For example, because

Where to Learn More

Books

Bennett, D.J. *Logic made Easy: How to Know When Language Deceives You.* New York: W.W. Norton & Company, 2004.

Gregg, J.R. *Ones and Zeros: Understanding Boolean Algebra, Digital Circuits, and The Logic of Sets.* New York: Wiley-IEEE Press, 1998.

Mukaidonon, M., and H. Kikuchi. *Fuzzy Logic for Beginners.* Singapore: World Scientific Publishing Company, 2001.

Web sites

Brain, M. "How Boolean Logic Works." <http://computer.howstuffworks.com/boolean.htm> (September 2, 2004).

Cohen, L. "Boolean Searching on the Internet: A Primer in Boolean Logic." *University Libraries-State University of New York.* <http://library.albany.edu/internet/boolean.html> (September 3, 2004).

Kemerling, G. "Arguments and Inference" <http://www.philosophypages.com/lg/e01.htm> (September 3, 2004).